D0984931

RESIDUE NUMBER SYSTEMS
Algorithms and Architectures

THE KLUWER INTERNATIONAL SERIES IN ENGINEERING AND COMPUTER SCIENCE

RESIDUE NUMBER SYSTEMS
Algorithms and Architectures

P.V. ANANDA MOHAN
I.T.I. Limited
Bangalore, India

KLUWER ACADEMIC PUBLISHERS
Boston / Dordrecht / London

Distributors for North, Central and South America:
Kluwer Academic Publishers
101 Philip Drive
Assinippi Park
Norwell, Massachusetts 02061 USA
Telephone (781) 871-6600 / Fax (781) 681-9045
E-Mail < kluwer@wkap.com>

Distributors for all other countries:
Kluwer Academic Publishers Group
Post Office Box 322
3300 AH Dordrecht, THE NETHERLANDS
Telephone 31 786 576 000 / Fax 31 786 576 254
E-Mail services@wkap.nl

Electronic Services < http://www.wkap.nl>

Library of Congress Cataloging-in-Publication Data

Mohan, P.V. Ananda, 1949-
Residue number systems: algorithms and architectures / P.V. Ananda Mohan.
p. cm.
Includes bibliographical references and index.
ISBN 1-4020-7031-4 (alk. paper)
1. Signal Processing—Digital techniques. 2. Algorithms. 3. Computer architecture. I. Title.

TK5102.5 .M638 2002
621.382'2—dc21

2002025487

Printed on acid-free paper.

Printed in the United States of America.

To

Srinivasa Kumar

CONTENTS

PREFACE

There has been continuing interest in the improvement of the speed of Digital Signal processing. The use of Residue Number Systems for the design of DSP systems has been extensively researched in literature. Szabo and Tanaka have popularized this approach through their book published in 1967. Subsequently, Jenkins and Leon have rekindled the interest of researchers in this area in 1978, from which time there have been several efforts to use RNS in practical system implementation. An IEEE Press book has been published in 1986 which was a collection of Papers.

It is very interesting to note that in the recent past since 1988, the research activity has received a new thrust with emphasis on VLSI design using non-ROM based designs as well as ROM based designs as evidenced by the increased publications in this area.

The main advantage in using RNS is that several small word-length Processors are used to perform operations such as addition, multiplication and accumulation, subtraction, thus needing less instruction execution time than that needed in conventional 16 bit/32 bit DSPs. However, the disadvantages of RNS have been the difficulty of detection of overflow, sign detection, comparison of two numbers, scaling, and division by arbitrary number, RNS to Binary conversion and Binary to RNS conversion. These operations, unfortunately, are computationally intensive and are time consuming. The RNS has an interesting feature that error detection and correction also is possible using redundant moduli which feature perhaps is not available with conventional arithmetic based Digital Signal Processors.

There has been continuous search for selecting the moduli for RNS so as to overcome these difficulties. A particular moduli set viz., $\{2^n-1, 2^n, 2^n+1\}$ has received considerable attention. The Binary to RNS conversion for this moduli set is rather simple and first investigated by Bi and Jones. The extension to general moduli set based on the periodic properties of $2^x \bmod m_i$

also has been described in literature. The topic of RNS to binary conversion has been investigated by Bi and Jones, Bernardson, Ibrahim and Saloum, Andraros and Ahmad, Dhurka Das, Taylor, Vinnakota and Rao, Piestrak, Dhurkadas, Ananda Mohan, Bharadwaj et al, Gallaher et al. It is very interesting to note that the RNS to binary conversion time today achievable with the fastest technique among these is comparable to that of instruction cycle time of commercial digital signal processors. Once the bottleneck of RNS to binary conversion is removed, high speed signal processors using RNS will be a reality and these may be quite competitive with commercial DSPs. We focus on this moduli set in Chapter III so as illustrate all the aspects of RNS based systems. Attention will be paid to other recently investigated moduli sets such as $\{2^n, 2^n\text{-}1, 2^{n-1}-1\}$, and $\{2n\text{-}1, 2n, 2n+1\}$.

In this book, we will be discussing the basics of RNS, techniques for RNS to binary conversion using Mixed Radix conversion (MRC) and Chinese Remainder Theorem (CRT), techniques of Binary to RNS conversion for general moduli set in Chapter II. We also deal with realizations of Fast RNS multipliers in Chapter IV. It may be noted that considerable work has been done on the realization of modulo multipliers for word-lengths as large as 1024 bits for application in Cryptographic Processors implementing RSA (Rivest, Shamir and Adleman) algorithm. This topic, hence, is of immense practical interest.

The topic of scaling is of immense interest in the implementation of IIR Filters, Adaptive filters etc. Scaling needs to be accompanied by base extension. These two topics as well as the problem of division are addressed in detail in Chapter V.

The topic of error detection and correction is considered in detail in Chapter VI. There is a need for complex signal processing in practical applications. This is realized in a Quadratic Residue Number System such that simplification of computation can be achieved. This topic will be explored in Chapter VII. Chapter VIII deals with several applications of RNS for implementation of FIR Filters, IIR filters, Adaptive Filters, Digital Frequency Synthesis, 2-D filters etc.

The aim of the book is to introduce the topic comprehensively to the designers of VLSI systems. This fills a gap of lack of a suitable text in this area. The audience could be senior graduate students, researchers in this area and Custom VLSI designers in the industry. Examples will be presented throughout the text in order to illustrate the concepts in a simple manner.

The author wishes to thank Jennifer Evans, Anne Murray of Kluwer Academic Publishers for their editorial guidance which has improved the quality of the presentation of this book and for their patience and understanding. The author wishes to thank I.T.I. Limited for providing a stimulating R&D environment which enabled the author to present this work. He also wishes to thank his wife Radha Nirmala and children Miss Sri Devi, Mr. Ramakrishna Kumar and Mr. Shiva Ram Kumar for their understanding and encouragement, love and affection which helped the author to focus on this effort.

P.V. Ananda Mohan

Bangalore
June 2001.

1

INTRODUCTION

1.1. HISTORICAL SURVEY

The origin of the topic RNS [Gar59] is credited to the Chinese scholar Sun Tzu of first Century AD and Greek Mathematician Nichomachus and Hsin-Tai-Wei of the Ming dynasty (1368AD-1643AD). Later, Euler presented a proof for the Chinese Remainder Theorem (CRT) in 1734. In the twentieth Century, Lehmer, Svoboda and Valach built hardware using RNS and much work was done at various laboratories during 1950's and 1960's. The text books by Szabo and Tanaka [Sza67] and Watson and Hastings [Wats67] in 1967 are the first documenting the results in this area. Subsequently, in 1977, resurgence of interest in this area was due to the pioneering work of Jenkins and Leon [Jenk77]. By 1980, interesting results were available in all areas of RNS such as residue Number scaling, RNS to Binary conversion, Binary to RNS conversion, scaling and error correction. These are largely due to the research of Soderstrand, Miller, Jullien, Jenkins, Leon among others. These early designs were mostly ROM based. The advent of VLSI technology has stimulated the work in the area of non-ROM based solutions. This led to work on special moduli sets. It is interesting to note that the computational speed has been increasing and excellent solutions have been evolving even as recently as December 2000.The work carried our till 1986 was compiled in an IEEE Press collection of papers [Sode86]. The reader is referred to an excellent historical introduction to this subject in this book. A book Chapter due to Jenkins was published in Handbook of Digital Signal Processing in 1993 [Jenk93].

Perhaps, the availability of inexpensive 16 bit and 32 bit digital signal processors in market for performing DSP intensive computation such as those in speech coding, Picture coding, and other Signal processing applications has obscured the need for RNS based signal processors. Special purpose applications e.g. in cryptography have the need for adding and multiplying numbers as large as 1024 bits in length, leading to extensive research in single modulus ALUs [Tay85b]. However, no general purpose Digital Signal Processor using RNS similar to DSP has appeared in market.

There has been recently extensive interest in low-power designs [Chan92] and high-speed computing engines. The hallmark of high-speed DSP designs is a high-speed multiplier accumulator and ALU. Conventional techniques such as Booth's algorithm using multi-bit recoding and fast multi-operand additions have decreased the multiplication time considerably. If further speeds are desired e.g in Radio modems where FIR filtering is needed at high sampling frequencies, conventional DSPs are not attractive. Designs based on distributed architecture e.g. Peled-Liu bit-slice approach [Pele74], or conventional multiplier-less designs using only few additions are preferred. An example for the latter case is the CSD (Canonical Signed Digit) representation of coefficients [Lim83].

RNS aims to increase the speed of computation using a divide and conquer approach, however, having its own problems as will be shown in the later chapters. Nevertheless, in applications where certain difficult operations such as scaling, division, comparison of two numbers are minimal, such as FIR filters, significant gains can be achieved. A typical RNS based signal processor will be as shown in Fig 1.1. The front-end is a binary to RNS converter whose n output words corresponding to n moduli will be processed by the n parallel processors in the RNS Signal Processor block to yield n output words. The last stage converts these n words to a conventional binary number representing the output of the Signal processor. Thus the architecture is SIMD (Single Instruction Multiple Data) type. This chapter introduces the basic concepts of RNS and the addition operation in the next few sections.

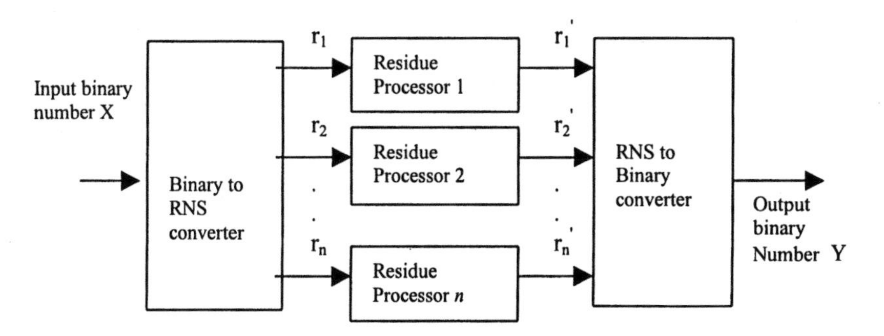

Fig 1.1 A typical RNS based Digital Signal Processor

1.2. BASIC DEFINITIONS OF RNS

An RNS is defined by a set of integers $\{m_1, m_2, m_3, \ldots m_L\}$ called " moduli " which are relatively prime to each other. Using such a set, any integer X between 0 ....(M-1) can be represented uniquely by the set of integers $\{x_1, x_2, x_3, \ldots x_L\}$ called the " residues " where $x_i = X \bmod m_i$ $i=1,2,\ldots L$. Note that $M= \pi_{i=1}^{L} m_i$. Such number system using residues is known as a Residue Number System. In this system, the operation $\{x_1, x_2, \ldots x_L\}$ □ $\{y_1, y_2, y_3 \ldots y_L\} = \{z_1, z_2, z_3, \ldots z_L\}$ holds where □ could be addition mod m_i, or subtraction mod m_i or multiplication mod m_i. Note that z_i is solely decided by x_i and y_i. Such fully decoupled system leads to carry-free arithmetic since the computations with respect to all the moduli are totally independent. It is unfortunate, however, that division cannot be easily accomplished in the RNS nor scaling defined as division by a constant. The size of the number also cannot be determined easily by looking at the residues. Fortunately, however, in certain DSP applications, only the operations of multiplications, additions and subtractions suffice for which RNS is a good choice compared to the large word length DSPs in conventional number systems.

The range 0 to M-1 available can be divided into positive and negative ranges, if desired. If M is odd, the dynamic range of the system is [-(M-1)/2 and (M-1)/2] resulting in a symmetric RNS. If M is even, the dynamic range is [-M/2, ((M/2)-1)]. Thus, residues of a positive number and a negative number are evaluated as follows:

$x_i = X \bmod m_i \quad X \geq 0$

$x_i = (m_i - X \bmod m_i) \quad X < 0.$

1.3. ADDITION OPERATION IN RNS

1.3.1. General addition

The fundamental operations in RNS signal processing are addition mod m_i and subtraction mod m_i. The sum [Bay87a] of two numbers x and y can exceed m_i and hence a conditional subtraction needs to be done as shown in Fig 1.2(a). This subtraction needs a two input adder whose one input is output of the first adder and the other is two's complement of m_i i.e.(2^n-m_i). Depending on whether the result is positive or negative the result (X+Y) or (X+Y+2^n-m_i) is selected using a multiplexer. Evidently, the computation time needed is that of two n bit adders and a 2:1 MUX delay. Simple variations of this technique [Dug92] are possible, as shown in Fig 1.2(b) and (c). In the second method shown in Fig 1.2(b), a comparator is used to check whether output of the first addition step is larger than m_i. In case, the result is less than m_i, the second addition step need not be performed. In the third method shown in Fig 1.2(c), one adder is multiplexed to perform both additions. The multiplexers at the top first allow X and Y to be added and then allow (2^n-m_i) and (X+Y).

A high-speed version of this modulo addition can be realized using the architecture of Fig 1.3. In this architecture, X+Y and (X+Y+2^n-m_i) are parallelly computed and based on the sign of the latter, one of the two outputs is selected. The three input addition needs an additional level of n full adders.

The adders can be realized by carry-propagate adders or carry-look-ahead adders or adders with regular VLSI layout. Evidently, the CPAs are slowest. Carry look ahead adders for small n can use one level CLA and block carry look ahead whereas designs for large n need more levels of Block carry look-ahead. The reader is referred to [Hwa79] for an interesting discussion on this subject. These adders feature constant delay. The adders using regular VLSI layout, due to Brent and Kung [Bren82] use carry computation blocks with the resulting adder delay given as $(\log_2 n+1)\Delta_{FA}$ and area $n.(\log_2 n+1).A_{FA}$ where Δ_{FA} and A_{FA} are the delay and area of a full-adder. The concepts of carry generation in Brent and Kung adders can be used even if regular layout is not used, since the achievable delays are still the same and lowest possible.

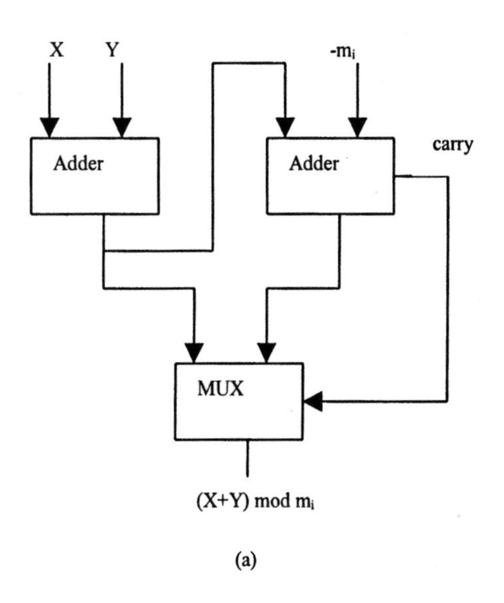

(a)

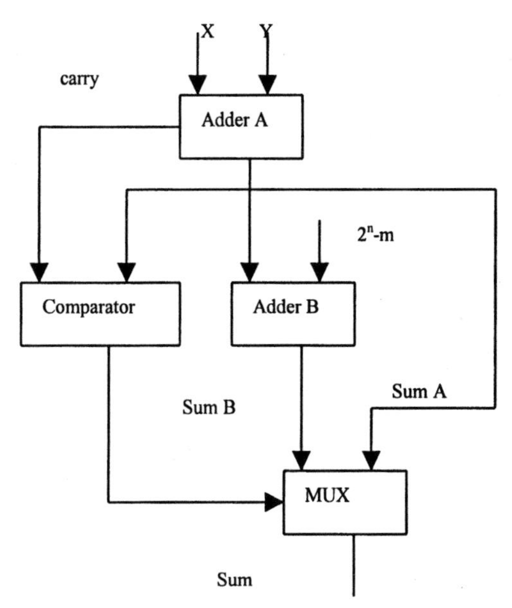

(b)

Fig 1.2 (contd)

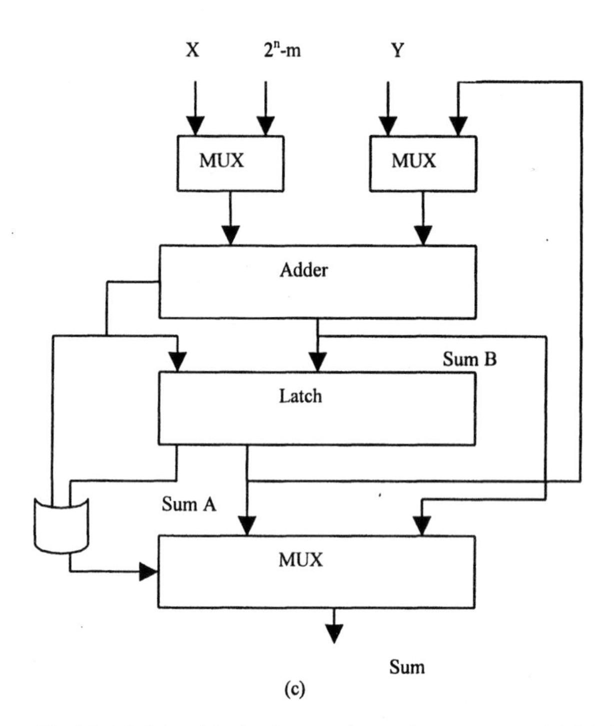

Fig 1.2 Modulo adder implementations using two stages (a)-(c) ((a) adapted From [Bay87a] 1987©IEEE and (b)and (c) adapted from [Dug92] 1992©IEEE)

It is thus seen that the cascade and parallel adders feature the following delay and area expressions:

Cascade Adder: Area = $2n(\log_2 n+1).A_{FA} + n.A_{INV} + n.A_{2:1Mux}$ (1.1a)

Delay = $2(\log_2 n+1).\Delta_{FA} + \Delta_{MUX} + \Delta_{INV}$. (1.1b)

Parallel Adder: Area = $2n(\log_2 n+1).A_{FA} + n.\ A_{FA} + n.A_{INV} + n.A_{2:1Mux}$ (1.2a)

Delay = $(\log_2 n+1).\Delta_{FA} + \Delta_{FA} + \Delta_{MUX} + \Delta_{INV}$. (1.2b)

Multiplexed Adder: Area = $n(\log_2 n+1).A_{FA}+n.A_{INV} +2n.A_{DFF} + 2n.A_{2:1Mux}$ (1.3a)

Delay=$2(\log_2 n+1).\Delta_{FA}+\Delta_{MUX}+\Delta_{DFF}$. (1.3b)

Thus, it can be seen that the multiplexed adder needs same delay as a cascade adder while needing smaller area. The parallel adder is also denoted as high-speed adder.

The realization of a modulo subtractor is next considered. This is similar to modulo adder except that first A-B is realized by adding A with two's

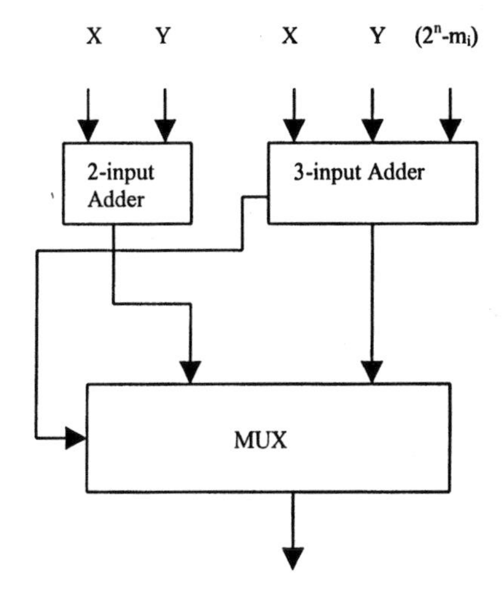

Fig 1.3 A modulo adder using parallel adders (adapted from [Ell90] 1990©IEEE)

complement of B. Since the result can be negative, A-B+m_i is evaluated to obtain a positive result. The area and delay are same as in the case of modulo adders described above except that the n inverters are needed to obtain twos complement of B in stead of m_i.

It may be mentioned that for some special cases of m_i such as 2^n, 2^n-1, and 2^n+1, the addition operation can be simplified. In the case of 2^n, modulo addition is normal addition ignoring the generated carry. The addition mod 2^n-1 needs an end-around-carry operation since the carry has a weight 2^n which mod (2^n-1) is 1. The addition mod (2^n+1) needs subtracting the carry from the sum since 2^n is -1 mod (2^n+1).

The case of modulo addition mod (2^n-1) has received considerable attention in literature. A particularly attractive technique due to Efsathiou et al [Efsta94] is considered next.

1.3.2. Addition mod (2^n-1)

Efsathiou, Nikolos and Kalamatianos [Efsta94] have described an area-efficient modulo (2^n-1) adder design. This uses the property that the carry of an n-bit adder has weight 2^n which mod (2^n-1) is unity. As such, the carry can be implicitly added to perform modulo reduction and addition in less time and area. This is as explained below. Consider the expression for carry of an n-bit adder obtained using the carry look-ahead expression,

$$c_{out} = c_{n-1} = G_{n-1} + P_{n-1}.G_{n-2} + ... + P_{n-1}.\ P_{n-2}.\ P_1.G_0 \quad (1.4)$$

where $G_i = a_i.b_i$ and $P_i = a_i \oplus b_i$. Normally, we would add the c_{out} in the next step as c_{-1} to obtain the final result. Interestingly, Eftasthiou et al [Efsta94] observe that since c_{-1} depends only on G_is and P_is, we can as well eliminate c_{-1} using c_{-1} computed using the expression for carry computation $c_i = G_i + P_i.c_{i-1}$ as follows:

$$c_0 = G_0 + P_0.c_{-1}$$
$$= G_0 + P_0.c_{n-1} = G_0 + P_0.G_{n-1} + P_0.P_{n-1}.G_{n-2} + ... + P_0.P_{n-1}.P_{n-2}......P_2.G_1 + P_0.P_{n-1}.P_{n-2}....P_1G_0. \quad (1.5a)$$

Noting that $G_0 + P_0.P_{n-1}.P_{n-2}....P_1G_0 = G_0$, (1.5a) gets simplified as

$$c_0 = G_0 + P_0.G_{n-1} + ...+ P_0.P_{n-1}.P_{n-2}.....P_2.G_1 \quad (1.5b)$$

The other carries can be next computed using

$$c_0 = G_0 + P_0G_2 + P_0P_2G_1. \quad (1.6a)$$

and

$$c_1 = G_1 + P_1G_0 + P_1P_0G_2. \quad (1.6b)$$

In general, the following formula holds good:

$$c_i = G_{(n+i)\bmod n} + \sum_{j=o}^{n-2} \left(\pi_{k=j+1}^{n-1} P_{(k+i+n)\bmod n} \right).G_{(j+i+1)\bmod n} \quad (1.7)$$

An implementation is as shown in Fig 1.4 (a). Note that this evidently shows us that the low order sum bits are dependent on the high order input bits unlike in an conventional adder.

It is interesting to note that in this adder described above, there is a possibility of double representation of zero. This can be eliminated using the modified circuit of Fig 1.4 (b). This is appreciated by noting that the double zero representation results when one of the inputs is ones complement of another or when both are all ones vectors. In a system where inputs are barred from being all ones, then the second case is possible. This case is sensed by the three input NAND function of all the propagate inputs and is used to block the all ones condition at the output.

The method can be extended to 2-level carry-look ahead adders as well, which are needed for large word length adder implementation using a two level CLA adder consisting of PG unit, Group Propagate and Generate unit (GPG), Between Groups Carry Look Ahead Unit (BGCLA), and Group Carry Look Ahead unit (GCLA) and summation unit. The reader is referred

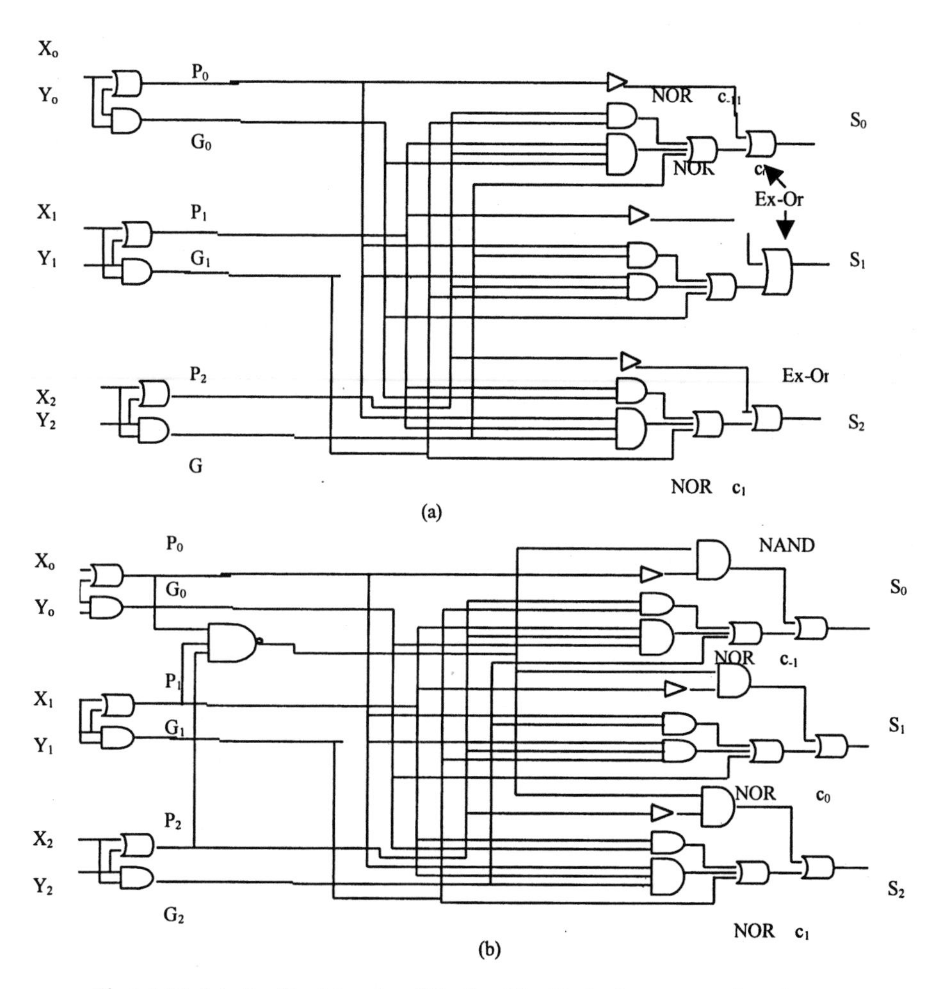

Fig 1.4 Modulo 7 adder (a) and modification (b) with single representation of zero. (adapted from [Efsta94] 1994©IEEE)

to Hwang [Hwa79] for an excellent treatment of this subject. Note that the only difference between conventional and mod (2^n-1) adders is that GP_{m-1} is remain the same. The reader is referred to [Kala00] for an interesting discussion on modulo (2^n-1) adders.

1.4. CONCLUSION

This Chapter gave an introduction to RNS and considered the simple operation of addition in some detail. The topics of Residue to binary conversion for general moduli set and specific moduli sets, the architectures for multiplication, division, scaling and error correction using redundant residues are considered in detail in the next five chapters. VLSI architectures using QRNS are considered together with applications of RNS in the next two Chapters.

2

FORWARD AND REVERSE CONVERTERS FOR GENERAL MODULI SETS

2.1. INTRODUCTION

In this Chapter, the topic of RNS to binary conversion for the general moduli set is considered in detail. There are basically two techniques for RNS to Binary Conversion, one based on Mixed Radix Conversion and another based on Chinese Remainder Theorem (CRT). Several architectures using ROMs as well as not using ROMs have been described in literature. The choice of these techniques is decided by the RNS to Binary and Binary to RNS conversion speeds as well as the area needed for implementation. The discussion in this Chapter excludes specialized three moduli sets which will be discussed in the next Chapter in detail. The binary to RNS conversion also can be realized in several ways which are also considered in detail in this Chapter.

2.2. MIXED RADIX CONVERSION BASED TECHNIQUES.

2.2.1. Szabo and Tanaka MRC technique

Consider the four moduli set $\{m_1, m_2, m_3, m_4\}$ in which the residues $\{r_1, r_2, r_3, r_4\}$ are given. It is needed to find the binary number B corresponding to this moduli set. In MRC, the desired number B is expressed as

$$B = r_1 + a_1.m_1 + a_2.m_1.m_2 + a_3.m_1.m_2.m_3 \qquad (2.1)$$

where $0 \le a_1 \le (m_2-1)$, $0 \le a_2 \le (m_3-1)$, $0 \le a_3 \le (m_4-1)$. The values of a_1, a_2 and a_3 are successively determined as follows. It can be first noted that $B-r_1$ is exactly divisible by m_1. Thus first $(B-r_1)$ is determined by subtracting r_1 from all the other residues. Next, division by m_1 is accomplished by multiplication of all residues of $(B-r_1)$ with $(1/m_1) \bmod m_2$, $(1/m_1) \bmod m_3$, $(1/m_1) \bmod m_4$ respectively. These quantities are known as multiplicative inverses of m_1 with respect to m_2, m_3 and m_4 respectively. This procedure continues till the last

MRC digit is obtained. The first MRC digit is $[(r_2-r_1) \bmod m_2].(1/m_1) \bmod m_2$. An example is presented in Fig 2.1. in order to illustrate this technique.

m_1	m_2	m_3	m_4
r_1	r_2	r_3	r_4
	$-r_1$	$-r_1$	$-r_1$
	$(r_2-r_1)_{m2}$	$(r_3-r_1)_{m3}$	$(r_4-r_1)_{m4}$
	$x(1/m_1)_{m2}$	$x(1/m_1)_{m3}$	$x(1/m_1)_{m4}$
	a_1	p	q
		$-a_1$	$-a_1$
		$(p-a_1)_{m3}$	$(q-a_1)_{m4}$
		$x(1/m_2)_{m3}$	$(1/m_2)_{m4}$
		a_2	r
			$-a_2$
			$(r-a_2)_{m4}$
			$x\ (1/m_3)_{m4}$
			a_3

Fig 2.1. Example for illustrating the Mixed Radix Conversion Technique

Evidently, the subtraction and multiplication by multiplicative inverse can be done using hardware modulo subtractors and modulo multipliers or using ROMs throughout. For a given n moduli system, n-1 steps are needed each involving several parallel modulo subtractions and modulo multiplications. The reader can compute easily the ROM size needed. Several simplifications / variations of Szabo and Tanaka Technique are feasible which are considered next.

2.2.2. Huang's RNS to binary conversion technique

Huang [Hua83] has suggested a RNS to binary conversion technique which is considered in this section. In this method, the given residues $\{r_1, r_2, r_3, \ldots\ r_n\}$ correspond to a binary number X which can also be equivalently obtained by considering the fact that

$$X = \{r_1,r_2,r_3, \quad \ldots\ldots r_n\} = X_1 + X_2 + X_3 + \ldots + X_n = \{ r_1,0,0\ldots\ldots,0\} +$$

$$(0,r_2,0\ldots\ldots\ldots,0\} + \ldots\ldots + \{ 0,0,0\ldots\ldots\ldots,r_n\} \qquad (2.2)$$

Using look-up tables (ROMs), each of the right hand terms can be parallelly computed to obtain the Mixed Radix Digits. It can be seen that the second residue set on the right hand side of (2.2), has residue 0 with respect to m_1, and similarly the third term on the right hand side of (2.2) has residues zero with respect to m_1 and m_2 and so on. Thus, the MRC digits available in the ROMs can be read and added in a mod m_i adder for each modulus (see Fig 2.2(a)).

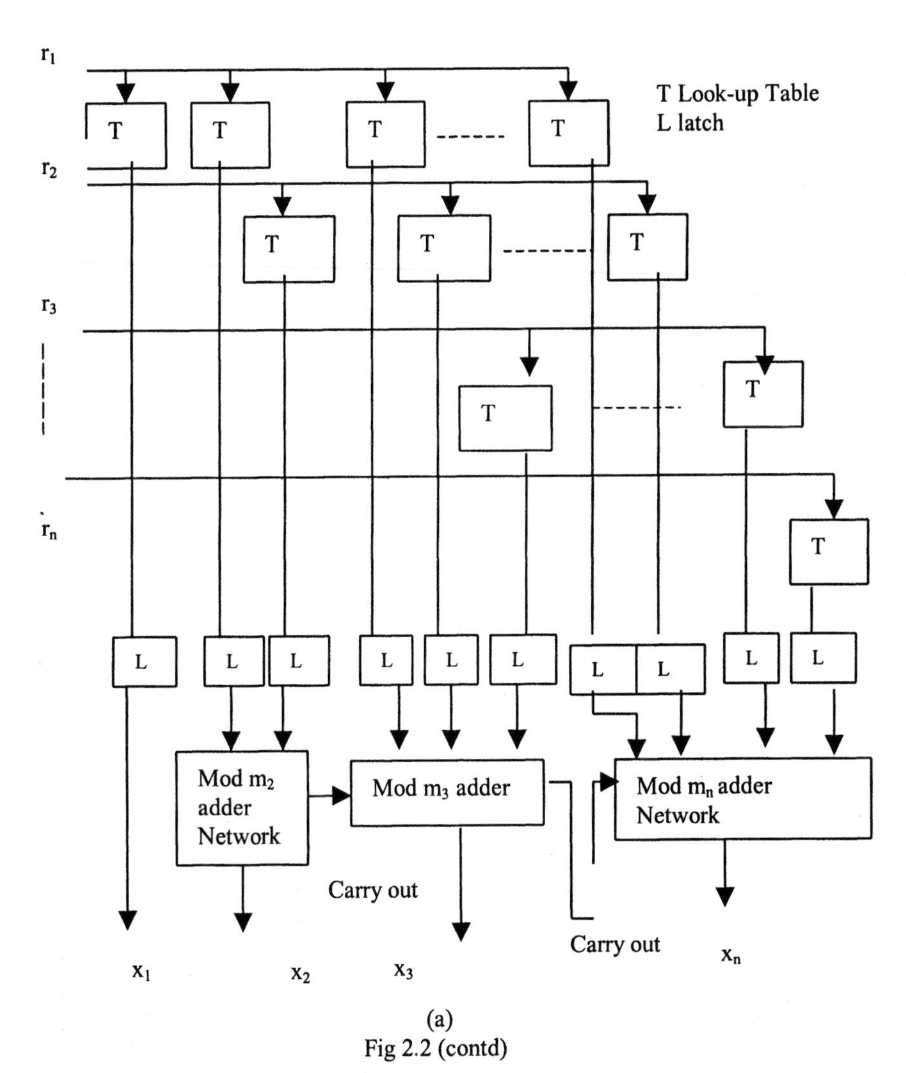

(a)
Fig 2.2 (contd)

Note, however, that the carry of the each adder mod m_i needs to be added to the subsequent modulo adder. The carry of the last adder, however, can be ignored. Thus, once the final MRC digits are obtained, the next step is to compute X using the well known expression for residue expansion of X (see e.g.(2.1)).

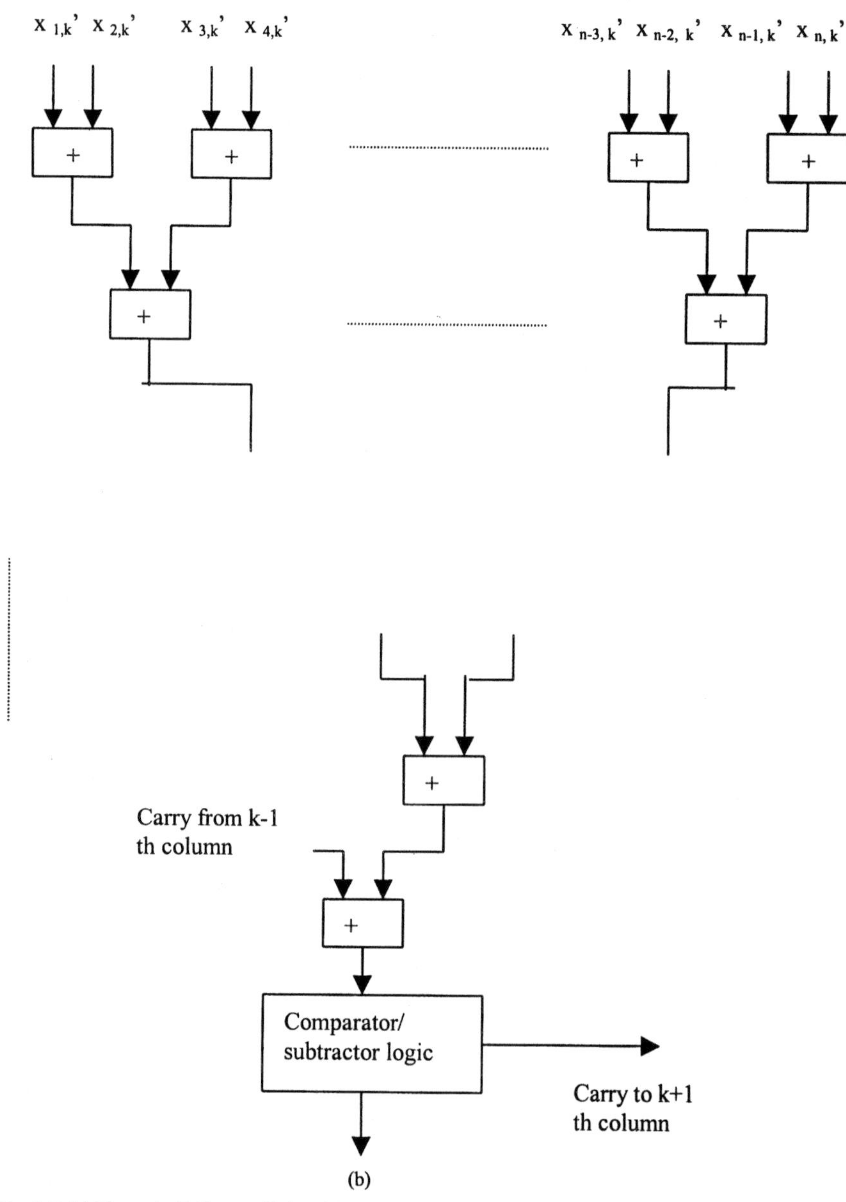

(b)

Fig 2.2 (a) Huang's Fully parallel residue to MRC architecture, (b) Adder network for column summation (adapted from [Hua83] 1983©IEEE)

Note that the addition of the mixed radix digits can be implemented using a $\log_2 n$ level adder tree as shown in the architecture of Fig 2.2(b).

An example is presented next to illustrate the procedure. Consider the moduli set P={3, 5,7}. Let the residues be {2, 2, 5} for which X has to be evaluated:

X={2, 2, 5}={2, 0, 0}+{ 0, 2, 0}+{0, 0, 5} (2.3)

The MRC digits corresponding to the three terms on the right hand side of (2.3) are respectively as follows:

	m_1=3	m_2=5	m_3=7		
{2,0,0}	2	1	2	⇒	2+1.3+ 2.15 =35
{0,2,0}	0	4	2	⇒	0+4.3+ 2.15 =42
{0,0,5}	0	0	5	⇒	0+4.3+ 5.15 =75
{2,2,5}	2	0	9		152
		⇒	1		
	2	0	3		47
				⇒ 1(discard)	

The addition of these together with the carry of the previous column gives the final MRC digits. Hence, the value of X can be seen to be

X = 2+0.3+ 3.15 = 47

2.2.3. Chakraborthi, Sundararajan and Reddy MRC technique

Chakraborthi et al [Chak86] consider two schemes viz., serial table look-up and parallel table look-up. ROMs are extensively used in this approach. The serial table look-up schemes for four and six moduli are presented in Fig 2.3(a) and (b) respectively. Here, two moduli are considered at each step. As an illustration of the four moduli scheme shown in Fig 2.3(a), LUT (Look-up Table) T_1 yields the mixed radix digits a_1, a_2 corresponding to residues r_1 ad r_2 in a straight forward manner. The LUT T_2 yields the residue corresponding to $(a_2.m_1 + a_1)$ with respect to moduli m_3 and m_4. Evidently, in the Szabo and Tanaka technique, it is known that these residues are to be subtracted from r_3 and r_4 to yield the new residues. The LUT T_3 reads the MRC digits corresponding to these residues and gives a_3 and a_4. The extension to six moduli set is straight forward as shown in Fig 2.3(b). An example will illustrate the procedure. Consider the moduli set {3, 5, 7, 11} and the residues {2, 3, 3, 0}.

	3	5	7	11	
	2	3	3	0	
Table Look-up		2	1	8	(Table T_1 outputs $a_2 = 2$)
r_3', r_4'			2	3	
Table Look-up			2	1	

Result X = 1.105 + 2.15 + 2.3 + 2 = 143.

The parallel look-up scheme for four moduli due to Chakraborthi et al [Chak86] is presented in Fig 2.3 (c). Note that this is another arrangement of the serial architecture so as to enhance the speed of conversion. The LUTs T_1 and T_2 handle r_1 and r_2, r_3 and r_4 respectively to yield MRC digits a_1,a_2,a_3', a_4' and a_3'' and a_4'. Note that instead of storing residues as in the serial technique, in this method MRC digits themselves are stored. The addition of MRC digits taking into account the carries in the modulo m_i adders yields the final MRC digits. As an illustration, consider again the same example as above of moduli set {3,5,7,11} and residues {2,3,3,0}. The procedure is as follows:

	3	5	7	11
Table Look-upT_1 {2,3,0,0}	2	2	6	2
Table Look-upT_2 {0,0,3,0}	0	0	3	9
	2	2	2	0
				$\Rightarrow$ 1
	2	2	2	1

The authors suggested some modifications of these techniques for which the reader is referred to their work. The hardware requirement for parallel look-up is larger than that of serial look-up, since the output entries in the first stage are large and requires large number of modulo additions with carry per modulus.

2.2.4. Barniecka and Jullien MRC technique

A somewhat similar technique has been suggested by Barniecka and Jullien earlier [Bar78], wherein first the given residues are used to compute the mixed radix digits following Szabo and Tanaka approach as shown in Fig 2.4(a). Next, using a bit slice approach, the MRC digits are used to look up in a ROM (see Fig 2.4 (b)) to give the weighted number corresponding to first LSB

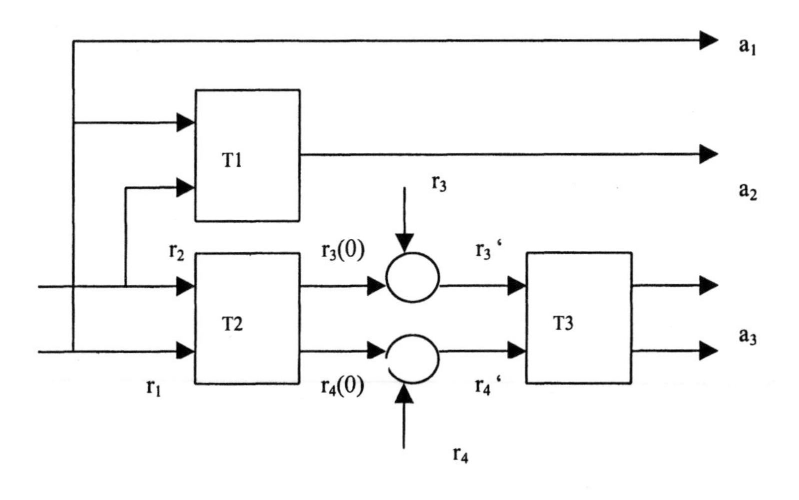

(a)

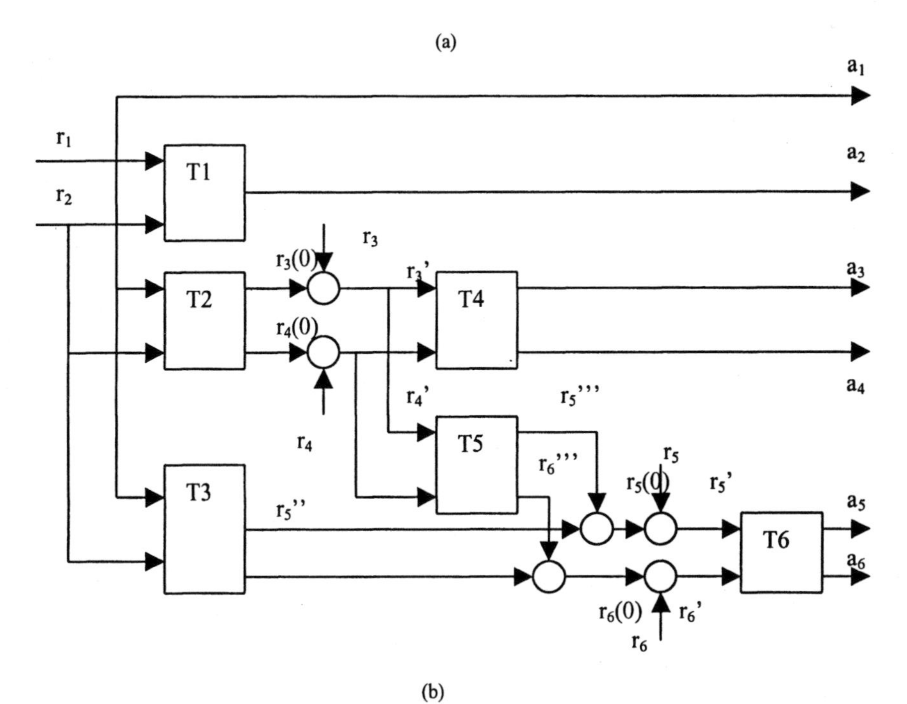

(b)

Fig 2.3 (contd)

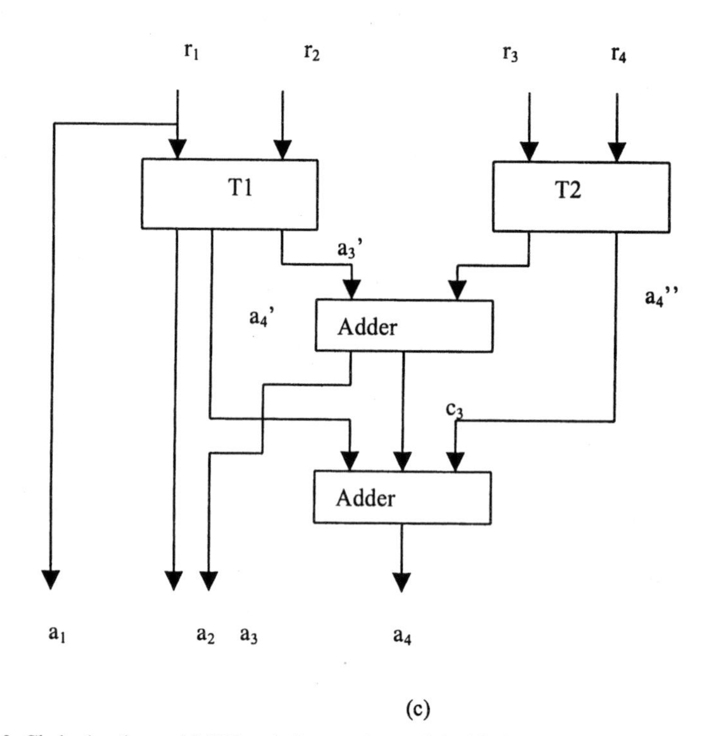

(c)

Fig 2.3. Chakraborthy et al MRC techniques using serial table look-up (a) for four moduli, (b) for six moduli and (c) using parallel table look-up for four moduli (adapted from [Chak86] 1986©IEEE)

r_{10}, next r_{11} etc. Then, these are accumulated with shifted value of the previous result to obtain X. An example is illustrated next.

Consider the moduli set {19,23,29,31}. Table 2.1 shows the partial number X corresponding to various bit values of the four mixed radix digits. Evidently, $P_1 = 19, P_2 = 19.23$, $P_3 = 19.23.29$, $P_0 = 1$ where we define X as

$$X = a_o + a_1.P_1 + a_2P_2 + a_3P_3 \tag{2.4}$$

and a_i is defined as $a_{i3}a_{i2}a_{i1}a_{io}$. The summation X is formed for this example with a_0=5, a_1=6, a_2=3, a_3=18 by first looking into memory, the word corresponding to MSBs and other bits next. The results are as follows:

$a_{34}a_{24}a_{14}a_{04}$	1	0	0	0	12673
$a_{33}a_{23}a_{13}a_{03}$	0	0	0	0	0
$a_{32}a_{22}a_{12}a_{02}$	0	0	1	1	20
$a_{31}a_{21}a_{11}a_{01}$	1	1	1	0	13129
$a_{30}a_{20}a_{10}a_{00}$	0	1	0	1	438

Then

$X = 2(2(2(2.12673)+0)+120)+13129)+438 = 229544$

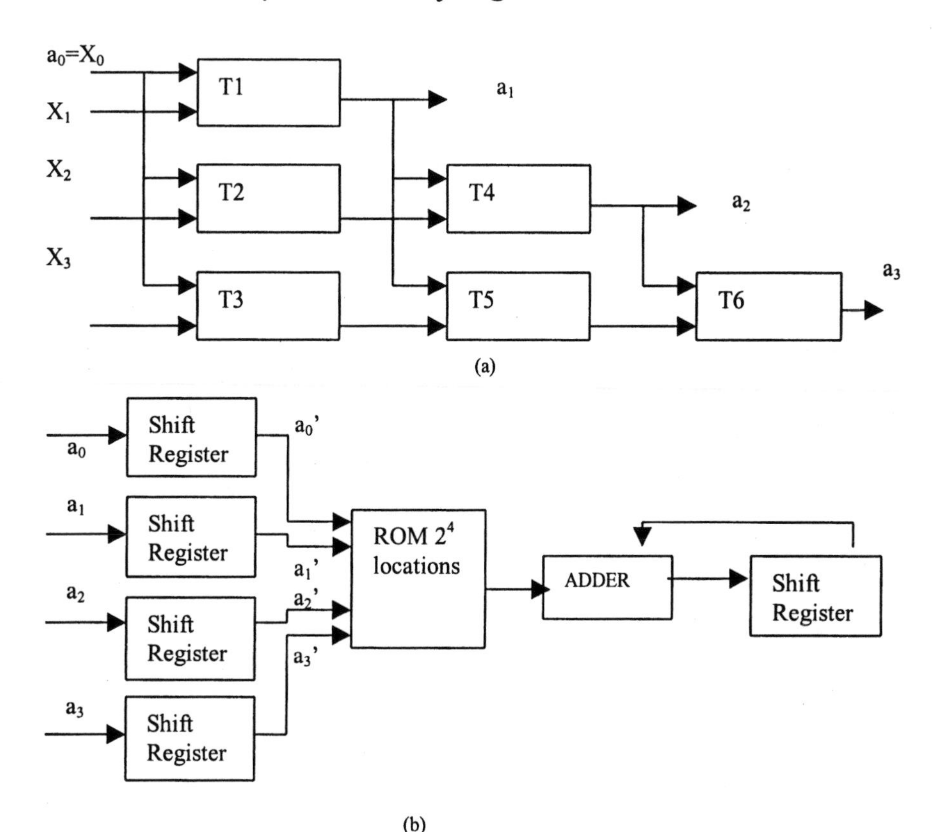

(b)

Fig 2.4. Barniecka and Jullien two-stage conversion Technique: (a) MRC digit calculation and (b) RNS to binary conversion using bit slice approach (adapted from [Bar78] 1978©IEEE)

An architecture for implementing this method is shown in Fig 2.4 (b). The a_i values enter serially and these bits address the ROM whose output is accumulated to the previous value shifted left by one bit using the shift register.

2.2.5. Soderstrand, Vernia and Chang technique of RNS to binary conversion

Soderstrand et al [Sode83] have described ROM based RNS to binary converters using MRC as well as CRT. These are considered briefly next. The MRC based architecture for a three moduli set is shown in Fig.2.5. Note that the first level of ROMs 1-3 perform the subtraction $(r_{13}-r_{11})$, $(r_{15}-r_{11})$, $(r_{16}-r_{11})$, and post multiplication by multiplicative inverse of m_1 with respect to m_2 , m_3 and m_4 respectively. Thus the second mixed radix digit is available. The next two PROMs viz., ROM4 and ROM5 perform the second level of MRC to obtain the third MRC digit. Next, the third level of ROMs 6 and 7 obtain the fourth MRC digit. Note that the architecture uses 256x8 bit memories. The last step is performing the calculations ($b_{11}+b_{13}.m_1$) in ROM8 and adding the result to that of ROM 6 and ROM7 which corresponds to ($b_{16}m_1m_2m_3+b_{15}m_1m_2$). The architecture has the advantage that depending on the desired precision of the result, the ROMs may be not equipped corresponding to LSBs. Evidently, this architecture has many levels. The use of CRT reduces the levels to one as will be shown later.

2.2.6. Yassine and Moore MRC technique

Yassine and Moore [Yass91] suggested simplification of Mixed Radix conversion technique for particular choice of moduli. Their approach starts with Szabo and Tanaka MRC technique, in which the desired binary number is expressed in terms of mixed radix digits γ_i as

$$X = W_1.\gamma_1+W_2\gamma_2+W_3\gamma_3 + \dots + W_n\gamma_n \quad (2.5)$$

in a n moduli RNS where $W_1=1, W_2=m_1, W_3=m_1.m_2$, ...and $W_n = m_1.m_2.m_3....m_{n-1}$. Yassine and Moore defined γ_i further as

$$\gamma_i=(V_iU_i)\bmod m_i \quad (2.6)$$

where

$V_1=1$

$V_2= (1/m_1) \bmod m_2$

$V_3=(1/(m_1m_2)) \bmod m_3$

....

$$V_n = (1/(m_1m_2...m_{n-1}))\bmod m_n \quad (2.7)$$

With the definition of γ_i and V_i in (2.6) and (2.7), we have

$$X = W_1.(U_1V_1) \bmod m_1 + W_2.(U_2V_2)\bmod m_2 + \dots + W_n(U_nV_n)\bmod m_n \quad (2.8)$$

By definition of residues r_i mod m_i, application to (2.8) yields

$r_1=U_1$

$r_2 = (U_1+U_2) \bmod m_2$

$r_3 = (U_1+W_2.(U_2V_2) \bmod m_2+U_3)\bmod m_3$

....

$$r_n = (U_1+W_2.(U_2V_2) \bmod m_2+ \dots + U_n)\bmod m_n \quad (2.9)$$

Iteratively, next U_i can be evaluated starting from the top equation in (2.9). Yassine and Moore observe that proper choice of V_i such that all are unity, leads to the fact that $\gamma_i = U_i$ and thus (2.6) can be used to yield γ_i. An example will be used to illustrate their technique.

a_{3j}	a_{2j}	a_{1j}	a_{0j}	Inner Summation
0	0	0	0	0
0	0	0	1	1
0	0	1	0	19
0	0	1	1	20
0	1	0	0	437
0	1	0	1	438
0	1	1	0	456
0	1	1	1	457
1	0	0	0	12673
1	0	0	1	12674
1	0	1	0	12692
1	0	1	1	12693
1	1	0	0	13110
1	1	0	1	13111
1	1	1	0	13123
1	1	1	1	13130

Table.2.1. Memory contents for stored output functions (adapted from [Bar78] 1978©IEEE)

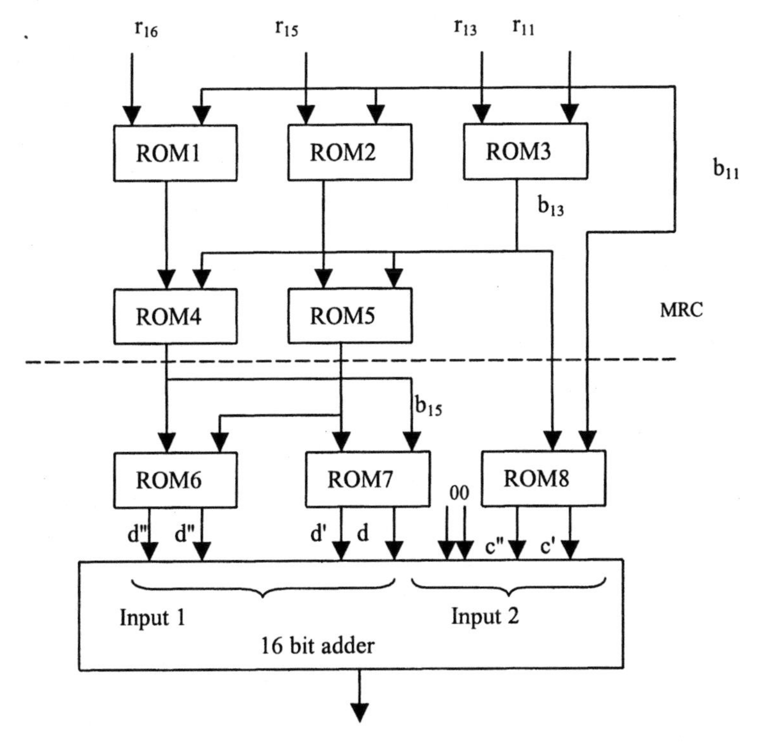

Fig 2.5. Soderstrand et al MRC architecture using ROMs (adapted from [Sode83] 1983©IEEE)

Consider the moduli set {127, 63, 50, 13} for which one can verify that $V_i = 1$ for all i. Let the residues be (78, 41, 47, 7) for which we need to evaluate X. We have by (2.9),

$\gamma_1 = U_1 = 78$

$\gamma_2 = |41-78|_{63} = 26$

$\gamma_3 = |47-78-127.26|_{50} = 17$

$\gamma_4 = |7-78-127.26-127.63.17|_{13} = 9.$

The desired X can then be found as

X=78+26.127+17.127.63+ 9.127.63.50 = 3739847

The advantage of this technique is that the subtractions are n(n-1)/2 as in the Szabo-Tanaka MRC technique. Only (n-2) modulo multiplications are needed.

2.2.7. Wang and Abd-el-barr MRC implementation

Wang and Abd-El-Barr [Wan96] have described a MRC algorithm which is presented in Fig 2.6. Note that pairs of moduli are considered parallelly in the first level to generate the binary numbers in the ring $(0, m_i.m_{i-1} - 1)$. The second level takes the outputs of the first level in pairs and generates the binary numbers in the rings $(0, m_i.m_{i-1}.m_{i-2}.m_{i-3} - 1)$. Thus, if there are n moduli, $\log_2 n$ levels of conversion are needed. Each level can be implemented by a look-up table. However, the sizes of the LUTs progressively increase and become extremely large in the last stages. For small moduli values, and fewer moduli, the method may be attractive.

Kumaresan and Shenoy [2.8] have described an interesting method of RNS to binary conversion based on MRC, which is considered in a later Chapter since this method needs the application of the base extension technique, which will be discussed therein.

2.2.8. Miller and McCormick MRC implementation

Miller and McCormick [Mill98] have described an interesting MRC technique. This technique has the advantage that the need for common wiring bus in each level for subtracting the residue from all other residues is avoided e.g.(r_2-r_1), (r_3-r_1), (r_4-r_1) in a four moduli system in the first level. Herein, only local interconnections are employed. It is interesting to note that in the MRC technique, linear Diophantine equations are solved. Consider a two- moduli RNS for which MRC is sketched in Fig 2.7(a). Note that the mixed radix digit k_1 is related to the residues r_1 and r_2 corresponding to moduli m_1, m_2 as follows:

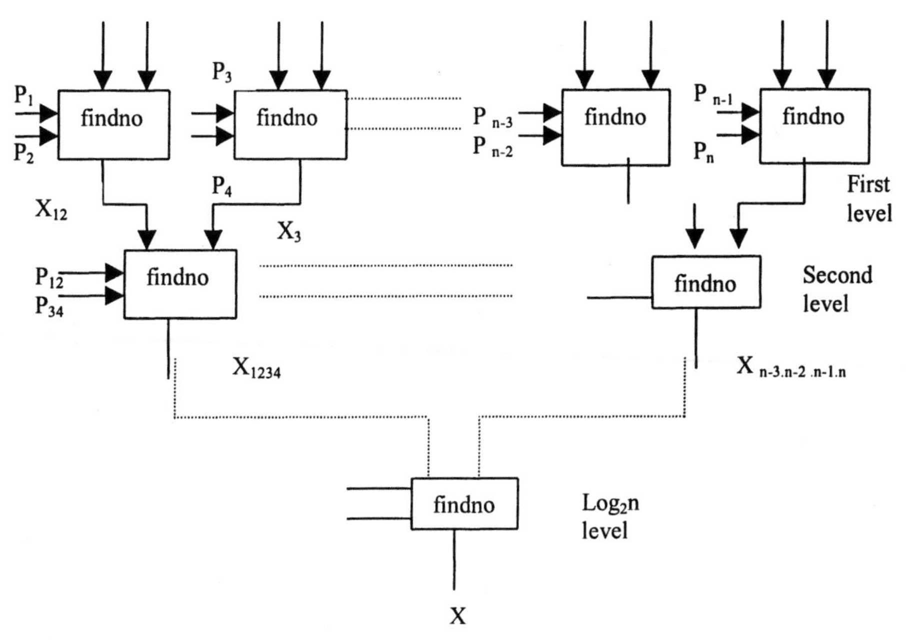

Fig 2.6. Wang and Abd-el-Barr MRC Architecture
(adapted from [Wan96] 1996©IEEE)

$X = k_1m_1+r_1 = l_1m_2+r_2$ (2.10)

However, l_1 is usually ignored and not calculated. Since X satisfies the residues with respect to m_1, m_2, X is the solution i.e. X mod $m_1 = r_1$ and X mod $m_2 = r_2$. In other words, k_1 shall be calculated by solving (2.10) by taking residue both sides with respect to m_2. Interestingly, Miller and McCormick observe that by solving for both k_i and l_i for adjacent moduli, MRC can be performed using only local interconnects.

We will not present the proof here, which can be found derived by induction in [Mill98]. The procedure is illustrated for a four moduli RNS {3, 5, 7, 11} and residues {2, 2, 5, 9} in Fig 2.7 (b). The top level uses adjacent moduli and generates digits k_1 and l_1 such that $r_1-r_2 = l_1m_2-k_1m_1$. As an illustration, for the left most center block, we have

$2-5 = 1.7 - 2.5 = -3.$

Thus, the output of each block (i,j) yields a number in the residue ring (0, $m_i.m_{i-1}$). The numbers indicated in the box i, i+1 are moduli m_i and m_{i+1} dealt by that block. The left most top block yields the equality of the right and left paths:

$7 = 2.3+1 = 1.5+2.$

The same can be verified for other blocks also. The next level adds m_1m_2 and m_2m_3 to the left and right paths. Here, the moduli considered are 3 and 7 since 5 happens to be common to both the blocks. It can be seen that the mixed radix digits are available as shown both catering for expansions viz.,

$$X = r_1 + a_1m_1 + a_2m_1m_2 + a_3m_1m_2m_3 \quad (2.11a)$$

and

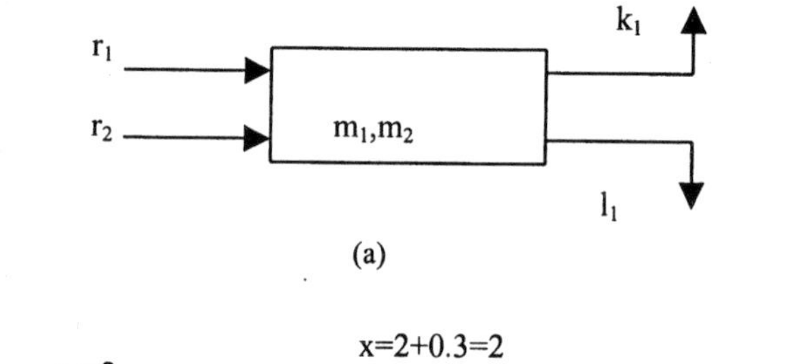

(a)

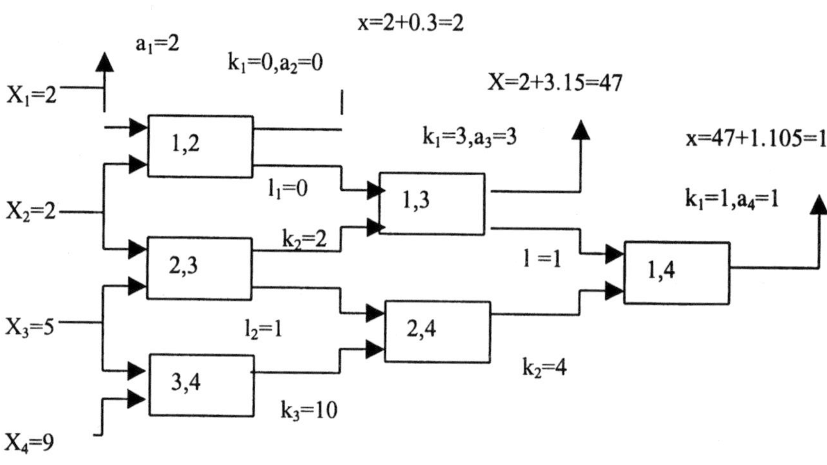

(b)

Fig 2.7. Miller and McCormick MRC architecture: (a) Two moduli case and (b) Four moduli case (adapted from [Mill98] 1998©IEEE)

$$X = r_4 + b_1m_4 + b_2m_4m_3 + m_3m_2m_3m_4 \quad (2.11b)$$

Even though, the structure is regular, the memory requirements are double that of MRC technique. Note also that the remaining paths available lead to other expansions for X.

2.2.9. Taylor and Ramnarayan RNS to binary conversion methods

An interesting technique has been described by Taylor and Ramnarayan [Tay81b], shown in Fig 2.8 (a), wherein three circulating shift registers of lengths $m_i.\log_2 m_i$ bits are employed. These contain the residues ranging from 0, 1, 2, ...(m_i-1) sequentially. It is evident that only m_i locations each of length $\log_2 m_i$ bits are needed. These shift registers are clocked by a regular clock of different frequencies so that the one corresponding to m_i is given $\log_2 m_i$ pulses and so on. The contents of the shift registers are compared with the given residues. When the comparison is sensed, the clock is stopped. The counter driven by the clock contains the binary decoded word corresponding to the given residue set. The method is extremely slow since $m_1.m_2...m_j.\lceil \log_2 m_i \rceil$ cycles are needed to complete the conversion in the worst case .
Ananda Mohan and Poornaiah [Anan91] suggested the use of modulo m_i counters (see Fig 2.8 (b)) in stead of shift registers which architecture consumes less area and needs only $m_1.m_2.....m_3$ clock cycles. The speed can be doubled by having two parallel hardware blocks which sense the matching of $\{r_1,r_2,....r_n\}$ as well as $\{m_1-r_1,m_2-r_2,.....m_n-r_n\}$. In other words one hardware block proceeds from comparison of numbers from 0 upwards, whereas the other proceeds from $(m_1.m_2....m_n - 1)$ downwards. Thus the speed of conversion can be doubled, since whichever hardware block senses comparison, stops the counters (see Fig 2.8(c)).

Anandamohan and Poornaiah suggest yet another improvement where MRC is indirectly employed. Herein, (m_1-m_2) mod m_2, (m_1-m_3) mod m_3 are added respectively in the first step to r_2, r_3 ...r_n till comparison of residue mod m_2 is achieved. The operation is repeated, however, with (m_3-m_2) mod m_3 being added to the current residue in the next column (m_3) till comparison is achieved with r_3. Evidently, the counters counting the number of times the constants are being added, contain the Mixed Radix Digits. The advantage of this method is that only m_2-1 + m_3-1 + ...m_n-1 cycles are needed at most. These methods may be attractive for RNS with small moduli.

Taylor and Ramnarayan [Tay81b] suggested two techniques for the moduli set $\{2^n-1, 2^n, 2^n+1\}$. The first shown in Fig 2.9 is based on MRC. The values (r_1-r_2) mod m_1 and (r_3-r_2) mod m_3 are used to address a ROM to obtain the value mod $(m_1.m_3)$. Then, the addition of r_2 is performed to yield the final result. They did not recognize that the adder is not needed since r_2 is LSB of the word to be determined and is already available as data. The ROM contents are 2n bit MSBs of the result. An example will be illustrative. Consider the moduli set {7, 8, 9} and given the residues {1, 2, 3}. Evidently, $(r_1\text{-}r_2) \bmod m_1 = 6$ and $(r_3\text{-}r_2) \bmod m_3 = 1$. The PROM contains 62 corresponding to these values which when used as MSB 6 bit word and appended with r_2 viz., 2 as LSBs yields (62.8+2) =498 which is the desired result.

Yet another technique due to Taylor and Ramnarayanan [Tay81b] uses a sub-cover decomposition. This also is indirectly a MRC technique and uses PROMs. This technique is presented in Fig 2.10. Here, J_1-J_3 denoted as c is computed where J_1 and J_3 are shown, same as in the previous technique. Then a ROM is looked into to obtain corresponding to c, a value Q(c) where $Q(c)= m_1m_2I_1 = S_1$. Then $x' = m_2J_1+S_1$ is computed and added to x_2 to yield the result. Note that in other words, X is realized as

$$X = r_2 + m_2.J_1 + m_1.m_2.I_1.$$

The example considered above is again used to illustrate the technique. For this case $J_1 = 6$ and $J_3=1$ and c=5. The PROM contains 15 locations corresponding to c=-8 to +6. In this case for c=5, $I_1 = 8$ corresponds to the result $498 = x_2 + m_2.J_1 + m_1.m_2.I_1 = 2+8.6+ 8.56$.

It can be appreciated that J_1 and I_1 are the mixed radix digits. In other words, $I_1 = -(J_3\text{-}J_1).2^{n-1} \bmod (2^n+1)$. The method can be explained also as follows: Once J_1 and J_3 are determined, the residues become $\{J_1, 0, J_3\}$. Then X* corresponding to $\{J_1, 0, J_3\}$ can be seen to be written in two ways as

$$X^* = (m_1.I_1+ J_1) \bmod m_2 \quad \text{or} \quad X^* = (m_3.I_3+J_3) \bmod m_2$$

Equating the two values of X*, we have

$$J_1\text{-}J_3 = m_3I_3 - m_1I_1 \qquad (2.12)$$

Thus, I_3 and I_1 are determined a priori and stored in PROMs corresponding to all possible J_1 and J_3 values. Note that $X = X^*.m_2 + r_2$.

2.3. CRT BASED CONVERSION TECHNIQUES

2.3.1. The basic CRT technique.

In this method, the desired binary number X corresponding to the residues $\{r_1, r_2, \ldots r_n\}$ in the moduli set $\{m_1, m_2, \ldots, m_n\}$ is determined using the formula

$$X=[\Sigma_{i=1}^{n} r_i.(M/m_i).(1/(M/m_i)) \bmod m_i] \bmod M \qquad (2.13)$$

where $M = m_1.m_2.\ldots.m_n$. Evidently, there are n terms which need to be added whose sum may exceed M and hence modulo reduction mod M needs to be done. The terms other than r_i are constants and can be stored in PROMs which can be read and multiplied by respective residues r_i and accumulated mod M or alternatively the product itself can be stored in PROM. An example will be illustrative.

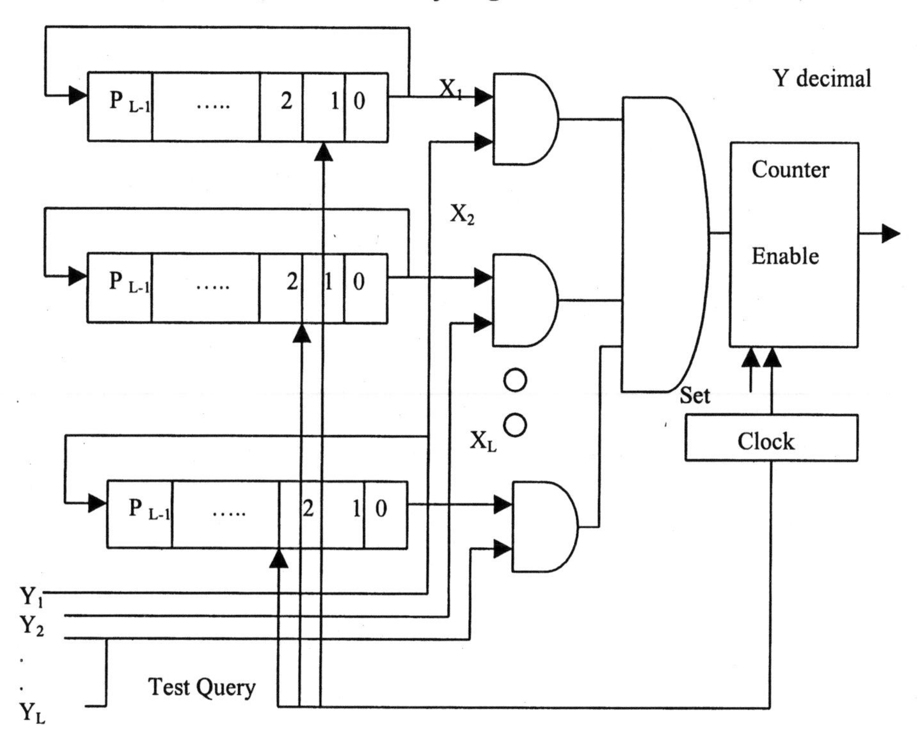

(a)

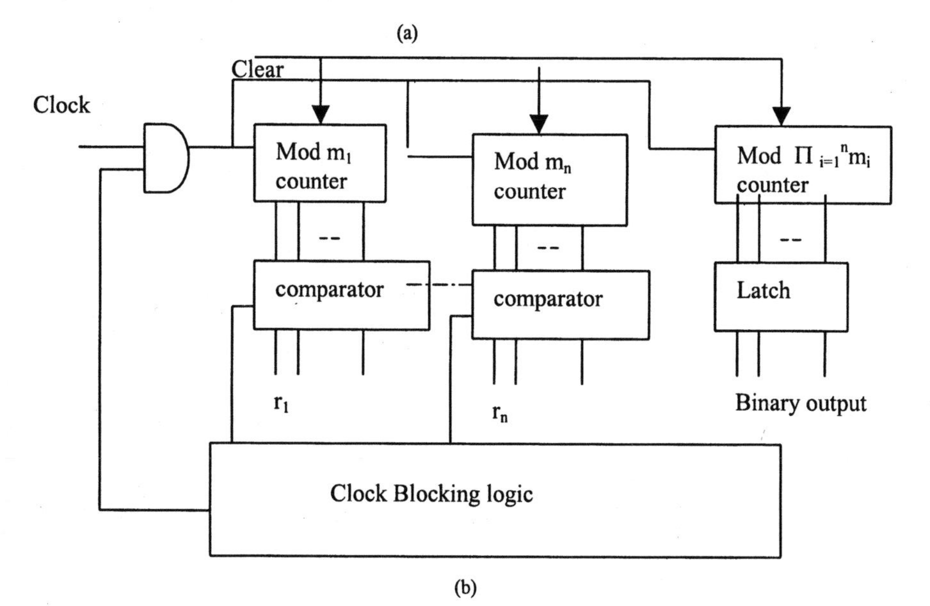

(b)

Fig 2.8 (contd)

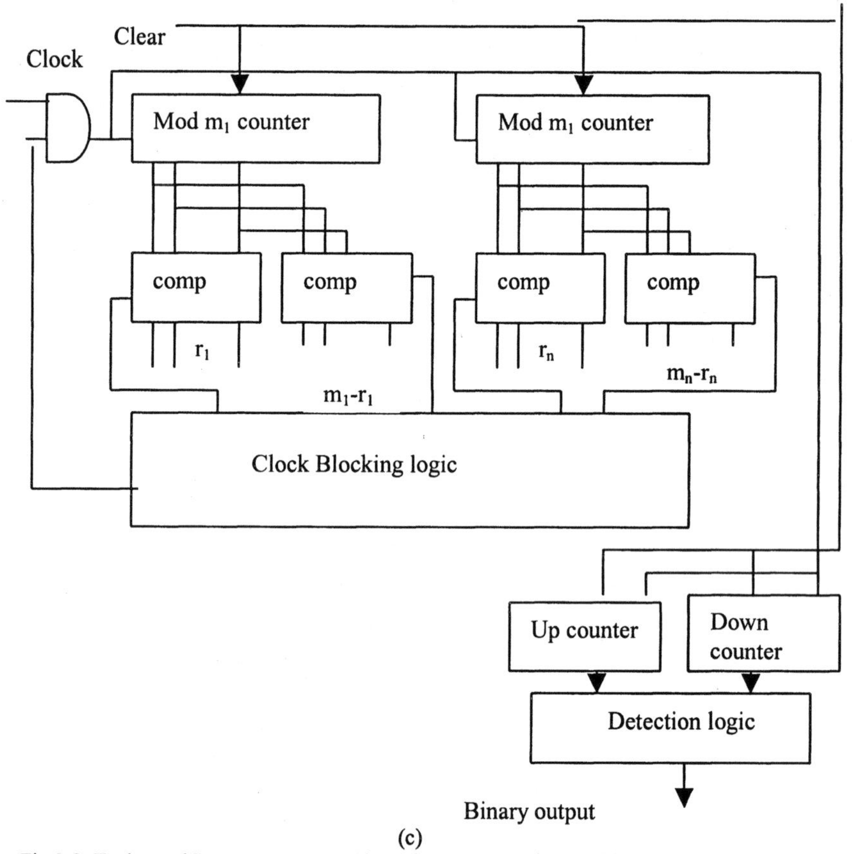

(c)

Fig 2.8. Taylor and Ramnarayanan RNS to binary Conversion Architecture (a), Ananda Mohan and Poornaiah's modified implementations (b) and (c) ((a) adapted from [Tay81b] 1981©IEEE and (b)and (c) adapted from [Anan91] 1991©IEEE)

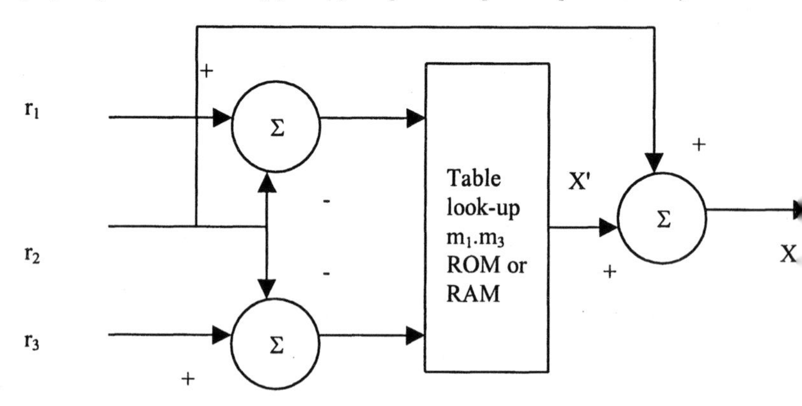

Fig 2.9. Taylor and Ramnarayanan architecture for RNS to binary conversion for the moduli set $\{2^n-1,2^n,2^n+1\}$.(adapted from [Tay81b] 1981©IEEE)

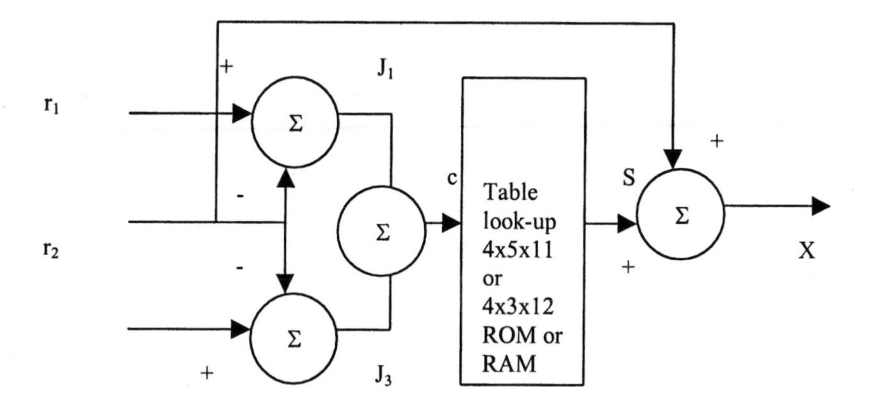

Fig 2.10. Taylor and Ramnarayanan architecture using sub-cover decomposition. (adapted from [Tay81b] 1981©IEEE)

Consider the modulo set {3, 5, 7, 11}. Evidently M=1155, M/m_1 = 385, M/m_2 =231, M/m_3=165, M/m_4=105.$(1/(M/m_1))$ modm_1=1, $(1/(M/m_2))$ modm_2=1, $(1/(M/m_3))$ mod m_3=2, $(1/(M/m_4))$ modm_4 = 2. Thus the coefficients in the expansion (2.13) become 385, 231, 330, 210 respectively. It can be easily interpreted that these numbers correspond to the residue sets {1, 0, 0, 0}, {0, 1, 0, 0}, {0, 0, 1, 0} and {0, 0, 0, 1} respectively. Thus the given residue set {2, 1, 5, 9} can be evaluated as {2, 1, 5, 9} ={2, 0, 0, 0} + {0, 1, 0, 0} + {0, 0, 5, 0} + {0, 0, 0, 9}. Each of these terms can be obtained by multiplying the coefficients derived above by 2, 1, 5 and 9 respectively. Thus, we have

$X = \{385.2 + 231.1 + 330.5 + 210.9\} \bmod 1155 = 1076.$

For this example it may be noted that the maximum of unreduced X could be {385.2+ 231.4+ 330.6+ 210.10} =5774 i.e.> 4.1155. Thus, in general, n.M atmost shall be subtracted for the general case. The additions can be done sequentially or parallelly in a tree of modulo adders in $\log_2 n$ steps. The reduction mod M can be done in the end or in the intermediate stags. The weighting can be done using ROMs or using hardware multipliers. Thus, a variety of structures have been suggested for implementation of CRT which will be considered next.

2.3.2. Soderstrand, Vernia and Chang technique

The basic CRT based architecture is shown in Fig 2.11 for which, a modification has been suggested by Soderstrand et al [Sode83]. They recommend computing a scaled version of the desired result:

$$X/M=\Sigma_{i=1}^{n}[(M/m_i).((1/(M/m_i)) \bmod m_i).r_i) \bmod m_i]/M \qquad (2.14)$$

Evidently, X/M can be greater than unity and hence only the fractional part is taken and integer value dropped at the output of the final adder. Further, it is possible to restrict each term in (2.14) to fewer bits than those needed to represent M which will lead however to reduced accuracy in the converted value of X. Soderstrand et al [Sode83] have conducted an exhaustive computer aided study for a particular moduli set { 16, 15, 13, 11}. They observe that for adders more than 11bit wide, the results are accurate. Reduction below 10 bits however reduced the dynamic range. Interestingly the sign bit of the result was found to be correct for over 75% of input values even if just two bits are considered.

2.3.3. Vu RNS to binary conversion technique

Van Vu [Vu85] has observed that the empirical results of Soderstrand et al [Sode83] are not correct. He has derived the number of bits needed to represent accurately the fractional form so that sign detection can be accurately accomplished. Assuming that the error e in the fractional representation of all the terms viz., $(2/M).((1/M_i) \bmod m_i).r_i) \bmod m_i$ is uniform and that there are n moduli, the total error is n.e. For a t bit representation of all the above values, the total error is therefore $< n.2^{-t}$. The resolution of the converter is 2/M since the total range {0, 2} is divided into M equal intervals. In case of M even and M odd, there is a little difference in that the gap between largest positive integer and 1 is 2/M in the M even case and 1/M in the M odd case. Accordingly, we have $n.2^{-t} \le 2/M$ for M even and $n.2^{-t} \le 1/M$ for M odd. For the example of Soderstrand et al [2.5], M =34320 and hence t = 17 bits is required for each of the components in (2.14).

Van Vu has presented an example to illustrate the theory above. For the moduli set {11, 13, 15, 16} corresponding to X_1=1 and X_2= -1, we have X_1 = {1, 1, 1, 1} and X_2 = {10, 12, 14, 15}. Then,

$\Sigma_{i=1}^{4} u_i$ = (2/11)(8x1) 11 +(2/13)(1x1)13+(2/15(2x1)15+(2/16)(1x1)16

and

$\Sigma_{i=1}^{4} u_i$= (2/11)(8x10)11+(2/13)(1x12)13+(2/15).(14x2)15+(2/16)(15x1)16

where 8, 1, 2, 1 are the multiplicative inverses $(1/(M/m_i)) \bmod m_i$.

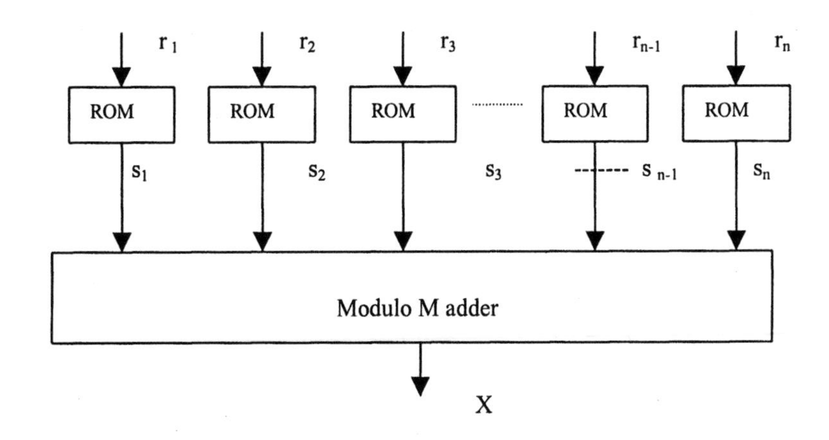

Fig 2.11. The basic CRT implementation.

The result in both the cases is

$X_s' = 0.0000\ 0000\ 0000\ 0101$

$X_s' = 1.1111\ 1111\ 11111101$

showing that the sign is positive in the former case and negative in the latter case. If we consider the 11 bit precision recommended by Soderstrand et al [Sode83], we have $X_s' = 0.0000\ 0000\ 0000\ 0000$, thus leading to a wrong sign.

2.3.4. Elleithy and Bayoumi architectures for CRT implementation

Elleithy and Bayoumi [Ell92] have considered the implementation of CRT in a fast and flexible manner. They focus on the addition problem of the n vectors in the metric expansion of CRT, which also needs modulo reduction. The carry-save-approach is used herein. The basic cell shown in Fig 2.12 (a) adds four inputs and generates two outputs. Such cells can be connected in a carry-save adder as shown in Fig 2.12 (b) where six inputs are being added. The reader is referred to Hwang [2.14] for an excellent discussion on the design of carry-save-adders. The outputs of the cell when summed may exceed M. However, the reduction mod M will be performed only at the end.

The four-input modulo M adder is shown in Fig 2.12 (a). Note that A is represented by two vectors A_S, A_C and B is represented by two vectors B_S, B_C. Herein, the three inputs are added in each CSA to yield two outputs and a carry. To the two outputs, the fourth remaining input is added which may yield another carry and two output carry vectors. If both the carries are 1, then

$2(2^n\text{-}M)$ is added which may result in another carry. If the carry is one, then $(2^n\text{-}M)$ is added to complete the addition. The output finally is available as two vectors. Elleithy and Bayoumi [Ell92] observe that five steps are needed at most to reduce the result. However, we observe that an additional step may be
needed. As an illustration, A=B=C=D=15 say with M of 9 is considered. We have the following:
A+B+C = 16+15+14 (16 is due to Carry)
15+14+D = 16+14+14 (D is added to result of step 1.)
14+14+14=16+14+12 (2.(16-9) is added since two carries have been generated in previous steps; again carry is generated.)
14+12+7 = 16+5+12 ((16-9) is added; Again, carry is generated.)
5+12+7= 0+14+10 (16-9) is added. Result.
It is evident that one more step is needed for final reduction mod 9.
A modulo adder tree may be used arranged in a $\log_2 n$ level architecture as shown in Fig 2.12 (c) to add all the inputs each with two inputs A and B each written as two sub-vectors as explained before. Elleithy and Bayoumi also suggested the use of a range determinator which reduces modulo M the sum of the carry-save-adders adding all partial metric vectors. The essential principle is to subtract a multiple of M to get the correct result. This follows the concept outlined for two input modulo adder. However, this technique appears not to be comprehensive and correct and hence omitted here.

2.3.5. Meehan et al RNS to binary conversion technique

Meehan et al [Mee90] have considered a variation of CRT so as to suit their RNS implementation. Herein, the processing is done on scaled residues throughout the signal processing. This means that the residues are multiplied by 2^s and fed as input to the signal processor. As will be shown later, scaled residues i.e. corresponding to $2^s.X$ are generated from the given binary input number using an unique algorithm. The final conversion to binary form shall effectively evaluate $2^{-s}.(X.2^s)$. Thus, the technique is applicable only in specialized cases where both I/O conversions are taking into account scaling as an implicit operation.

In this technique we show that scaling residues as well as weights in the CRT expansion yields the advantage of reduction of the final metric sum mod M easily.
The CRT actual expansion is
$X=\Sigma y_i.x_i=\Sigma(M/m_i).(w_i.x_i)\text{mod } m_i$ (2.15a)
where $w_i = (1/(M/m_i)) \bmod m_i$. Assuming that x_i is scaled to $x_i' = (x_i.c_i) \bmod m_i$ with $c_i=2^s$, w_i needs to be scaled as $w_i.c^{-1}$ where $c^{-1}=2^{-s}$ thus yielding

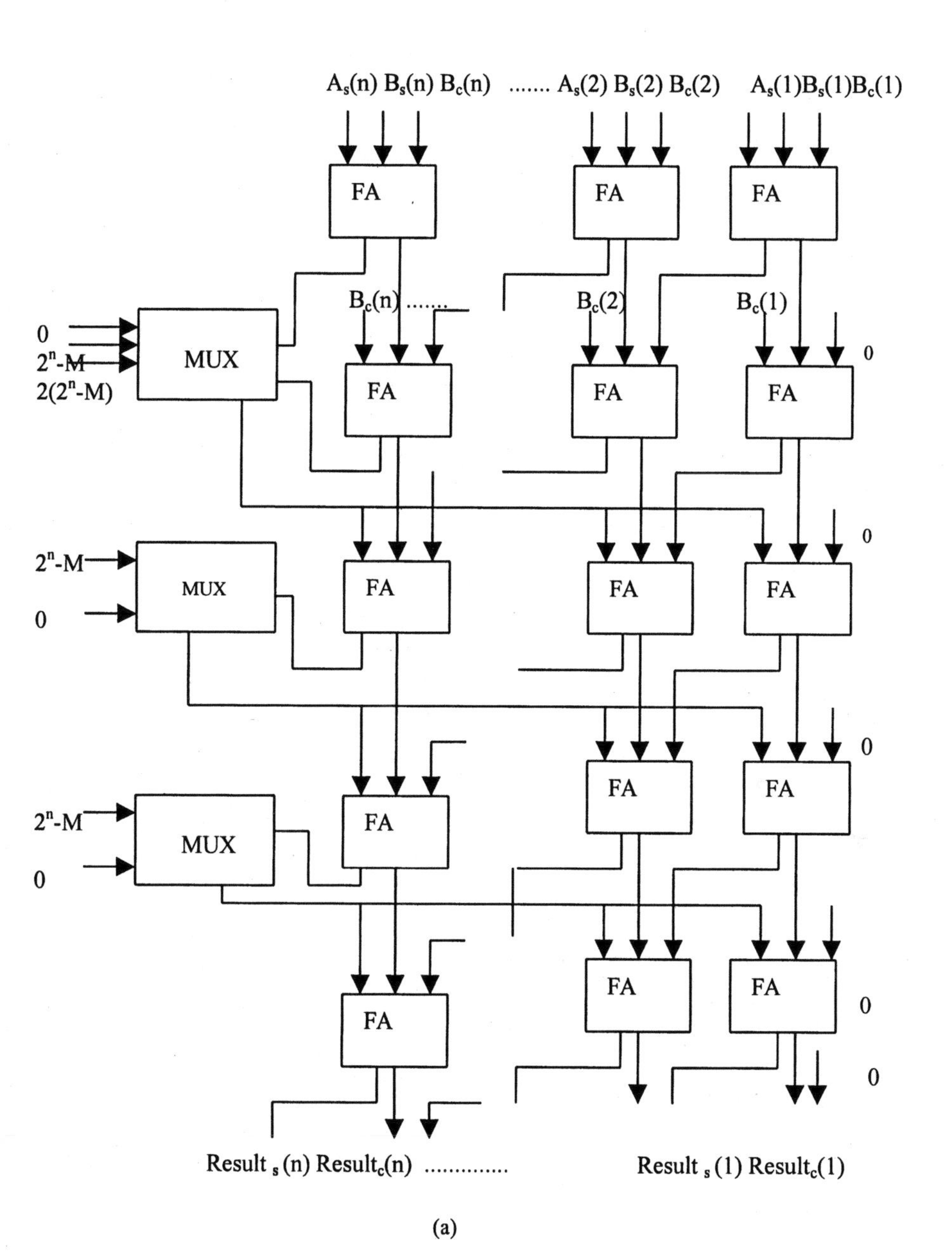

(a)

Fig 2.12 (contd)

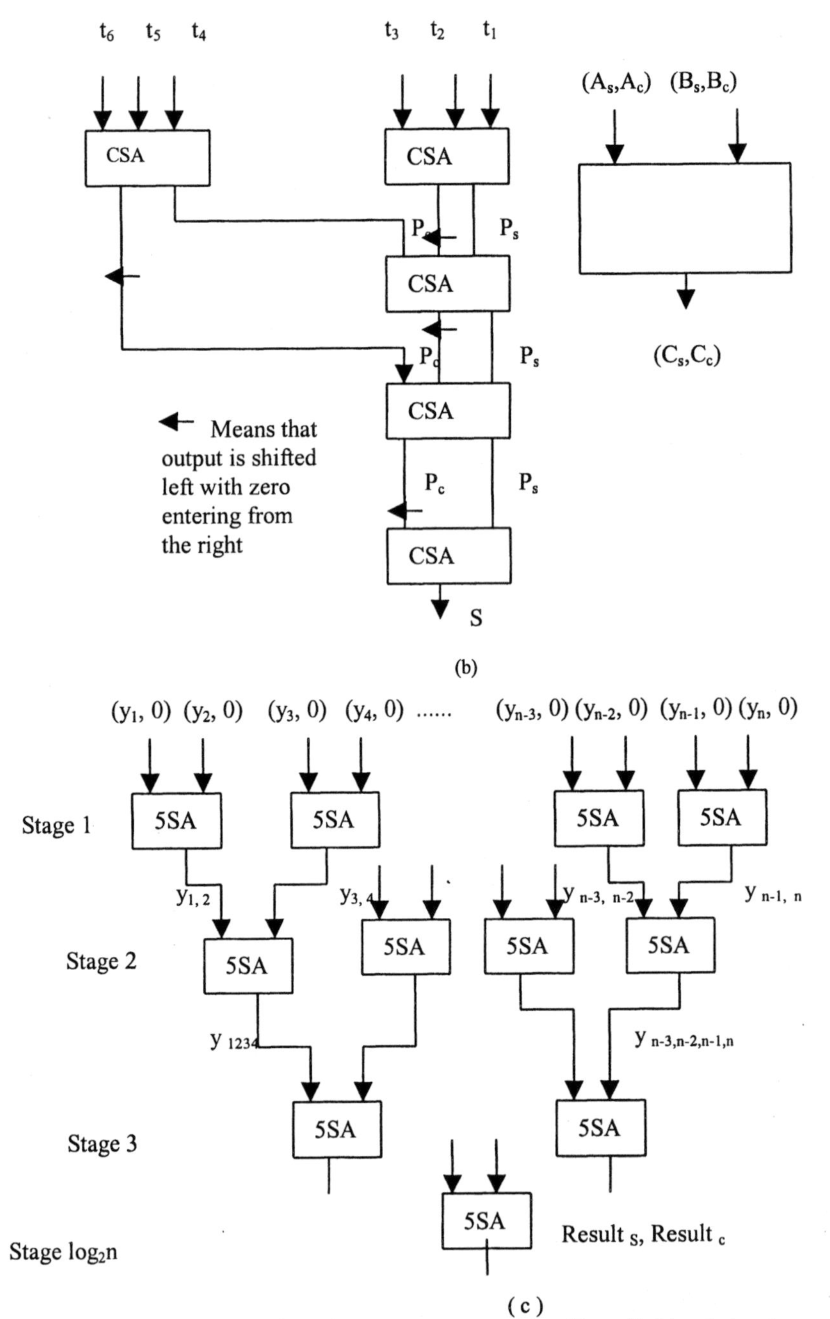

Fig 2.12 (a) Elleithy and Bayoumi four input two output modulo adder cell, (b) a six input carry-save adder and (c) a $\log_2 n$ level modulo adder tree (adapted from [Ell92] 1992©IEEE)

$X_s' = \Sigma x_i'.y_i' = \Sigma(M/m_i).x_i'.w_i'$ (2.15b)

Next X is multiplied by (2/M) so that X.(2/M) is between 0 and 2:

$X.(2/M) = \Sigma(2/m_i).\ x_i'.w_i'$

Thus, looking at the MSB one can determine the sign. This is illustrated by an example.

Assuming a scale factor of 2^5 for a moduli set {3,5,7} and residues corresponding to 92, the results are presented below. Choosing a 11 bit precision for the fractional part, exact results can be obtained.

$x = 92 = \{2,2,1\}$, $c = 2^5 = 32 = \{2,2,4\}$, $x' = x.c = 92.32 = \{1,4,4\}$

$w = \{2,1,1\}$

$c^{-1} = 2^{-5} = \{2,3,2\}$

$w_i' = \{1,3,2\}$

$x_i' = \{1,4,4\}$

$y_1' = \lceil (2/3).1.2^{11} \rceil .2^{-11} = 0.10101010110$

$y_2' = \lceil (2/5).3.2^{11}/2 \rceil 2^{-11} = 1.00110011010$

$y_3' = \lceil (2/7).2.2^{11}/2 \rceil 2^{-11} = 0.10010010011$

The ceiling brackets indicate that least integer greater than or equal to the enclosed quantity is used.

$y_1'x_1' = 0.10101010110$

$y_2'x_2' = 100.11001101000$

$y_3'x_3' = 10.01001001100$

$X_s' = xx1.11000001010 = 1.7548828\ldots$

$X = \lfloor (M/2).X_s' \rfloor = \lfloor 92.13 \rfloor = 92.$

It can be seen that the number of bits (s+1) needed to do the computation are given by $s \geq \lceil \log_2 (M\ \Sigma_{i=1}^{n} (m_i - 1) \rceil - 1$.

2.3.6. Wang's RNS to binary converters

Wang [Wan00] has described another technique known as New CRT1 which is briefly considered next. Given the residues $(x_1, x_2, x_3, \ldots x_n)$ corresponding to moduli $(P_1, P_2, \ldots P_n)$ the decoded binary number has been shown to be

$X = \{x_1 + (x_2 - x_1).k_1P_1 + (x_3 - x_2).k_2.P_1.P_2 + \ldots\ldots + (x_n - x_{n-1}).k_{n-1}.P_1.P_2\ldots.P_{n-1}\}$ mod $(P_1.P_2\ldots..P_{n-1})$ (2.16a)

where $k_1.P_1 = 1 \bmod (P_2\ldots.P_{n-1})$, $k_2.P_1.P_2 = 1 \bmod (P_3.P_4.P_5\ldots.P_{n-1})$, etc.

The proof for this follows by verifying that

$X \bmod P_1 = x_1 \quad X \bmod P_2 = x_2$

and so on. Next, Wang rewrites X next by combining $x_1, x_2, \ldots x_n$ terms as

$X = [\ x_1.\{(1-k_1.P_1)\} + x_2.\{(k_1 - k_2.P_2).P_1\} + \ldots + \{(k_{n-2} - k_{n-1}.P_{n-1}).P_1.P_2\ldots P_{n-2}\}.x_{n-1} + k_{n-1}.x_n.P_1.P_2\ldots P_{n-1}]$ mod $(P_1.P_2\ldots..P_{n-1}.P_n)$ (2.16b)

The coefficients in { } bracketed terms in (2.16b) are denoted as $a_0, a_1, a_2, \ldots, a_{n-1}$ i.e. $a_0 = (1-k_1.P_1)$, $a_1=(k_1-k_2.P_2)$ and so on. Next, Wang suggests writing these a_i s in Mixed Radix digit form viz.,

$$a_0 = a_{0,0}+a_{0,1}.P_1+a_{0,2}.P_1.P_2 + \ldots\ldots + a_{0,n-1}.P_1.P_2\ldots P_{n-1} \quad (2.17a)$$

$$a_1.P_1 = a_{1,1}.P_1+a_{1,2}.P_1.P_2 + \ldots\ldots + a_{1,n-1}.P_1.P_2\ldots P_{n-1} \quad (2.17b)$$

$$a_2.P_1.P_2 = a_{2,2}.P_1.P_2+a_{2,3}.P_1.P_2.P_3\ldots + a_{2,n-1}.P_1.P_2\ldots P_{n-1} \quad (2.17c)$$

....

$$a_{n-1}.P_1.P_2.P_3\ldots P_{n-1} = a_{n-1,n-1}.P_1.P_2\ldots P_{n-1} \quad (2.17d)$$

Next, using (2.17) in (2.16), X can be written in the form

$$X=[B_0 + B_1.P_1 + B_2.P_1.P_2 + . . + B_{n-1}.P_1.P_2.P_3\ldots P_{n-1}] \bmod (P_1.P_2\ldots.P_{n-1}.P_n) \quad (2.18)$$

B_i are termed as the first-order Mixed radix digits by Wang. It is evident that the various Mixed Radix Digits $a_{i,j}$ weighted by x_i and added together yield the B_i values. Note, however, that Wang suggests an architecture comprising of two stages. The first stage shown in Fig 2.13 (a) evaluates B_i based on stored $a_{i,j}$ and input x_i values. Then the second stage performs implementation of (2.17) using the architecture of Fig 2.13 (b).

It is interesting to note that [Anan00c] the B_i values turn out to be same as in the original CRT expansion given in $X = [\,\Sigma_{i=1}^{n} s_i.(x_i//s_i) \bmod P_i\,] \bmod M$ where $s_i = M/P_i$ and $1//s_i$ stands for multiplicative inverse of s_i with respect to mod P_i. Thus, even though Wang started the derivation based on his new CRT1 given by (2.16a) above, he finally obtained the well-known CRT expansion given in (2.18) above. Wang suggests calculating B_i using the MRC representation of coefficients of x_i which is precisely the technique described by Huang [Hua83]. Huang's derivation is based on the fact that X is the sum of the decoded values corresponding to $\{x_1, 0, 0, \ldots, 0\}$, $\{0, x_2, 0, \ldots, 0\}$, …. , $\{0, 0, \ldots, 0, x_n\}$ reduced mod M, which is the essence of CRT.

We also wish to state that the result in (2.18) can be much larger than M and hence in the architecture of Fig 2.13, the subtraction of M is not sufficient and a large multiple of M may need to be subtracted as in the well-known original CRT implementations. As an illustration, for the example moduli set {3, 5, 7, 11} considered by Wang, considering the residue set {2, 4, 6, 10}, the evaluation of

$$X = [B_0+B_1.P_1+B_2.P_1.P_2+B_3.P_1.P_2.P_3] \bmod (P_1.P_2.P_3.P_4)$$
$$= [x_1+(3x_1+2x_2).P_1 + (4x_1+x_2+x_3).P_1.P_2$$
$$+(3x_1+2x_2+3x_3+2x_4).P_1.P_2.P_3] \bmod (P_1.P_2.P_3.P_4)$$

yields $B_0 + P_1.B_1 = 44$ and $P_1.P_2.(B_2 + P_3.B_3) = 5730$ which when summed yields 5774. Thus, four times the dynamic range M (=1155) needs to be subtracted to reduce the result mod M. As against this approach, Huang [Hua83] recommends the use of modulo P_i addition of the Mixed Radix digits with carry (which can be larger than unity as well), to obtain the final MRC form. Then, Huang evaluates the result in the conventional manner, using $\log_2 n$ level binary adder tree (see Fig 2.2 (a) and (b)). Since addition of MRC digits needs small length adders, only the last MRC digit to binary conversion needs large adders. In essence, the approach of Wang is same as that of Huang in principle.

2.3.7. Cardiralli et al Scaled CRT implementation

Cardiralli et al [Card00] have described an alternative implementation of scaled CRT based residue to binary conversion. In this technique, the CRT result divided by M is evaluated:

$X/M = \left(\Sigma_{i=1}^{N} (x_i/m_i) \right) \bmod 1$ (2.19a)

where $x_i = ((1/M_i).r_i) \bmod m_i$ (2.19b)

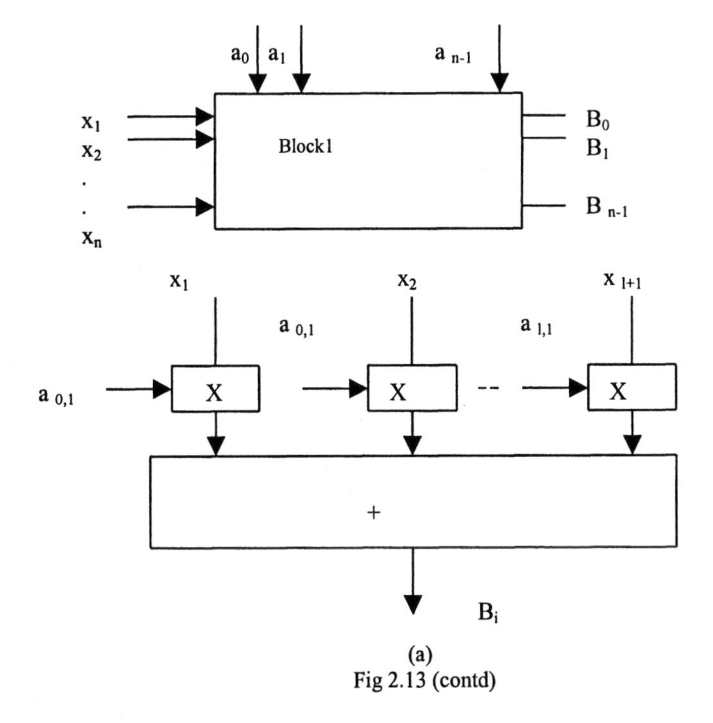

(a)
Fig 2.13 (contd)

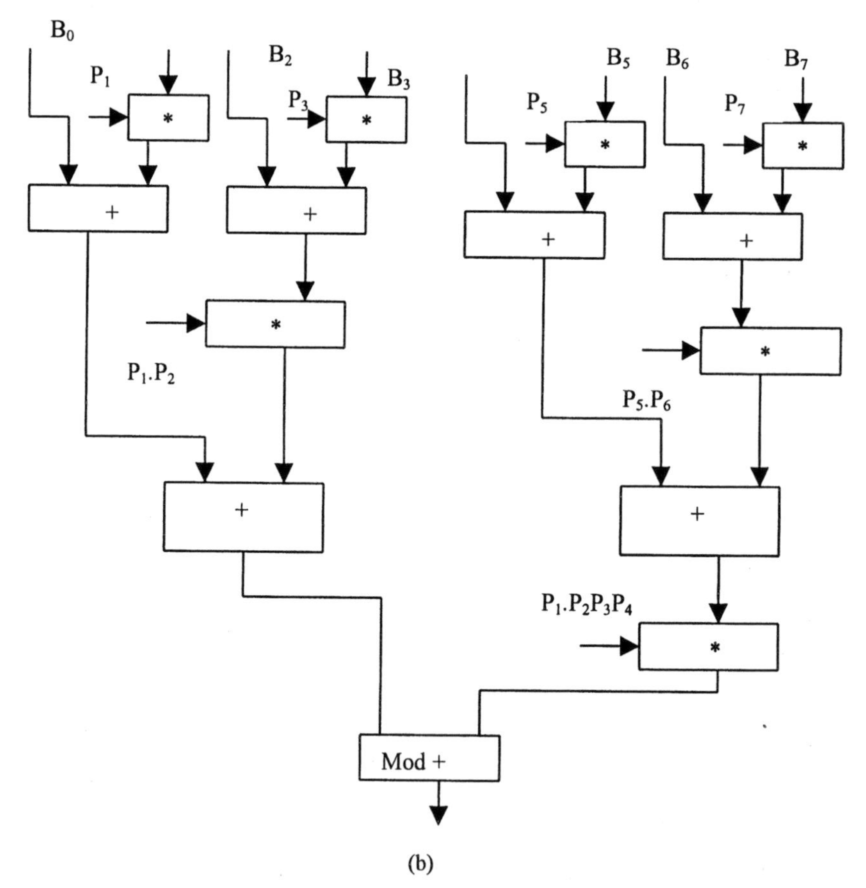

Fig 2.13. Wang's Two-stage New CRT implementation: (a) Architecture for obtaining B_i and (b) Second stage to obtain the final result (adapted from [Wan00a] 2000©IEEE)

Evidently, this method is similar to Vu's method where multiplication by 2/M is performed. The aim is to eliminate the mod M reduction in CRT. The result in (2.19a) can be computed by evaluating all x_i/m_i and adding them. Cardiralli et al suggest the use of small Look-up-tables (LUT) to enable computation of x_i/m_i. For this purpose, we write x_i/m_i as

$$x_i/m_i = (H_{1i}/2) + (H_{2i}/2^2) + \ldots(H_{Li}/2^L) + (\varepsilon_{Li}/(m_i.2^L)) \quad (2.20)$$

where L is such that $L=\lceil \log_2(N.M) + 1 \rceil$. Next, we wish to determine iteratively, H_{ji} j=1,2,...L while ε_{Li} are stored in memory for each x_i. ε_{Li} is estimated as follows:

From (2.20), multiplying both sides by $2^L.m_i$ and taking mod 2 both sides, we have

m_i	K	1	2	3	4	5	6	7	8	9	10	11
m_1=	Δ_k^1	-	1	2	1	2	1	2	1	2	1	2
3	H_k^1	1	0	1	0	1	0	1	0	1	0	
m_2=	Δ_k^2	-	2	4	3	1	2	4	3	1	2	4
5	H_k^2	0	0	1	1	0	0	1	1	0	0	
m_3=	Δ_k^3	-	0	0	0	0	0	0	0	0	0	0
7	H_k^3	0	0	0	0	0	0	0	0	0	0	

Table 2.2. Terms of (2.23 c, d) for different moduli (adapted from [Card00] 2000©IEEE)

$0 = (H_{Li}.m_i+\varepsilon_{Li}) \bmod 2$ (2.21)

Note that ε_{Li} is defined as follows;

$\varepsilon_{Li} = \{ (x.2^L/m_i) - \lfloor x.2^L/m_i \rfloor \} .m_i = \Delta_{L+1}$ (2.22)

In other words, ε_{Li} represents the truncation error. As an illustration, for the moduli set {3, 5, 7} for a residue set {2, 1, 0}, we have the ε_{Li} as $\varepsilon_{L1}=2$,$\varepsilon_{L2}=4$ and $\varepsilon_{L3}=0$. This is as explained as follows: $(x_i.2^L/m_i)$ for $m_i=3$,$x_i=2$ and L=10 yields

$\varepsilon_{L1} = ((2.2^{10}/3) - \lfloor 2.2^{10}/3 \rfloor).m_1 = (682 + 2/3 - 682).3 = 2$

Next, with the knowledge of ε_{L1} we determine the various H_{ji}. We already know that $H_L=(-\varepsilon_{L1}) \bmod 2 = 0$. Next, to compute H_{L-1}, we multiply (2.20) by $2^{L-1}.m$ to obtain

$(H_{L-1}) \bmod 2 = (-((H_L.m_i+\varepsilon_{L1})/2).m_i^{-1}) \bmod 2$

Note that m_i^{-1} here is unity for odd moduli. Thus, for x=2, $m_1=3$ considered above, using the already known ε_L and H_L, we have

$H_{L-1} = (-1) \bmod 2 = 1$.

In a similar manner, the other coefficients can be obtained as shown in Table.2.2. Note that the following general equations can be employed:

For k=L: $\varepsilon_L = \{ (x.2^L/m) - \lfloor x.2^L/m \rfloor \} .m = \Delta_{L+1}$ (2.23a)

$(H_L)_2 = H_L = (-\varepsilon_L.m^{-1})_2$ (2.23b)

for k ε (L-1,...1) : $\Delta_{k+1} = ((H_{k+1}.m+\Delta_{L+2})/2$ (2.23c)

$(H_k)_2 = H_k = (-\Delta_{k+1}.m^{-1}) \bmod 2$ (2.23d)

Note that the mod 2 operators in (2.23b) and (2.23d) mean taking just the LSB. We also observe that all Δ_{k+1} are less than m_i.

An architecture for implementing Cardiralli et al technique is presented in Fig 2.14 (a). The individual hardware for evaluating the x_i/m_i term can be seen in the top whose outputs are fed to a CSA slice to yield the final output. The CSA adder tree follows conventional designs see e.g Hwang [2.14]. The CSA cell for four inputs is shown in Fig 2.14 (b). The x_i/m_i computation architecture

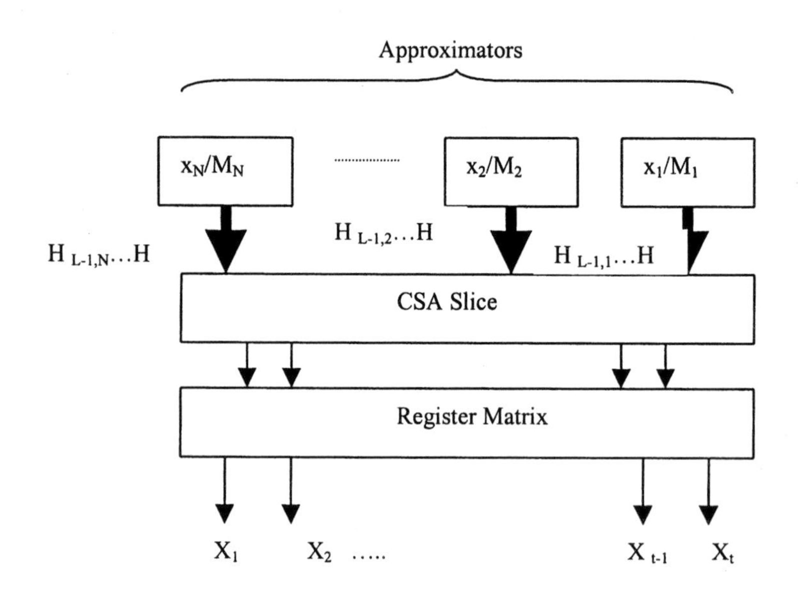

(a)

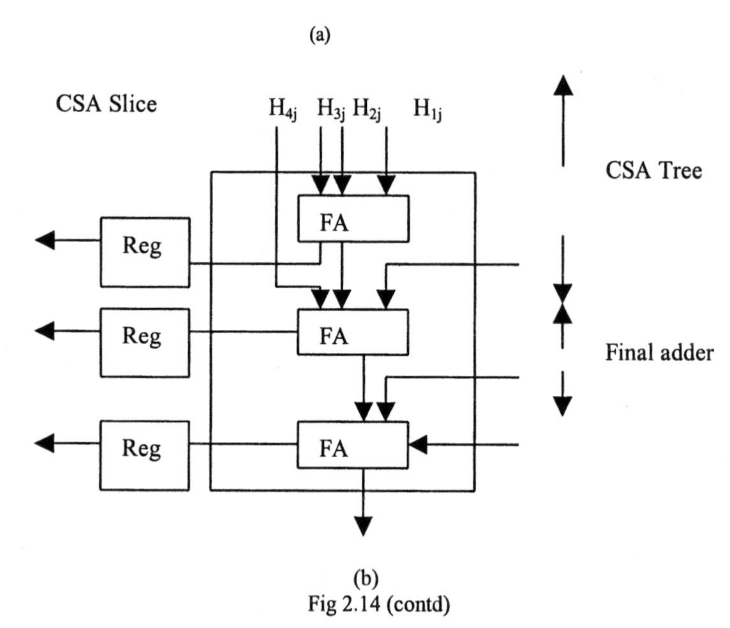

(b)
Fig 2.14 (contd)

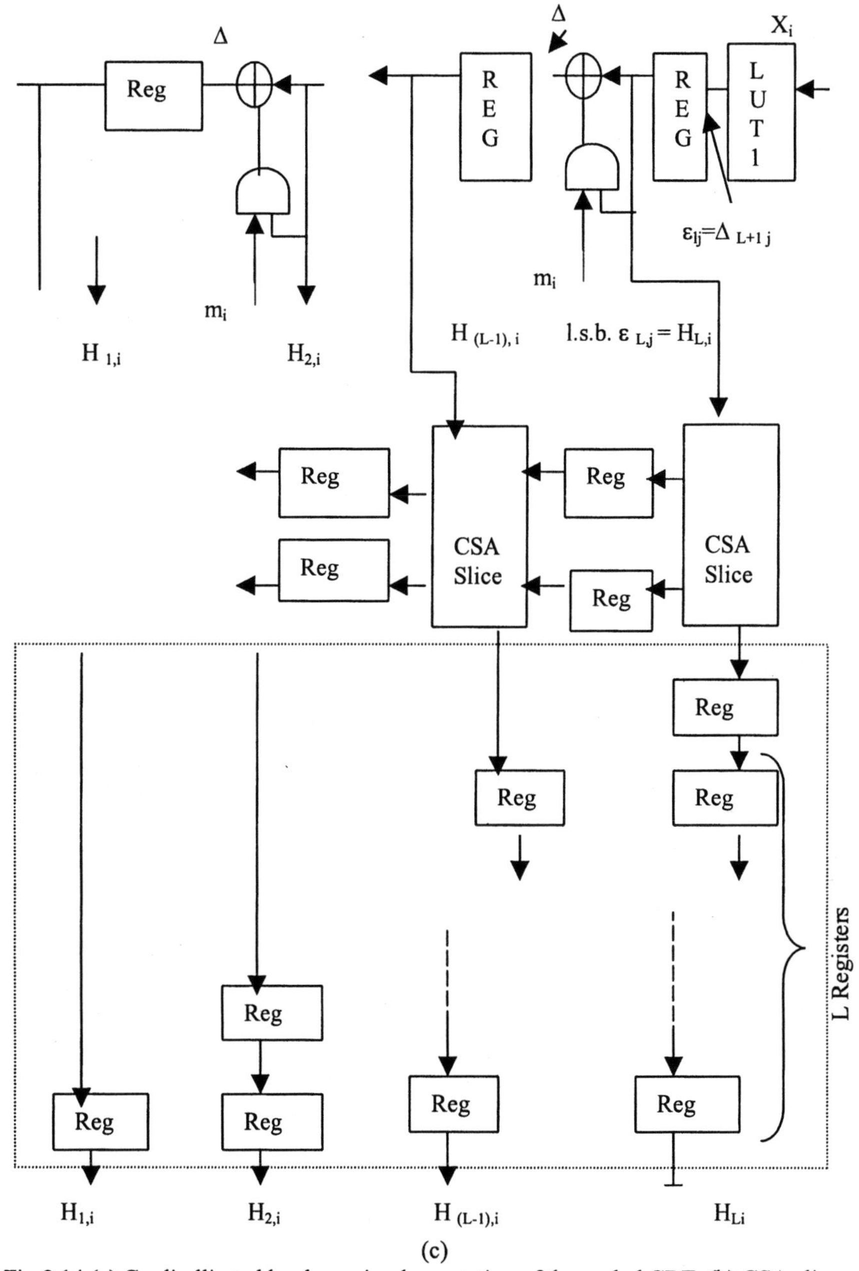

(c)

Fig 2.14 (a) Cardiralli et al hardware implementation of the scaled CRT, (b) CSA slice architecture for four moduli, and (c) hardware implementation of (2.20) (adapted from [Card00] 2000©IEEE)

is sketched in Fig 2.14 (c) wherein e_i are stored in LUTs and Δ_is and H_is are computed. The fully parallel systolic architecture needs several registers as shown.

2.3.8. Alia and Martinelli implementation

A bit slice implementation of the RNS to binary converter using CRT is presented by Alia and Martinelli [Ali84]. It is shown in (2.2) that X is the sum of n binary numbers each corresponding to one of the residue sets
$\{ r_1, 0, 0, 0\ldots\}$, $\{ 0, r_2, 0, 0, \ldots..0\}$, ... $\{ 0, 0, 0, \ldots.., r_{n-1}\}$. Thus, having n processor elements (PE), $X_1, X_2, \ldots, X_n$ can be calculated. Each of the r consists of b_i bits. Noting that

$$r_j = b_{i-1}.2^{i-1} + b_{i-2}\, 2^{i-2} + \ldots.. + b_1.2 + b_o \qquad (2.24)$$

By storing X_i corresponding to all 2^i mod m_i, then successively adding them conditionally based on the bit value of b_i, the individual expressions can be realized. Alia and Martinelli suggest that two each of the bits can be fed to one PE (see Fig 2.15 (a)) to yield the result. Having $b_i/2$ PEs in the first step, $b_i/2$ individual items can be computed. Then, the PES are interconnected in such a way that their results are added mod mi in $\log_2 (b_i/2)$ steps. Each modulus will have its own dedicated i/2 PEs. The final reduction can be done by the PES of any one modulus. The hardware evidently needs a bussed architecture with numerous interconnections. Each PE typically consists of two registers to store the value corresponding to 2^kmod m_i and 2^{k-1} mod m_i, two sets of AND gates to permit the input to the adder only when b_i=1, and in addition one modulo M adder. It also accepts an external input to perform the summation of the results of each modulus associated hardware.

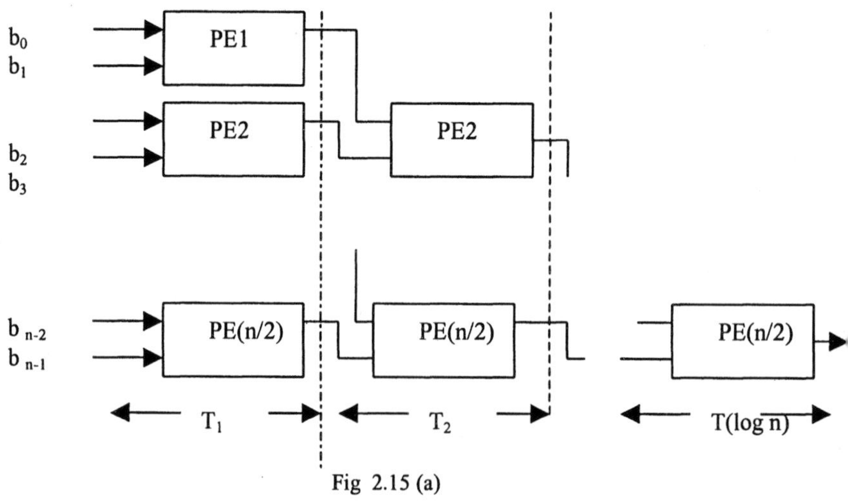

Fig 2.15 (a)

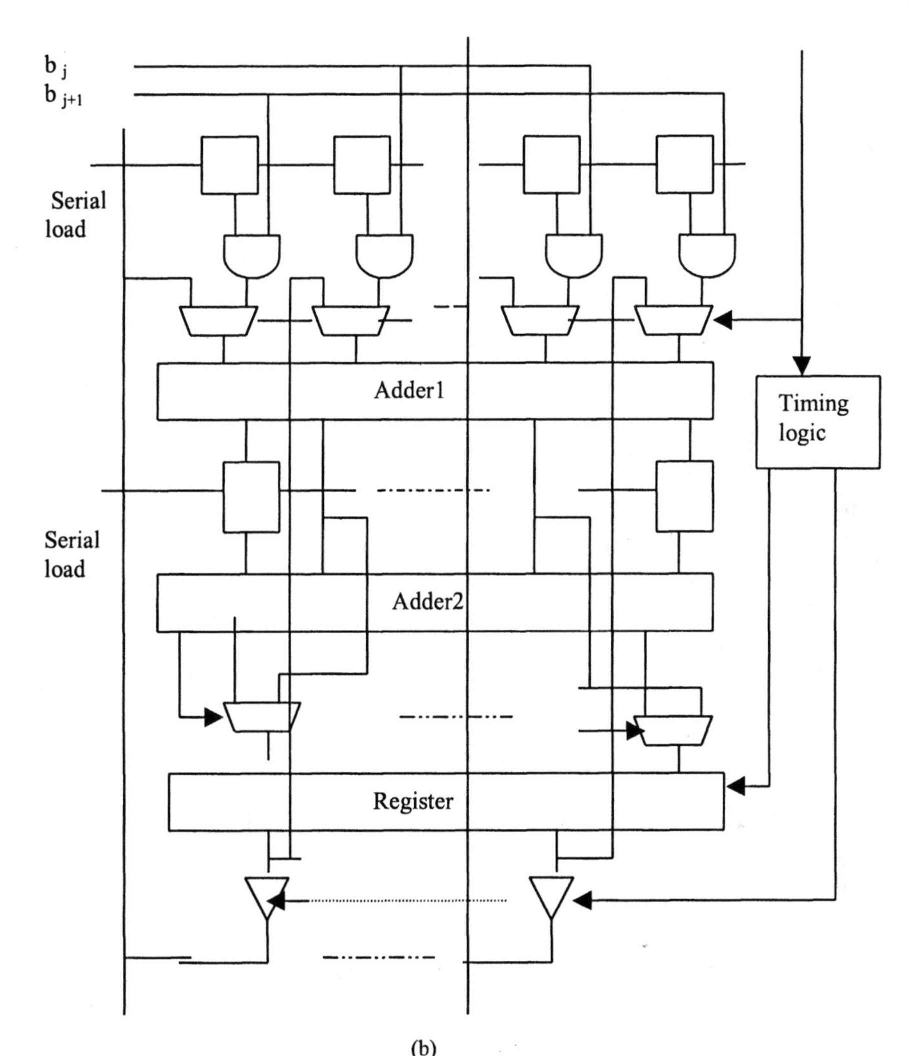

(b)

Fig 2.15. Alia and Martinelli CRT based RNS to Binary Conversion architecture (a) and Architecture of each PE (b) (adapted from [Ali84] 1984©IEEE)

2.3.9. Capocelli and Giancarlo implementations

Capocelli and Giancarlo [Capo88] suggested a method of reducing the PEs from n/2 to n/log n. Here, expressing l=log n and defining t=n/l and assuming that l and t are integers, the CRT expansion is realized using t PEs each catering for weighing of one modulus. After accumulating mod M the result, the PEs are used to compute the result corresponding to next t moduli and so on. Evidently, n registers for storing $(M/mi).(M/mi)^{-1}$ mod m_i are needed. Hardware multipliers have been recommended by Capocelli and Giancarlo in their architecture presented in Fig 2.16. Note that the queue is a ROM, storing the weights mentioned above which is circularly addressed to cater for various moduli m_i, m_{i+t}, m_{i+2t} etc.

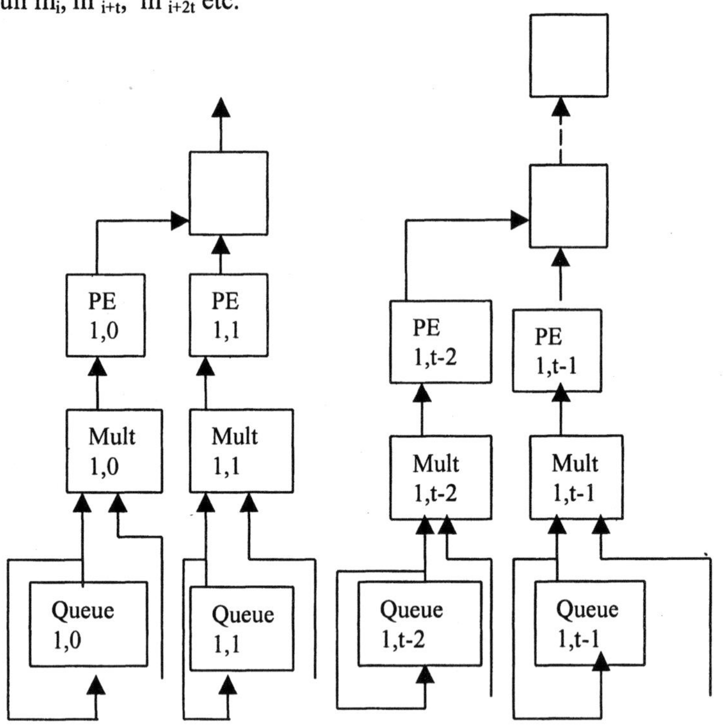

Fig 2.16. Hardware architecture of Capocelli et al RNS to Binary converter (adapted from [Capo88] 1988©IEEE)

2.3.10. RNS to binary converters due to Jenkins

Jenkins [Jenk78a] suggested the use of commercial microprocessors for RNS to binary conversion for certain special two moduli RNS using twin primes e.g. (239, 241), (59, 61), (107, 109), (29, 31). The dynamic range for these example moduli sets is roughly 15.8, 11.8, 13.5 and 9.8 bits respectively. Jenkins observes that the CRT application in these cases yields expressions of the form

$$X= \left| m_2.y_1' + m_1 y_2' \right|_M \qquad (2.25a)$$

where y_1' and y_2' are $((m_2^*)_{m1}.r_1) \bmod m_1$ and $((m_1^*)_{m2}.r_2) \bmod m_2$ and m_2 is the large modulus and m_1 is the small modulus. Knowing that $m_2 = m_1+2$, we have from (2.19a),

$$X = |(m_1+2)y_1'+m_1y_2'|_M \quad (2.25b)$$

Certain approximations can be made if a scaled result is required. This helps to reduce the ROM size. For instance, scaling by m_1 is thus approximated as

$$X_s = [X/m_1]_Q = (y_1'+y_2') \bmod m_2 \quad (2.26)$$

where Q is the quantized approximate result. The error is bounded by $2(1 - (1/m_1))$.

They also considered scaling in the three moduli system where the scaled result can be of the form

$$X_s = [X/(m_1m_3)]_Q = (f_1(y_1') + y_2' + f_3(y_3')) \bmod m_2 . \quad (2.27)$$

where $f_1(y_1') = [(m_3/m_2).y_1']_Q$ and $f_3(y_3') = [(m_1/m_2).y_3']_Q$.

Jenkins also considers a two moduli system $(2^k-1, 2^k)$ for which scaling by (2^k-1) yields similar relation as in (2.20). Similar results can be found for moduli system $\{2^k, 2^{k+n} - 1\}$ for which the result of scaling by 2^k is

$$X_s = (2^n.y_1'+y_2') \bmod (2^{k+n} - 1) \quad (2.28)$$

Finally the interesting moduli set, which will be studied in detail in the next Chapter viz., $\{2^n-1, 2^n, 2^n+1\}$ has scaled output given as follows, when scaled by $m_1.m_3 = 2^{2n}-1$:

$$X_s = [X/(2^{2n}-1)]_Q = (y_1'+y_2'+y_3') \bmod 2^k \quad (2.29)$$

where the error is $(y_1'/(2^k-1)) - (y_3'/(2^k+1))$.

2.3.11. Huang et al CRT implementation

Huang et al [Hua81] have described a CRT implementation which used some interesting techniques for modulo M reduction. They considered a five moduli RNS {16, 15, 13, 11, 7}. The application of CRT to this 18 bit dynamic range system (M = 240,240), yields a 20 bit result which is reduced directly to two's complement form by looking at the MSBs and then adding suitable correction factor. First, $2^{18} - M$ is added to the five operands Q_1, Q_2, Q_3, Q_4 and Q_5, yielding Q:

$$Q = (Q_1 + Q_2) \bmod M + (Q_3+Q_4) \bmod M + (Q_5 + 2^{18} - M) \bmod M$$

in three adders first to get the three terms and then another adder to obtain Q. The following logic is applied to reduce Q directly to twos complement result since Q lies between 2^{18}-M and $2M + 2^{18}$. The following cases are analyzed based on the MSBs b_{18} and b_{19} of Q: (a) If $b_{18} = 0, b_{19}=0$, then Q is compared with $2^{18} - (M/2)$. If $Q \geq 2^{18} - (M/2)$, no correction is required. Else, M is added mod 2^{18} to Q. (b) If $b_{18} = 1$, $b_{19} = 0$, then Q is compared to M/2. If Q >M/2, $2^{18} - M$ will be added mod 2^{18} to Q. Else, no correction is required. (c) If $b_{19} = 1$ and $b_{18}=0$, then Q is compared to $(3/2).M - 2^{18}$. If $Q \geq (3/2)M - 2^{18}$,

then $-2M$ is added and else, $-M$ will be added to Q. (d) The case $b_{18}=b_{19}=1$ is not possible since $Q < 3M+2^{18}-M < (2^{19}+2^{18})$.
It is interesting to note that the method uses all 2^{18} modulo adders and the result is directly in two's complement form.

2.4. BINARY TO RNS CONVERSION TECHNIQUES

The simplest technique of obtaining residues of a given number X is to divide X by each of the moduli and take the remainder. As is well known division is a quite complex operation and time consuming. Hence faster techniques of Binary to RNS conversion are of interest.

Jenkins and Leon [Jenk77] suggested binary to RNS conversion by splitting the given binary word into n smaller segments each segment addressing n ROMs to get the residues and adding these n residues mod m_i. Several variations of this technique have been developed later which will be studied in detail in what follows.

2.4.1. Alia and Martinelli technique

In this method the residue of the given binary word B mod m_i is written as

$$B \bmod m_i = b_0 \bmod m_i + (2.b_1) \bmod m_i + (2^2.b_2).\bmod m_i + \ldots + (2^{n-1}.b_{n-1}) \bmod m_i \tag{2.30}$$

Alia and Martinelli [Ali84] suggest the use of the same architecture described for RNS to binary conversion shown in Fig 2.15. Each PE stores residues mod m_i of two adjacent powers of two i.e.2^j and 2^{j+1} mod m_i and these are weighted by the b_j and b_{j+1} terms respectively and added mod m_i. Thus n/2 values are obtained. Next, using the bus structure these n/2 values are added mod m_i in $\log_2(n/2)$ steps to obtain the final residue. Note that similar hardware needs to be used for the other moduli also. Note also that the residues of pairs of powers of two are stored in ROM in the PEs.

2.4.2. Capocelli and Giancarlo technique

In this technique, there are only t PEs each computing the residue of a $\alpha = \log_2 n$ bit word where $t = n/\log_2 n$. A PE is shown in Fig 2.17 (a). The results of all these PEs are added in h steps where $h = \log_2 t$ to yield the final result.

Capocelli and Giancarlo [Capo88] suggest the online computation of the residues of powers of two rather than storing in the ROM. Note, however, the residue corresponding to 2^0, 2^α, $2^{2\alpha}$ etc need to be stored from which multiplication by 2 and reduction mod m_i yields the successive residues as shown in Fig 2.17 (b). Evidently $\alpha + h$ steps are needed.

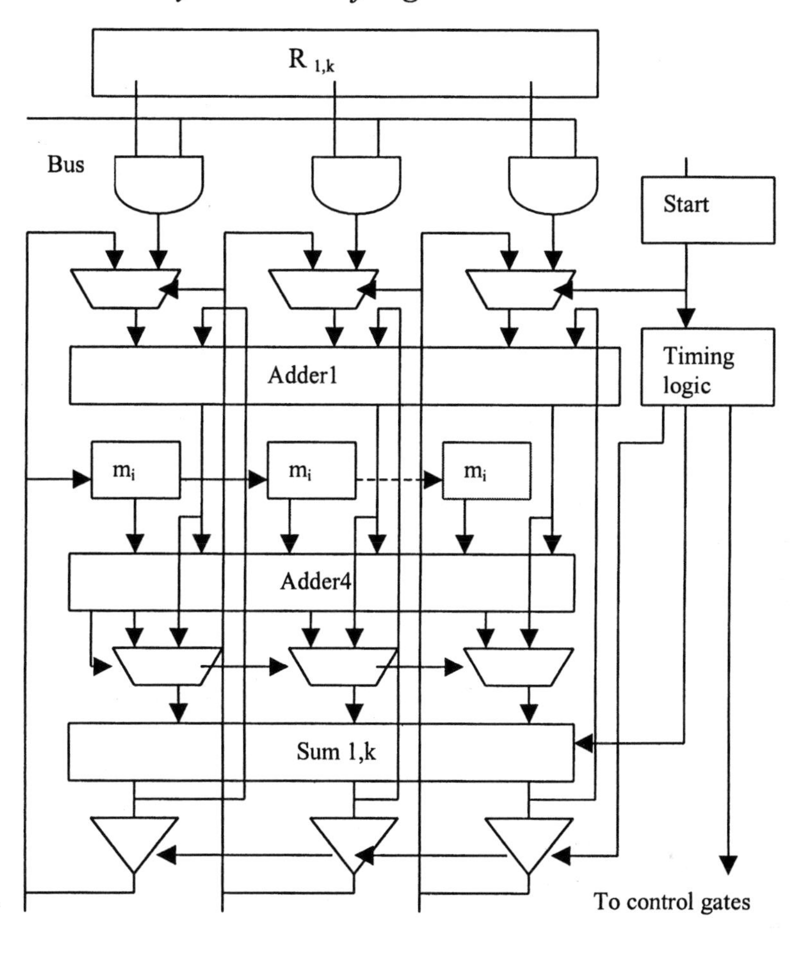

(a)

Fig 2.17 (contd)

2.4.3. Piestrak and Ananda Mohan technique

Piestrak [Pies91, Pies94] and Ananda Mohan [Anan94, Anan99] used the periodic property of powers of two residues mod m_i to simplify the hardware as well as ROM storage needed. In this method, first moduli are suggested to be chosen as to have an "order" which is small. The order for a number is defined as P_i such that 2^{Pi} mod m_i =1. As an illustration, for m_i=89, P_i = 11 since 2^{Pi} mod 89=1. In the case of certain moduli, we also have the property that $2^{\phi i}$ mod m_i=-1. As an example, for m_i =13, ϕ_i = 6. These two properties can be used to perform binary to RNS conversion efficiently.

We will first consider the case of m_i such that 2^{Pi} mod m_i=1. We denote the given binary word N as b_{n-1} b_{n-2} .... $b_2b_1b_0$. Due to the fact that 2^x, 2^{x+Pi}, 2^{x+2Pi}

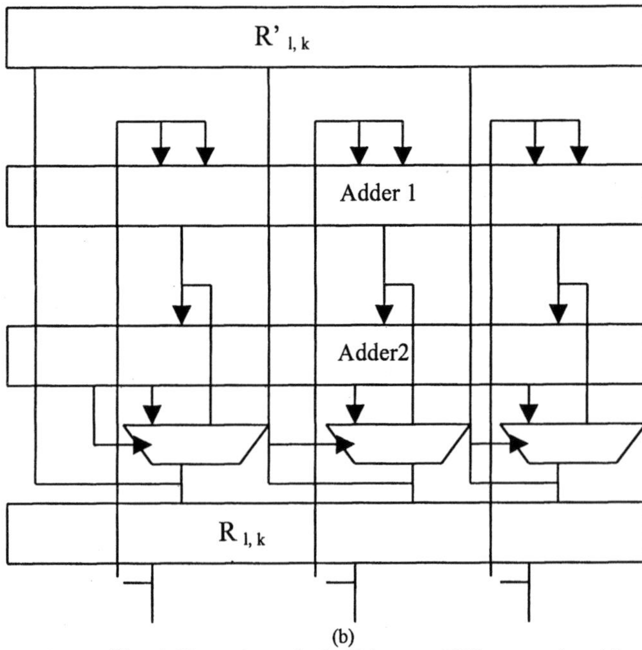

(b)

Fig 2.17. Capocelli and Giancarlo method of binary to RNS conversion: (a) Architecture of PE and (b) Architecture of 2^i mod m_i evaluation (adapted from [Capo88] 1988©IEEE)

have the same residue, the P_i bit words $b_{P(i-1)}......b_1b_0$, $b_{2Pi-1}.......b_{Pi+1}b_{Pi}$, etc can be first added using a carry save adder tree followed by a carry propagate adder with end around carry to yield a P_i bit word as shown in Fig 2.18 (a). The hardware needed is $P_i.([n/P_i] -1)$ full adders and the delay involved is $(n/P_i + 2P_i - 3)\ \Delta_{FA}$ where Δ_{FA} is a full-adder delay, taking into account the end-around carry addition in the last CPA stage. The residue of the P_i bit word W can be found using Alia and Martinelli technique. Note that if the period P_i is larger than $\log_2 m_i$ i.e. the number of bits needed to represent m_i, only residues of P_i-$\log_2 m_i$ number of 2^x mod m_i need to be stored. As an illustration, for the example m_i=89, P_i=11 whereas $\log_2 89$=7. Thus, expressing W as $w_{10}w_9w_8....w_1w_0$, only 2^7 mod m_i, 2^8 mod m_i, 2^9 mod m_i, 2^{10}mod m_i need to be stored and weighted and added modulo m_i.

Ananda Mohan [Anan94] also suggested that composite moduli i.e. moduli which have several small prime factors can be chosen with a small order i.e. P_i facilitating large dynamic range RNS simplifying at the same time Binary to RNS conversion. As an illustration a 128 bit dynamic range system can comprise of eight moduli viz., m_1=151.31.7, m_2 = 257.17.5, m_3 = 73.19.27, m_4 = 524287, m_5= 337.127, m_6=683.89.23, m_7=241.13, m_8= 331.468. The respective P_i values are 15, 16, 18, 19, 21, 22, 24 and 30 respectively. In this example, the word lengths of the moduli are uniformly distributed. As an illustration for a 64 bit RNS, the architecture of binary to RNS conversion corresponding to modulus 21845 (=257.17.5) is presented in Fig 2.18 (b).

The architecture for the second case with the property $2^{\phi i}$ mod m_i = -1 is next considered. Note that due to the definition of ϕ_i, alternate ϕ_i bit fields have to be added and sum of even fields shall be subtracted from sum of odd fields:

$$W=\Sigma W_U - \Sigma W_V \quad \text{for } U=0, 2, 4, .. \text{ and } V=1,3,5... \tag{2.31}$$

Note that we have defined the given binary word as $W_x\ W_{x-1}....W_2W_1W_0$ where all W_i are ϕ_i bit words. The value of W obtained is a (ϕ_i+$\log_2$s+1) bit word. The $\log_2$s term arises since we added s number of W_i terms and one extra bit is needed to take into account the sign of the result. The MSB w_y of this word $w_y....w_1w_0$ has a weight $-2^y = -2^{(\phi i+\log s)} = 2^{\log s}$ since $-2^{\phi i}$ mod m_i =1. Hence, if w_y is 1, we need to add a 1 to the (ϕ_i+$\log_2$s) bit word obtained in (2.31) excluding the sign bit at the bit position by where y = $\log_2$s+1. The next step is to find the residue mod m_i of this (ϕ_i+$\log_2$s) bit word W.
As an illustration, consider the evaluation of 16519104 mod 13. The binary number corresponding to 16519104 is
111111 000000 111111 000000
W_3 W_2 W_1 W_0
The various steps involved in the computation are shown in Fig 2.18 (c).
We next consider the extension to the case of even moduli. Note that one even modulus can be permitted in any RNS which is mutually prime to all other moduli. Consider that the even modulus is represented in general as $2^{\beta}.m_i'$. Denoting the period of m_i' as P_i', we have $2^{Pi'}$mod m_i'= 1 or $2^{Pi'}$-1 = Q .m_i'.
Multiplying both sides by 2^{β}, we have
$2^{(\beta+Pi')} - 2^{\beta}$ = Q. m_i'
or $2^{(\beta+Pi')} = 2^{\beta}$mod m_i.

As an illustration, m_i = 72 corresponds to the case m_i'=9 and β =3. Thus, the period of 72 is same as that of 9 but however the starting point of this periodic behaviour is from 8. Thus, the reduction of the word-length from n to P_i'proceeds from the β th bit position and to the resulting word, β LSBs are appended to get a P_I' +β bit word, which can be reduced as in the case of odd moduli. As an illustration, consider 27189 whose binary equivalent is 0110101000110101. Consider m_i =52 whence β = 2 and m_i =13. Partitioning the above word excluding the two LSBs into Q_i bit fields as
01 101010 001101 01
W_3 W_2 W_1 W_0
We have W_1-W_2+W_3 = 13 –42+1 =-28 =-128+ 100 =100+2 =102.

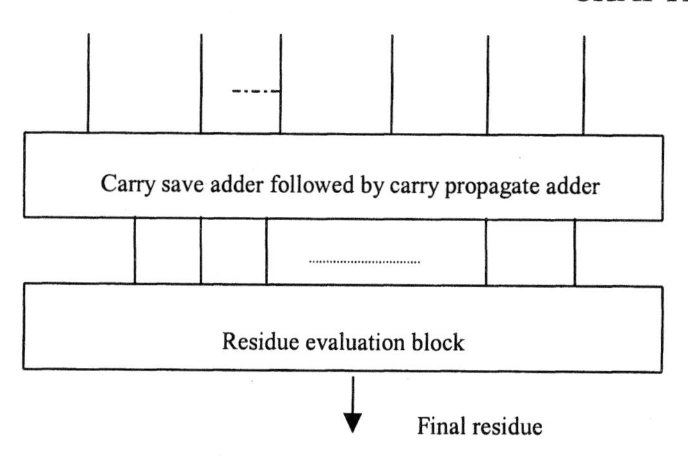

(a)

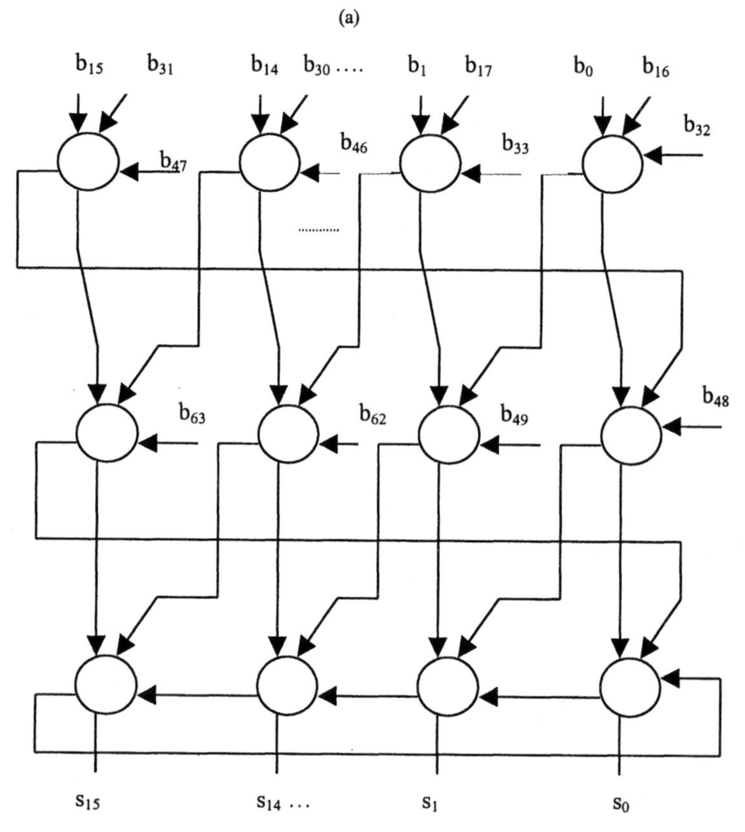

Fig 2.18. (contd)

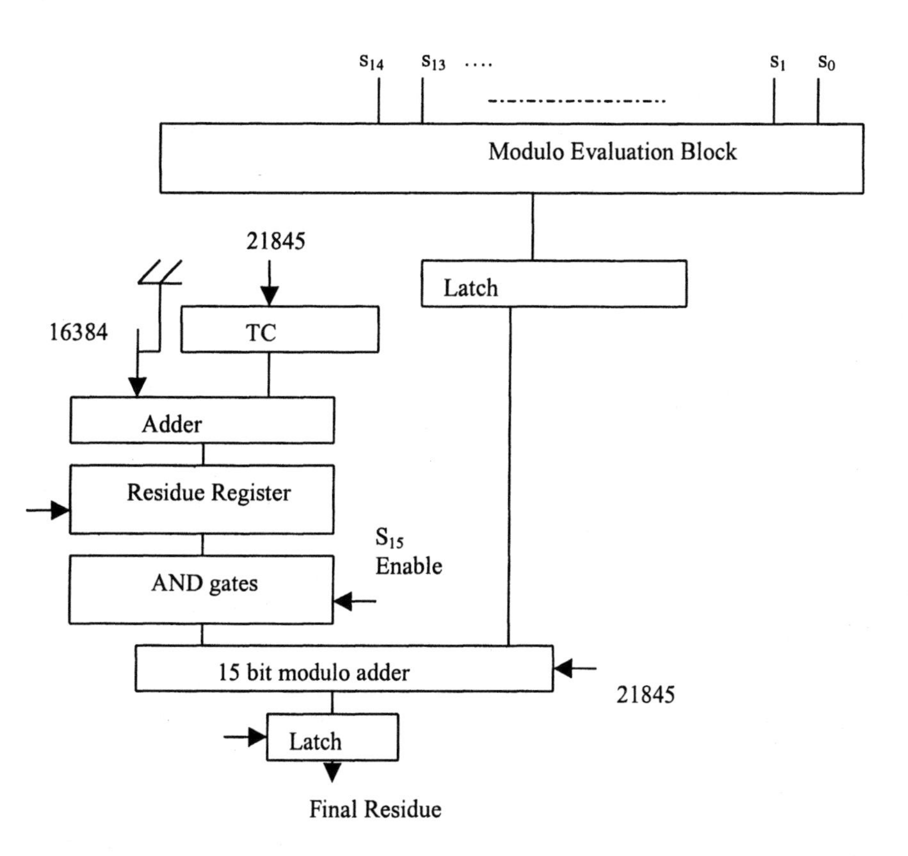

S_{14}
S_{13} ....
S_1
S_0
Modulo Evaluation Block
21845
Latch
16384
TC
Adder
Residue Register
S_{15}
Enable
AND gates
15 bit modulo adder
21845
Latch
Final Residue

Fig 2.18 (b)

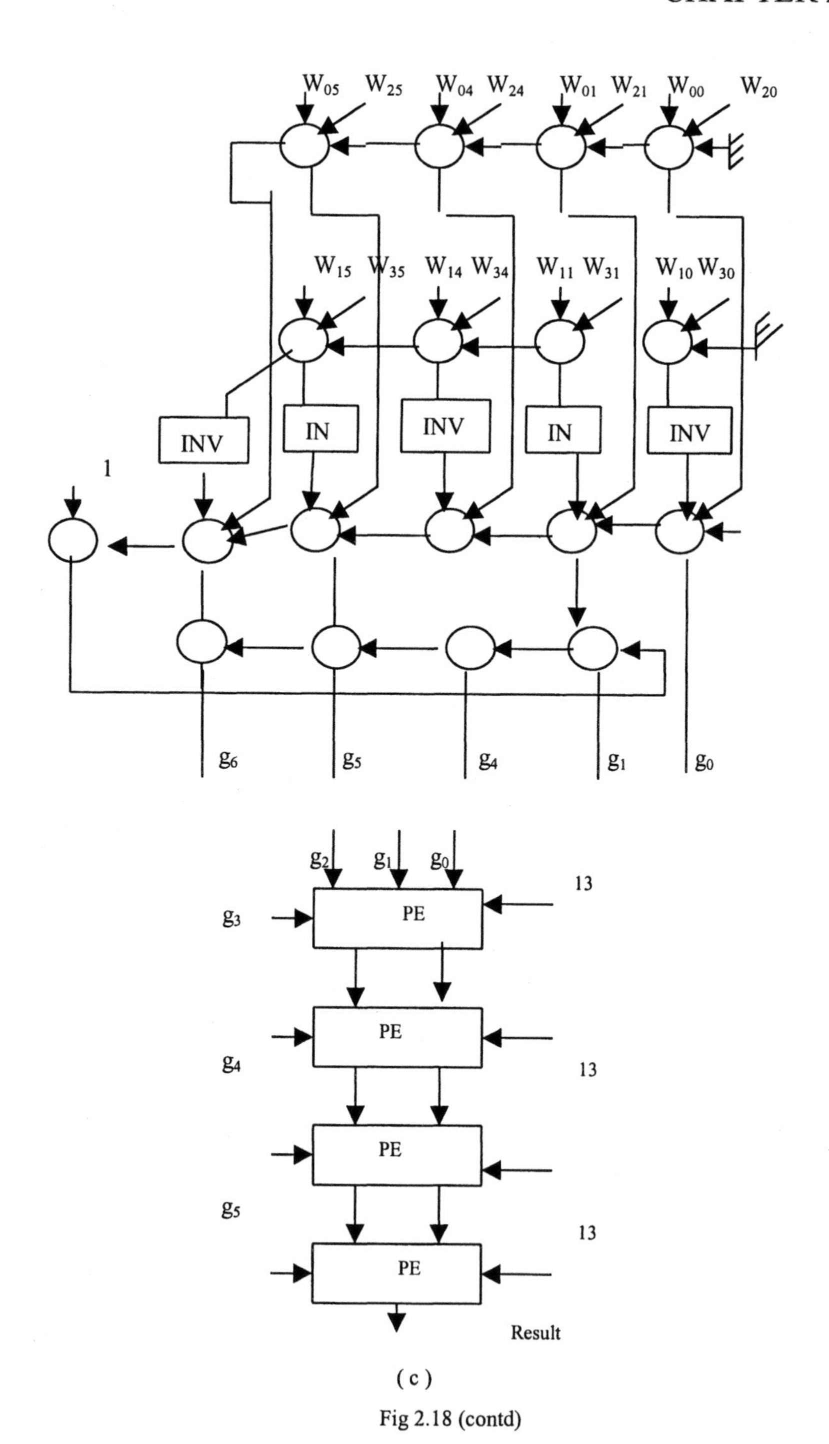

(c)

Fig 2.18 (contd)

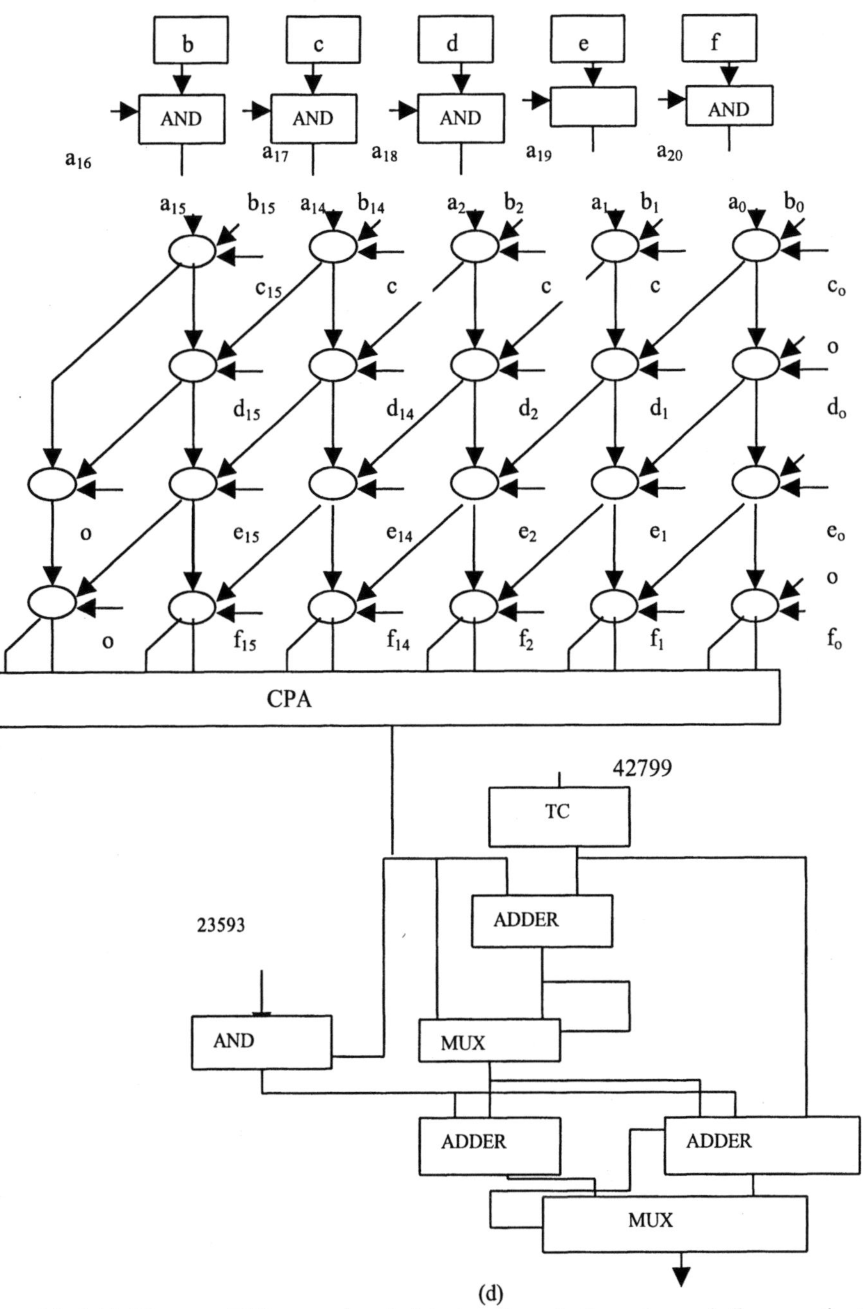

(d)

Fig 2.18. Binary to RNS conversion Architecture for reduction using periodic properties for odd moduli (a), for composite modulus (b), for modulus with half period (c) and (d) using Stouraitis et al architecture in the second stage

(adapted from [Anan94] 1994©World Scientific)

Next, the two bit word W_0 is appended to yield the reduced 9 bit word viz. 4.102+ 1 = 409. Next, the residue of this 9 bit word is found in six steps.
Ananda Mohan also suggested that Stouraitis et al technique [Stou92, Stou93] can be used to reduce the n bit word mod m_i which is considered next. In this method, the residues of the higher bit positions are stored in memory and added to the LSBs without modulo reduction to get a smaller length word than the original. This can be repeated at most few times to achieve the final reduction. As an illustration, for a 64 bit dynamic range system, considering a modulus 42779 = 337.127 with a period 21, first using the periodic property we obtain a 21 bit word. Note that the modulus 42779 is a 16 bit word. Hence the residues of 2^{16}, 2^{17}, 2^{18}, 2^{19} and 2^{20} are stored in five memory locations. Next based on the bit values of the 21 bit word at the output of the first stage of converter, we obtain using a Stouraitis converter, the reduced word. This can be at most 128397, a 17 bit word corresponding to the case all the bits are 1 at the output of the first stage converter. The use of Stouraitis converter on this further reduces the word length to a 16 bit word. The architecture is presented in Fig 2.18 (d). Note however, that alternatively, the successive stages can be based on Alia and Martinelli technique as well, where addition of moduli of higher powers of two is reduced mod m_i immediately. The reader is referred to Stouraitis et al [Stou92, Stou93] for more information on the stages needed and the full adder count needed. Note that some of the ideas developed in this section will be used in the next Chapter for the case of some special moduli set.

2.4.4. Meehan et al binary to RNS converter

Meehan et al [Mee90] have presented an interesting technique for binary to residue conversion with scaling inherent in the conversion process. This means that the residue obtained is not the exact value but a scaled version as discussed in Section 2.3.5. The proof for this technique is as follows. At every step we divide the current result by 2. If the previous result is even, no addition need be done. If it is odd, then m_i needs to be added. The procedure is repeated b times where b is the number of bits in X.

As an illustration, the evaluation of 23 mod 5 is presented. In the first step X=23 and hence (X/2) mod m_i = 11+(1/2)mod m_i. The value of (1/2) mod m_i can be obtained as $(m_i+1)/2$. Thus, for m_i=5, (1/2) mod 5 = 3 yielding X/2 = 11+3 = 14. This procedure is repeated as follows:

X(0)	10111	add modulus	23
	+ 101		5
	----------		-----
X(1)	1110	add nothing	14

```
        +  0000                        0
           -----                     ------
X(2)        111   add modulus          7
        +  101                         5
          ------                     ------
X(3)        110   add nothing          6
        +  000                         0
          -------                    ------
X(4)        11                         3
              101             add modulus     5
        -------------                ------
X(5)       100                        4
```

In other words, we have computed $(2^5)^{-1}.23 \bmod 5 = 4$. The scaled residues can be processed in the RNS processor and only at the time of output conversion, the scaled factor can be taken into account as explained before in Section 2.3.5. Note that this procedure can be simplified as below for actual implementation.

```
         0
         1        XOR =1, add modulus
   + 101
     110
       1          XOR =0, add nothing
  + 0000
    1000
       1          XOR=1, add modulus
 + 10100
  100000
       0          XOR =0, add nothing

+ 000000
  100000
       1          XOR=1, add modulus
+ 1010000
 10000000
```

In each step, the modulus is conditionally added to the b most significant bits output of the previous stage and to the j th LSB of X which comes through the carry in. The modulus is added if the exclusive OR of the j th bit from input X and the LSB of the result from the previous stage is 1. This method is same as the previous one and can be verified by the reader. The architecture for

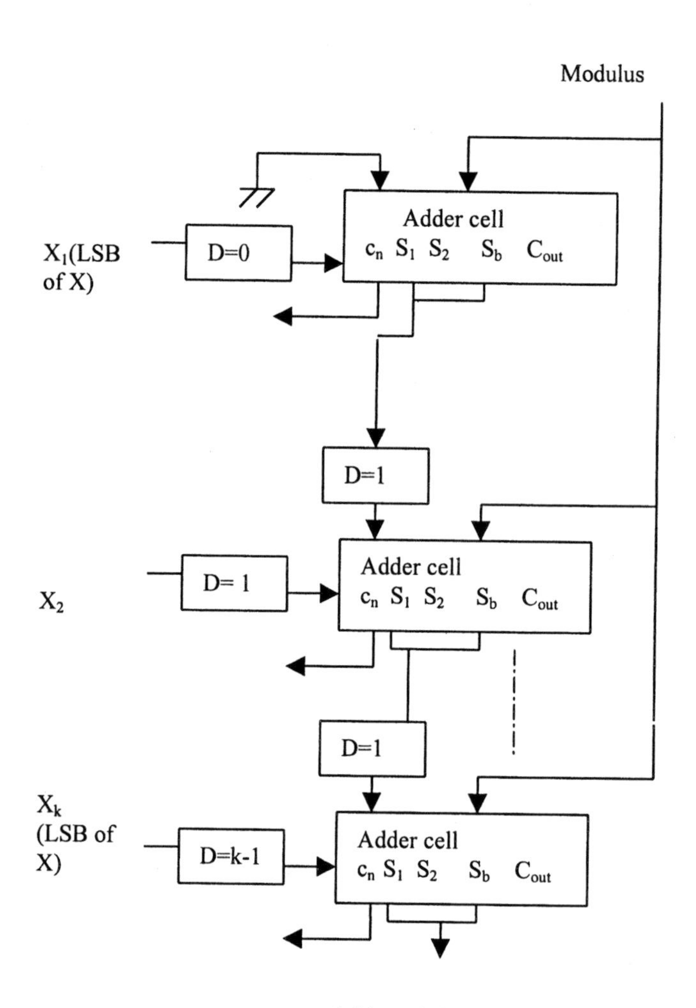

Fig 2.19. Meehan et al architecture for Binary to RNS conversion with scaling (adapted from [Mee90] 1990©IEEE)

implementation is as shown in Fig 2.19. Meehan et al [Mee90] have suggested the use of the same architecture for input/output conversion for which information the reader is referred to their work.

2.5. CONCLUSION

In this Chapter techniques for RNS to binary and Binary to RNS conversion have been studied in detail. ROM based implementations are attractive for fast implementations, whereas non-ROM based implementations are attractive when ROM is not available in certain custom VLSI designs. We have considered the general moduli sets in this Chapter. However, some particular moduli sets have received considerable attention recently and enable sufficiently fast forward and reverse converters to encourage the use of RNS in VLSI computing engines. This topic will be discussed in detail in the next Chapter.

3

FORWARD AND REVERSE CONVERTERS FOR THE MODULI SET $\{2^k-1, 2^k, 2^k+1\}$

3.1. INTRODUCTION

The particular three moduli set $\{2^k-1, 2^k, 2^k+1\}$ has received considerable attention in literature, since this moduli set offers considerable ease in forward and reverse conversion, scaling and other operations [Anan00b]. In this Chapter, VLSI architectures for forward and reverse conversion are discussed in detail. Related moduli sets $\{2^k-1, 2^k, 2^{k-1}-1\}$, $\{2n, 2n+1, 2n-1\}$ and $\{2n, 2n+1, 2n+2\}$ also will be considered.

3.2. FORWARD CONVERSION ARCHITECTURES FOR $\{2^K-1, 2^K, 2^K+1\}$ MODULI SET

The dynamic range of this particular moduli set is evidently $2^k.(2^{2k}-1)$ i.e. corresponding to 3k bits. Thus, given a 3k bit binary number, the residues corresponding to the three moduli need to be determined. The residue corresponding to 2^k is just the k bit LSB word ignoring the MSBs. The residues corresponding to 2^k-1 and 2^k+1 can be determined using the periodic property of $2^x \bmod m_i$ where m_i can be 2^k-1 or 2^k+1 as explained below [Bi88, Anan91].

It is easy to see that the given 3k bit binary word W can be split as three fields W_2, W_1, W_0 where

$$W=W_2.2^{2k}+W_1.2^k+W_0 \qquad (3.1)$$

Noting that $2^k \bmod (2^k-1)=1$ and $2^{2k} \bmod (2^k-1)=1$, we have from (3.1),

$$W' = W \bmod (2^k-1) = (W_2+W_1+W_0) \bmod (2^k-1) \quad (3.2)$$

Thus, all the three k bit words W_0, W_1 and W_2 can be added using a 3-input k bit adder with end-around-carry as shown in Fig 3.1. The hardware needed is k full-adders followed by a k bit carry- propagate-adder. The conversion time is at most 2k full adder delays noting that the end-around-carry adder delay needs (k-1) full-adder delays.

The residue evaluation corresponding to the modulus (2^k+1) of the 3k-bit word

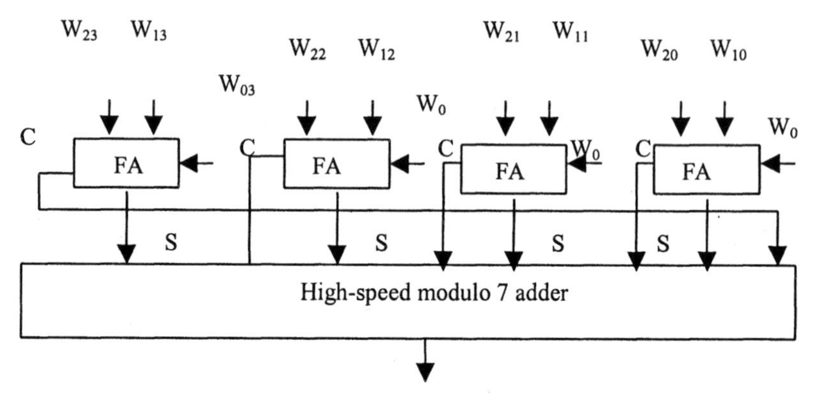

Fig 3.1. Binary to RNS converter for modulus (2^k-1)

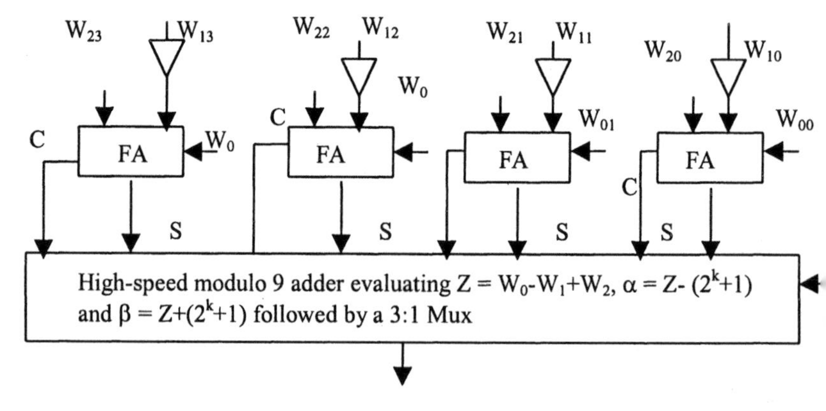

Fig 3.2. Binary to RNS converter for modulus (2^k+1)

described by (3.1) is given by

$W'' = W \bmod (2^k+1) = (W_2 - W_1 + W_0) \bmod (2^k+1)$ (3.3)

This can be seen to be true due to the fact that that $2^k \bmod (2^k+1) = -1$ and $2^{2k} \bmod (2^k+1) = 1$. The evaluation in (3.3) can be carried out by first adding W_2 and W_0 to the two's complement of W_1 and then evaluating the residue mod (2^k+1). We note for this purpose that the sum in (3.3) before residue evaluation can be between $2.(2^k-1)$ and $-(2^k-1)$. Hence, by evaluating $\alpha = W'' - (2^k+1)$ and $\beta = W''+(2^k+1)$ in parallel, either W" or α or β can be selected based on the sign of α and β. The resulting architecture is presented in Fig 3.2.

As an illustration for a given binary word 110 111 101 corresponding to the moduli set {7, 8, 9}, we obtain using (3.2) and (3.3), the residues corresponding to 7 and 9 as (6+7+5) mod 7=4 and (6-7+5) mod 9=4 respectively. The residue corresponding to 8 is the 3-bit LSB field i.e. 5. Note that, in general, for any large word length binary word W (even >3k bits), expressing the k bit fields as W_a, W_{a-1}, ………, W_2, W_1, W_0, the residues corresponding to (2^k-1) and (2^k+1) can be obtained [Anan99] as $(W_a+W_{a-1}+\ldots\ldots.W_2+W_1+W_0) \bmod (2^k-1)$ and $(\pm W_a+W_{a-1}-\ldots..+W_2-W_1+W_0) \bmod (2^k+1)$ where + sign applies for even a and –sign applies for odd a.

3.3. REVERSE CONVERTERS FOR THE MODULI SET $\{2^k-1, 2^k, 2^k+1\}$

Several approaches for RNS to Binary conversion for this particular moduli set have been described in literature. Some of these are based on Mixed Radix Conversion and some are based on Chinese Remainder Theorem. Some used combination of these two techniques. All these techniques will be described in detail in this section.

3.3.1. Mixed Radix Conversion technique [Vinn94, Anan98]

The MRC Technique described in Chapter 2 for the general moduli set can be applied to this moduli set as described next, using Fig 3.3 (a). Note that in Fig 3.3 (a), the three multiplicative inverses a, b and c exist, which respectively can be easily evaluated as a=1, $b=2^k$ and $c=2^k-1$. Thus, it can be seen that $a_1 = (r_1-r_2) \bmod (2^k-1)$ needing no multiplication. Next, the multiplication $p.2^k \bmod (2^k+1)$ can be seen to be $-p \bmod (2^k+1) = (r_2-r_3) \bmod (2^k+1)$ thus leading to further simplification i.e. instead of evaluating $(r_3\text{-}r_2)$ mod 2^k -1 an multiplying by b mod (2^k+1), we can instead just compute a_2 as $(r_2$ - $r_3)$mod(2^k+1). The last simplification arises due to the interesting property o multiplication by 2^{k-1} mod $(2^k\text{-}1)$. Given a k bit number $N = b_{k-1}b_{k-2}\ldots b_2b_1b_0$ the value of $N.2^{k-1}$ mod $(2^k\text{-}1)$ can be obtained by just circular shift of N b (k-1) bits to the left. As an illustration, 13.8 mod 15 can be seen to be 14 a follows:

$(1101)_2 \times (1000)_2$ mod $15 = (110\ 1000)_2 = (1110)_2$.

	m_2	m_3	m_1	8	9	7
B	r_2	r_3	r_1	5	7	2
B-q		$-r_2$	$-r_2$		-5	-5
		$(r_3\text{-}r_2)_{m3}$	$(r_1\text{-}r_2)_{m1}=\alpha$		2	$4=\alpha$
$(B\text{-}q)/m_2$	b=	$x(1/m_2)_{m3}$	$x(1/m_2)_{m1}$ =a		x8	x1
		a_2	a_1		$a_2=7$	$4=a_1$
			$-a_2$			-7
			$(a_1\text{-}a_2)_{m1}$			4=c
			$x(1/m_3)m_1$ =c			x4
			a_3			2=d

(a)

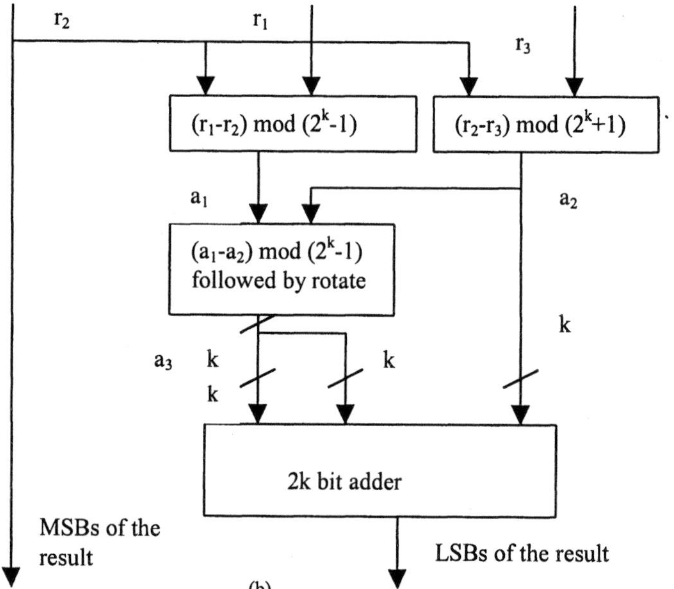

(b)

Fig 3.3 (a) MRC Technique for the moduli set $\{2^k\text{-}1, 2^k, 2^k+1\}$, (b) Architecture for implementation of (a).

Once thus, the mixed radix digits are obtained, the required binary word can be found to be

$$W= a_3.\ 2^k.(2^k+1) +a_2.2^k+r_2. \qquad (3.4)$$

Note, however, that the last k bits of W are r_2 itself. Thus, the 2k most significant bits of W can be obtained as $(W-r_2)/2^k= a_3.(2^k+1)+a_2$. Thus, a 2k bit adder is needed to evaluate the MSBs of W, to which appending r_2 yields the desired W. The architecture for implementation of this procedure is presented in Fig 3.3 (b). The two modulo subtractions for evaluating a_2 and a_1 can be performed in parallel followed by another modulo subtraction. This is followed by another 2-input 2k-bit addition.

This converter needs three n-bit modulo subtractors – two of them operating in parallel followed by one 2n-bit adder. The n-bit modulo m_i subtractors can be realized using cascade or parallel architectures [Bay87b] as described in Section.1.3.1. The area and delay of these converters are as follows:

Vinnakota and Rao Cascade (VRC) type design:

$$Area_{VRC} = 2n(\log_2 n+1)A_{FA} +6n (\log_2 n+1) A_{FA} + 3A_{MUX} +3n A_{INV} \qquad (3.5a)$$

$$Delay_{VRC} =(\log_2 2n+1)\Delta_{FA} +4(\log_2 n+1)\Delta_{FA} + 2\Delta_{INV} + 2\Delta_{MUX} \qquad (3.5b)$$

Vinnakota and Rao Parallel (VRP) type Design:

$$Area_{VRP} = 2n(\log_2 2n+1)A_{FA} +6n(\log_2 n+1)A_{FA} +3nA_{FA}+3A_{MUX} +3n A_{INV} \qquad (3.6a)$$

$$Delay_{VRP} =(\log_2 2n+1)\Delta_{FA} +2(\log_2 n+1)\Delta_{FA}+ 2\Delta_{MUX}+2\Delta_{INV} +2\Delta_{FA} \qquad (3.6b)$$

Note that A_{FA}, A_{MUX}, A_{INV}, A_{AND} are the areas of a full adder, an n-bit 2:1 Multiplexer, an inverter and an AND gate respectively whereas Δ_{FA}, Δ_{MUX}, Δ_{INV}, Δ_{AND} are the areas of a full adder, a 2:1 Multiplexer, an inverter and an AND gate respectively.

As an illustration for n=16RNS (i.e.48bit dynamic range), we obtain the following results:

Vinnakota and Rao Cascade (VRC) type design:

$$Area_{VRC} = 672.A_{FA} + 3A_{MUX} +48.\ A_{INV} \qquad (3.7a)$$

$$Delay_{VRC} =26.\Delta_{FA} + 2\Delta_{INV} + 2\Delta_{MUX} \qquad (3.7b)$$

Vinnakota and Rao Parallel (VRP) type Design:

$$Area_{VRP} = 720.A_{FA} +3A_{MUX} +48.\ A_{INV} \qquad (3.8a)$$

$$Delay_{VRP} =18.\Delta_{FA} + 2\Delta_{MUX}+2\Delta_{INV} \qquad (3.8b)$$

A design example is also presented in Fig 3.3 (a) in order to illustrate the technique.

3.3.2. Bernardson's RNS to binary conversion technique

Bernardson [Ber85] was the first to describe a non-ROM based design for RNS to binary conversion for the particular moduli set under consideration. His approach led to extensive research on non-ROM based implementations. In this method, CRT is first used on the residues corresponding to (2^k-1) and (2^k+1) to obtain the intermediate result P. Next, the moduli set $\{2^{2k}-1, 2^k\}$ is considered and MRC technique is applied to obtain the desired binary number W. It can be seen using CRT that

$$P = \{ [(1/(2^k+1)) \bmod (2^k-1)] .(2^k+1)r_1 + [(1/(2^k-1)) \bmod (2^k+1)].(2^k-1)r_3\} \bmod (2^{2k}-1) \quad (3.9)$$

Both the multiplicative inverses in (3.9) can be seen to be 2^{k-1}, thus yielding

$$P = \{2^{2k-1}.(r_1 + r_3) + 2^{k-1}(r_1 - r_3)\} \bmod (2^{2k}-1) \quad (3.10)$$

Next, the desired binary number W can be found to be

$$W=[(r_2-P \bmod 2^k) \bmod 2^k].[(1/(2^{2k}-1)) \bmod 2^k].(2^{2k}-1)+P \quad (3.11)$$

or

$$W=[(P'-r_2) \bmod (2^k)] .(2^{2k}-1))+P \quad (3.12)$$

where P' is the k bit LSB word of P. Note that the multiplicative inverse in (3.11) has been written as –1 yielding (3.12).

The architecture suggested by Bernardson is presented in Fig 3.4 (a). This architecture implements (3.9) using a ROM1 with a (2k+1) bit address and 2k-bit wide memory. The k LSBs of the ROM output yield P mod 2^k and using a subtractor and ROM2, the first term of (3.11) is implemented which when added to P yields the final output. An alternative implementation is presented in Fig 3.4 (b) wherein ROM2 is avoided and replaced by a 3k bit modulo 2^k subtractor while rest of the hardware is same as in Fig 3.4 (a). Another solution which does not need ROMs at all is presented in Fig 3.4 (c) wherein the modulo adder shown in Fig 3.4 (d) is employed.

The proof for this architecture is as follows. Given (a + b), we can write (a +b) mod $(2^{2k}-1)$ approximately as

$$(a+b) \bmod (2^{2k}-1) = \{ (a + b) - (2^{2k}-1)[(a + b)/2^{2k}]\} - \alpha. (2^{2k}-1) \quad (3.13)$$

where [ ] indicates the integer part. Denoting the { } part as R, α can be shown to be 0 for $R< 2^{2k}-1$ and 1 for $R>2^{2k}-1$. Note that R can have a maximum value $2^{2k}-2$ corresponding to $(a + b)_{max}$ of $2^{3k} - 2^{2k} - 2^k$. This architecture of Bernardson of Fig 3.4 (c) needs a (k+2) bit adder, a (k+1) bit adder, a (3k+2) bit adder, two (3k+2) bit subtractors, a 2k bit subtractor, a k bit subtractor, a 3k bit adder, a 3k bit subtractor and a 2k bit comparator.

3.3.3. Bi and Jones RNS to binary conversion technique

Bi and Jones [Bi88] described a simplification which needs four k bit adders and four (k+1) bit adders as shown in Fig 3.4 (e). Using the periodic properties of (a+b) mod $(2^{2k}-1)$, the value can be evaluated easily as shown in the architecture of Fig 3.4 (e) which fact was not recognized by Bernardson. Note further that $2^{k-1}.x \bmod (2^k-1)$ can be obtained by k-bit left circular shift as explained before. Denoting the desired W as

$$W = y_o + y_1.2^k + y_2.2^{2k} \quad (3.14)$$
$$= P + (2^{2k}-1).[P \bmod 2^k - r_2] \bmod 2^k$$

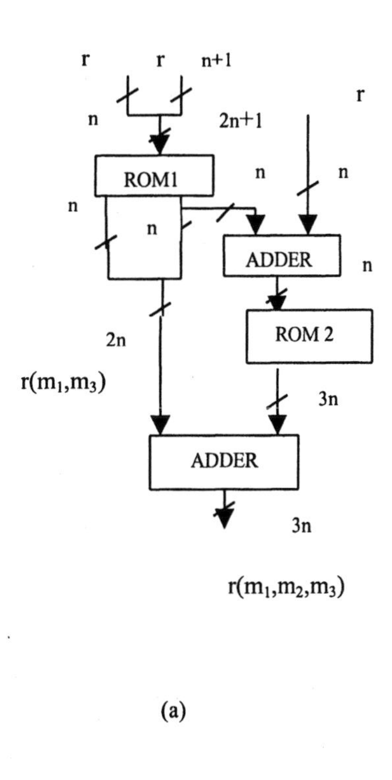

(a)

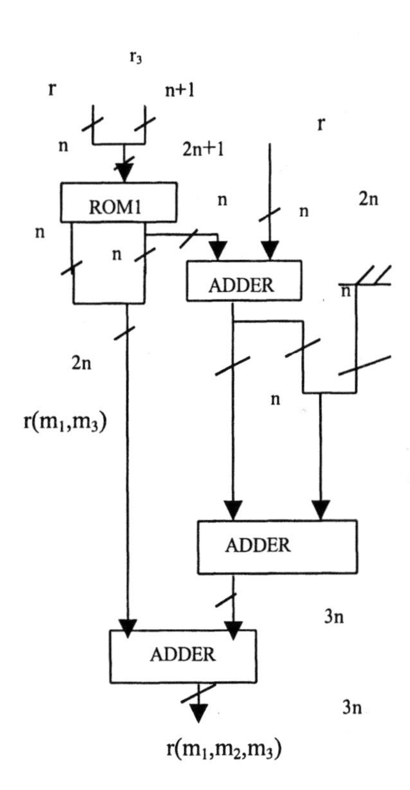

(b)

Fig 3.4 (contd)

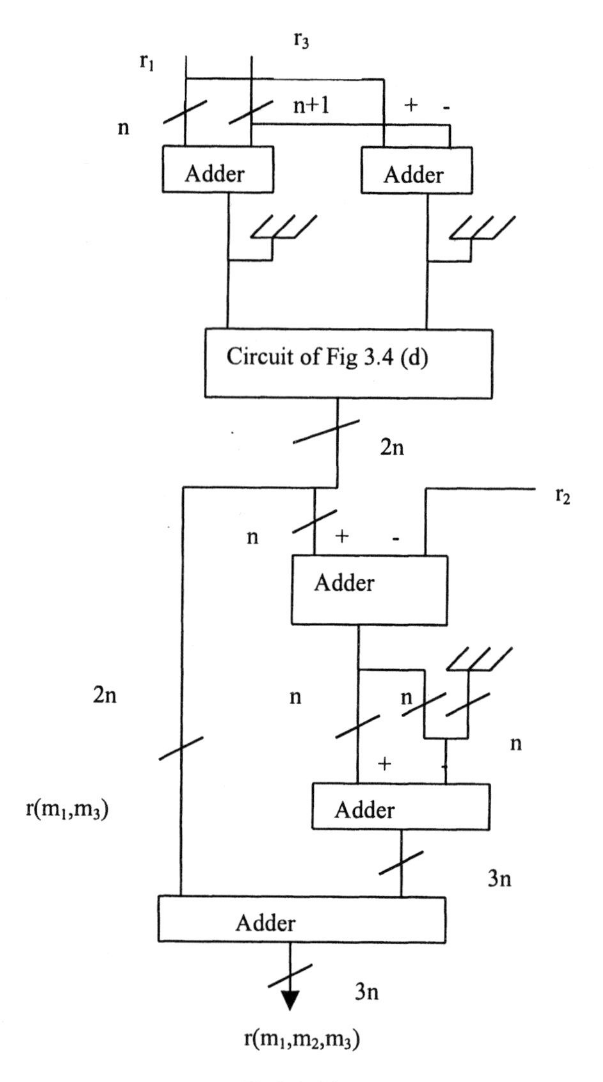

Fig 3.4 (c)

and writing the most and least significant fields of P in (3.10) as B and A, then y_o, y_1 and y_2 can be easily shown to be

$y_o = A - (A - r_2)$ (3.15a)

$y_1 = B -$ borrow obtained in (a) (3.15b)

$y_2 = (A - r_2) -$ borrow obtained in (b) (3.15c)

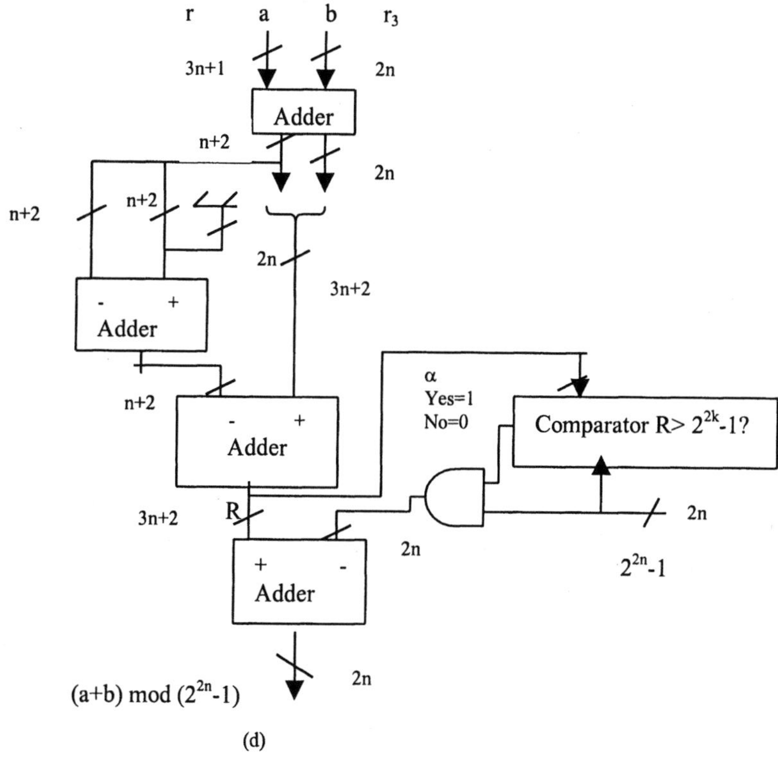

(d)

Fig 3.4 (contd)

The implementation is straight forward as shown in Fig 3.4 (e) needing only 8 adders. The left three adders produce $2^k.(r_1+r_3) + (r_1-r_3)$. The fourth adder realizes P where circular shifting and (n+1) bit addition is performed. The fifth adder evaluates $(P \bmod 2^k - r_2) \bmod 2^k$. The last three adders implement (3.15).

3.3.4. Ibrahim and Saloum RNS to binary conversion technique

Ibrahim and Saloum [Ibra88] have described another approach for RNS to binary conversion which also first determines r (m_1, m_3) =P and evaluates K next where

$$K = (P \bmod 2^k - r_2) \bmod 2^k \quad (3.16)$$

to yield

$$X = K(2^k-1)+P \quad (3.17)$$

They evaluate P as

$$P=D(2^k+1)/2 + r_3 \quad (3.18a)$$

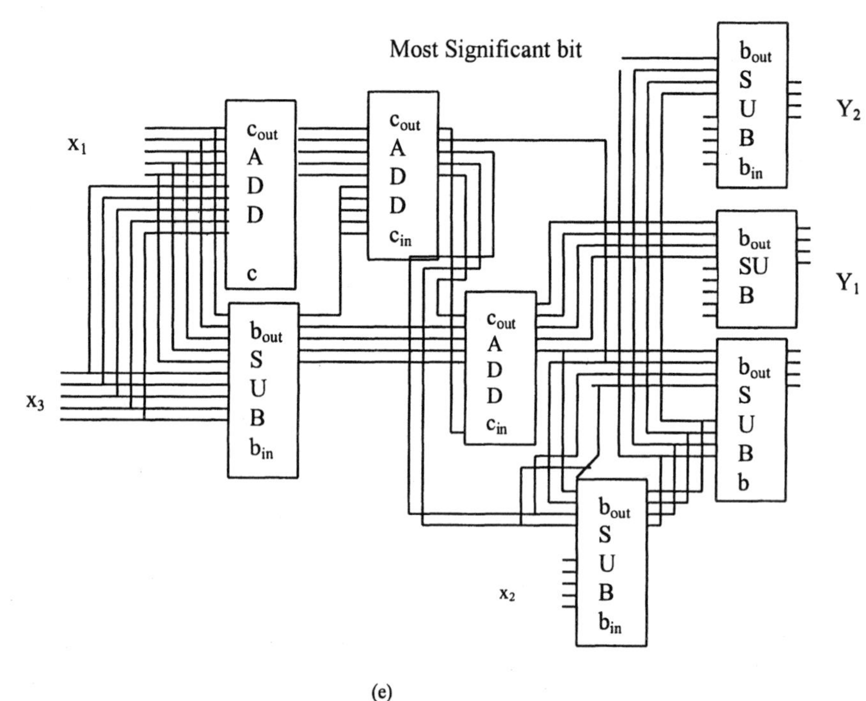

(e)

Fig 3.4. Reverse Converter Architectures Due to Bernardson: (a) Using ROMs only, (b) Using ROMs and Subtractors, (c) Using no ROMs, (d) Modulo Adder Architecture used in (c), and (e) Reverse Converter Architecture due to Bi and Jones ((a)-(d) adapted from [Ber85] 1985©IEEE and (e) adapted from [Bi88] 1988©IEEE)

$$P = (D + 2^k - 1).(2^k+1)/2 + r_3 \qquad (3.18b)$$
$$P = (D + 2(2^k - 1).(2^k+1)/2 + r_3 \qquad (3.18c)$$

depending on whether $D = (r_1 - r_3)$ is a nonnegative even integer, or an odd integer or negative even integer respectively. It can be seen that as against CRT for two moduli in the previous method due to Bernardson, Bi and Jones, here MRC is employed. However, the multiplicative inverse calculation is avoided. An architecture for evaluating P is shown in Fig 3.5 (a). Note that $(r_1 - r_3)$ is first evaluated in a full subtractor FS1. The LSB of the output indicates whether D is odd or even. The sign of the result is used to evaluate B. The proof is left to the reader and can be found in [Ibra88]. The next step is to evaluate B. $(2^k + 1)$ which can be done using a mapping of B as LSB word and MSB word by concatenation to yield a 2n bit word. The value B. $(2^k + 1)$ thus obtained is next added to r_3 in the 2k bit full adder FA1 to obtain P. P mod 2^k is simply the LSBs from which using a full subtractor FS2 (P mod $2^k - r_2$) mod 2^k is determined using additional exclusive OR gates and another full adder FA2 as shown in Fig 3.5 (b). If $(P - r_2)$ is positive, the sign bit is zero and the result is $(P - r_2)$. In the case of sign bit being one, the result is two's complement achieved by adding one to the ones complement of $(P-r_2)$ value realized by using the exclusive OR gates.

The last stage using a 3k bit full subtractor (see Fig 3.5 (c)) uses the two inputs one being K and the other $(K.2^k + P)$ a 3k bit word obtained by concatenating k bit K with 2k bit P as LSB. Thus the total hardware requirement is a (k +1) bit FS, a 2k bit FA, k bit FS, k bit FA, and 3k bit FS together with 2k exclusive OR gates.

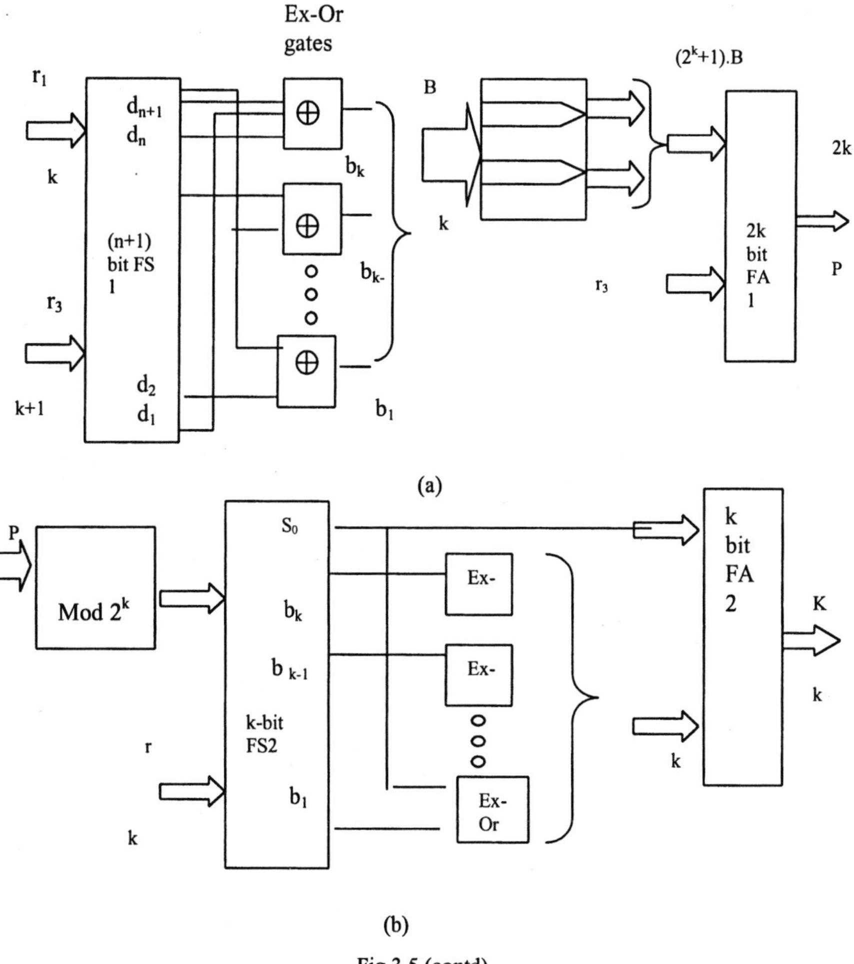

Fig 3.5 (contd)

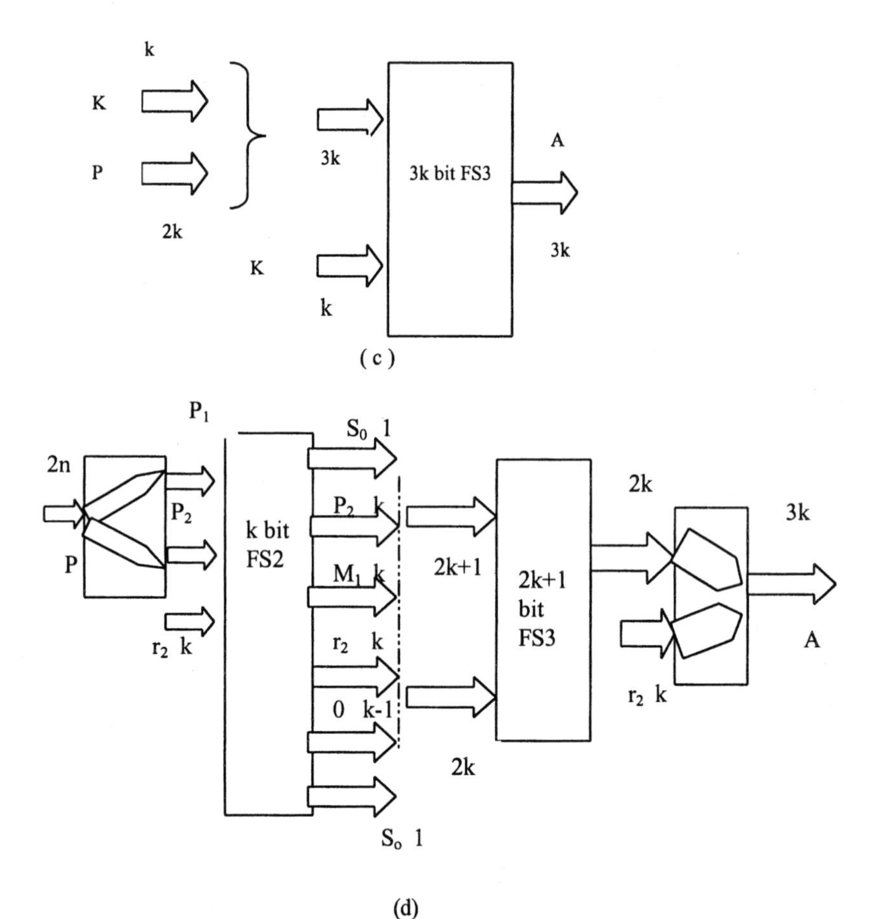

Fig 3.5. (a)-(c) Various blocks in Ibrahim and Saloum Reverse Converter Architecture and (d) Dhurkadas's modification of (a) ((a)-(c) adapted from [Ibra88] 1988©IEEE and (d) adapted from [Dhur90] 1990©IEEE)

3.3.5. Dhurkadas's modified Ibrahim and Saloum converter

Dhurkadas [Dhur90] suggested an interesting simplification of the above which is considered next. In this technique, once P is evaluated using the scheme of Fig 3.5 (a), then directly W can be evaluated without needing evaluation of K. This can be seen as follows. We know that

$$W = K\,(2^{2k} - 1) + P \tag{3.19a}$$

Denoting $P = P_1{}^{2k} + P_2$, and noting that $K = P_2 - r_2 + s_o.2^k$, where s_o can be 0 or 1 when P_2-r_2 is positive or negative respectively, we have from (3.19a)

$$A = (s_o2^{3k} + P_22^{2k} + P_12^k + r_2) - (r_2.\,2^{2k} + s_o2^k) = B - C \tag{3.19b}$$

We next observe that B can be easily formed as a 3k bit word by concatenating s_o, P_2, P_1, r_2 each of length 1, k, k and k bits as shown below:

$s_o p_k p_{(k-1)} \ldots\ldots p_1 p_{2k} p_{(2k-1)} \ldots\ldots p_{(k+1)} a_k \ldots\ldots a_2 a_1$.

Note that r_2 is defined as $a_k \ldots\ldots a_2 a_1$. Next C can be seen to be the 3n bit word

$a_k a_{(k-1)} \ldots\ldots a_1 0000 \ldots 0 s_o 00 \ldots\ldots 0$.

Noting that n LSBs are zero, B – C is evidently does not need a 3k-bit subtraction and can be performed using only the (2k + 1) MSBs as shown in the final architecture. Concatenating the result of the subtractor realizing (B – C) with r_2 yields A. Thus the hardware of Dhurkadas realization is much less than that of Ibrahim and Saloum architecture, as shown in Fig 3.5 (d).

3.3.6. Andraros and Ahmad architecture

Andraros and Ahmad [And88] have described a RNS to binary conversion technique based on CRT. Straight forward application of CRT for three moduli set $\{2^k+1, 2^k, 2^k-1\}$ yields corresponding to residues r_1, r_2 and r_3, the final result as

$$X = [2^k(2^k-1)(2^{k-1} - 1)r_1 + (2^{2k}-1)(2^k-1)r_2 + 2^k(2^k+1)2^{k-1}.r_3] \bmod \{2^k(2^{2k}-1)\} \quad (3.20)$$

where $M_i = M/m_i$ can be identified together with the multiplicative inverses of M_i with respect to m_i. Next $(X - r_2)$ can be computed and seen to be divisible by 2^k. This is advantageous, since the LSBs of the final result is r_2 as X mod $2^k = r_2$, given as data already. Thus $(X/2^k) \bmod (2^{2k} - 1)$ only needs to be evaluated:

$$(X/2^k) \bmod (2^{2k-1}) = [\ \{(2^{2k-1}+2^{k-1})r_1\} \bmod (2^{2k}-1) - r_1 + \{(2^{2k}-2^k-1)r_2\} \bmod (2^{2k-1}) + \{(2^{2k-1}+2^{k-1})r_3\} \bmod (2^{2k}-1)\ ] \bmod (2^{2k}-1)$$

$$=(C-r_1+B+A) \bmod (2^{2k}-1) \quad (3.21)$$

where we define

$$A =\{(2^{2k-1}+2^{k-1})r_3\} \bmod (2^{2k}-1) \quad (3.\ 22a)$$

$$B = \{(2^{2k}-2^k-1)r_2\} \bmod (2^{2k}-1) \quad (3.22b)$$

$$C = \{(2^{2k-1}+2^{k-1})r_1\} \bmod (2^{2k}-1) \quad (3.\ 22c)$$

Andraros and Ahmad next observe that A, B and C can be easily determined by looking at r_1, r_2 and r_3 and mapping their bits to new 2k bit words. Defining first r_1, r_2 and r_3 as $b_{1k} b_{1(k-1)} \ldots\ldots b_{11} b_{1o}$, $b_{2(k-1)} \ldots b_{21} b_{2o}$, and $b_{3(k-1)} b_{3(k-2)} \ldots b_{31} b_{30}$, we have

$A = b_{30} b_{3(k-1)} \ldots b_{32} b_{31} b_{30} b_{3(k-1)} \ldots\ldots b_{32} b_{31}$

$B = b'_{2(k-1)} b'_{2(k-2)} \ldots\ldots b'_{22} b'_{21} b'_{2o} 1111 \ldots 11$

$C = b_x b_{1(k-1)} \ldots b_{12} b_{11} b_x b_{1(k-1)} \ldots\ldots b_{11}$

where the prime indicates inverted bit and $b_x = b_{l0}$ or b_{lk} since both can neve be 1 at the same time. Note that we have used the property of $2^x.p \bmod (2^{2k}-1$ for evaluation by proper rotation of p as explained before.

Andraros and Ahmad suggest the architecture of Fig 3.6 (a), wherein parallell (A + B) and (C-r_l) are first computed. The carry of (A + B) since it has sam value as the LSB ($2^{2k} \bmod (2^{2k} - 1) = 1$), is added next in another 3 input adde with (A + B) and (C-r_l). The result and carry of this second stage are added i another adder with y_o to yield the final output. Note that y_o is obtained b AND ing all the bits of the output of the second adder. The addition of this y to the last stage facilitates modulo (2^{2k}-1) reduction of the output of the secon level adder.

The area and time requirements of this approach can be found to be as follows Andraros and Ahmad (AA) design:

$\text{Area}_{AA} = 8k(\log_2 k+1)A_{FA} + (2k-1)A_{AND} + 2kA_{INV}$ (3.23a)

$\text{Delay}_{AA} = 3(\log_2 2k+1)\Delta_{FA} + \Delta_{INV} + (\log_2 2k)\Delta_{AND}$ (3.23b)

As an illustration for k=16RNS (i.e.48bit dynamic range), we obtain the following results:

Andraros and Ahmad (AA) design:

$\text{Area}_{AA} = 768.A_{FA} + 32.A_{AND}$ (3.24a)

$\text{Delay}_{AA} = 18.\Delta_{FA}$ (3.24b)

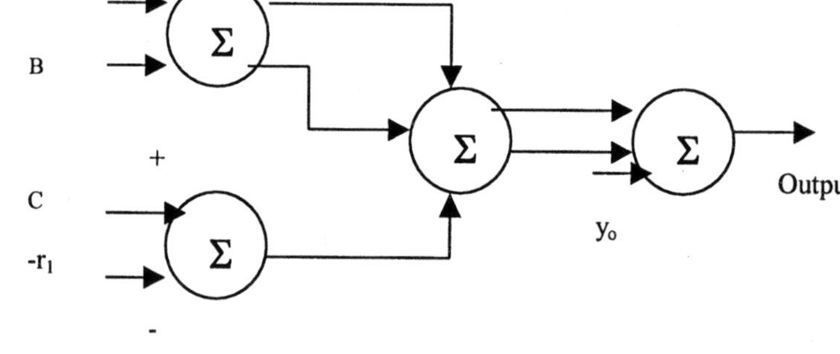

(a)

Fig 3.6 (contd)

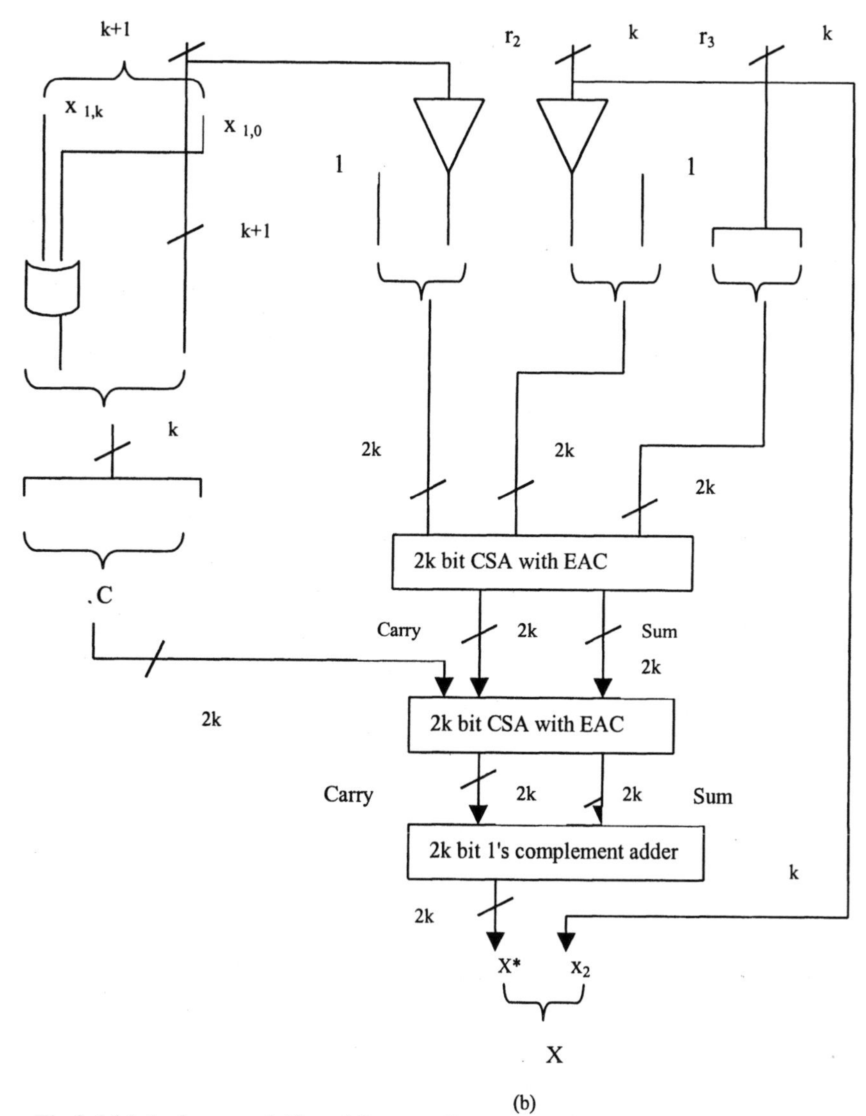

(b)

Fig 3.6 (a) Andraros and Ahmad Reverse Converter architecture, (b) Piestrak's modification of (a) ((a) adapted from [And88] 1988©IEEE and (b) adapted from [Pies95] 1995©IEEE)

3.3.7. Piestrak's modified Andraos - Ahmad architecture

Piestrak [Pies95] has suggested a fast implementation of Andraros and Ahmad implementation by observing that the four operands in (3.21) need not be added in three stages but can be added much faster using a carry save adder concept. For this purpose, first $(-r_1) \bmod (2^{2k}-1)$ is also recognized to be written easily as

$\{ 1111\ldots1r'_{1k}r'_{1(k-1)} \ldots\ldots r'_{11}r'_{10} \}$

Next the four operands A,B,C and $-r_1$ can be added using the architecture of Fig 3.6 (b). Note that EAC (end around carry) concept can be used since the carry having the weight $2^{2k} \bmod (2^{2k}-1)$ is 1. Note that full adder cells are needed (2k for each level of carry save addition) and a final Carry Propagate adder with EAC is needed. Note that the delay for computing the result is $2\Delta_{FA} + (4k-1)\Delta_{FA}$ assuming a CPA. An alternative version to reduce the delay at the expense of hardware can use two parallel adders calculating x+y and x+y+1 and select the former or latter depending on whether x+y resulted in a carry or not as mentioned earlier. The effective delay is thus reduced.

It may also be noted that the full adder cells having logic one as one input can be simplified to an exclusive-NOR gate and an OR gate as pointed out by Piestrak. Thus, 2k Full adder cells can be reduced to 2k Exclusive NOR gates and 2k OR gates, thus saving hardware.

The area and time requirements of this approach using cascade adders (cost-effective design) and using parallel adders (high-speed design) are evaluated below.

Piestrak Cost-effective design:

$\text{Area}_{PCE} = 2k(\log_2 2k+1)A_{FA} + 4k.A_{FA}$ (3.25a)

$\text{Delay}_{PCE} = 2(\log_2 2k+1)\Delta_{FA} + 2\Delta_{FA}$ (3.25b)

Piestrak High Sped (PHS) design:

$\text{Area}_{PHS} = 4k(\log_2 2k+1)A_{FA} + 4k.A_{FA} + 2\,A_{MUX}$ (3.26a)

$\text{Delay}_{PHS} = (\log_2 2k+1)\Delta_{FA} + 2\Delta_{FA} + \Delta_{MUX}$ (3.26b)

Operand	2^{2k-1}	2^{2k-2}	.....	2^{k+1}	2^k	2^{k-1}	2^{k-2}	2^0
A	$x_{3,0}$	$x_{3,k-1}$		$x_{3,2}$	$x_{3,1}$	$x_{3,0}$	$x_{3,k-1}$	$x_{3,1}$
B	$x_{2,k-1}'$	$x_{2,k-2}'$		$x_{2,1}'$	$x_{2,0}'$	1	1	1
C	$(x_{1,0} + x_{1,1})$	$x_{1,k-1}$....		$x_{1,2}$	$x_{1,1}$	$(x_{1,0} + x_{1,k})$	$x_{1,k-1}$ ...	$x_{1,1}$
$-X_1$	1	1		1	$x_{1,k}'$	$x_{1,k-1}'$	$x_{1,k-2}'$....	$x_{1,0}'$

(a)

2^{2k-1}	2^{2k-2}	.....	2^{k+1}	2^k	2^{k-1}	2^{k-2}	2^0
$x_{3,0}$	$x_{3,k-1}$		$x_{3,2}$	$x_{3,1}$	$x_{3,0}$	$x_{3,k-1}$	$x_{3,1}$
$x_{2,k-1}'$	$x_{2,k-2}'$		$x_{2,1}'$	$x_{2,0}'$	$(x_{1,k}' + x_{1,k-1}')$	$x_{1,k-2}'$....	$x_{1,0}'$
$(x_{1,k} + x_{1,k-1})$	$x_{1,k-1}$		$x_{1,2}$	$x_{1,1}$	$x_{1,0}$	$x_{1,k-1}$	$x_{1,1}$

(b)

Fig 3.7 (contd)

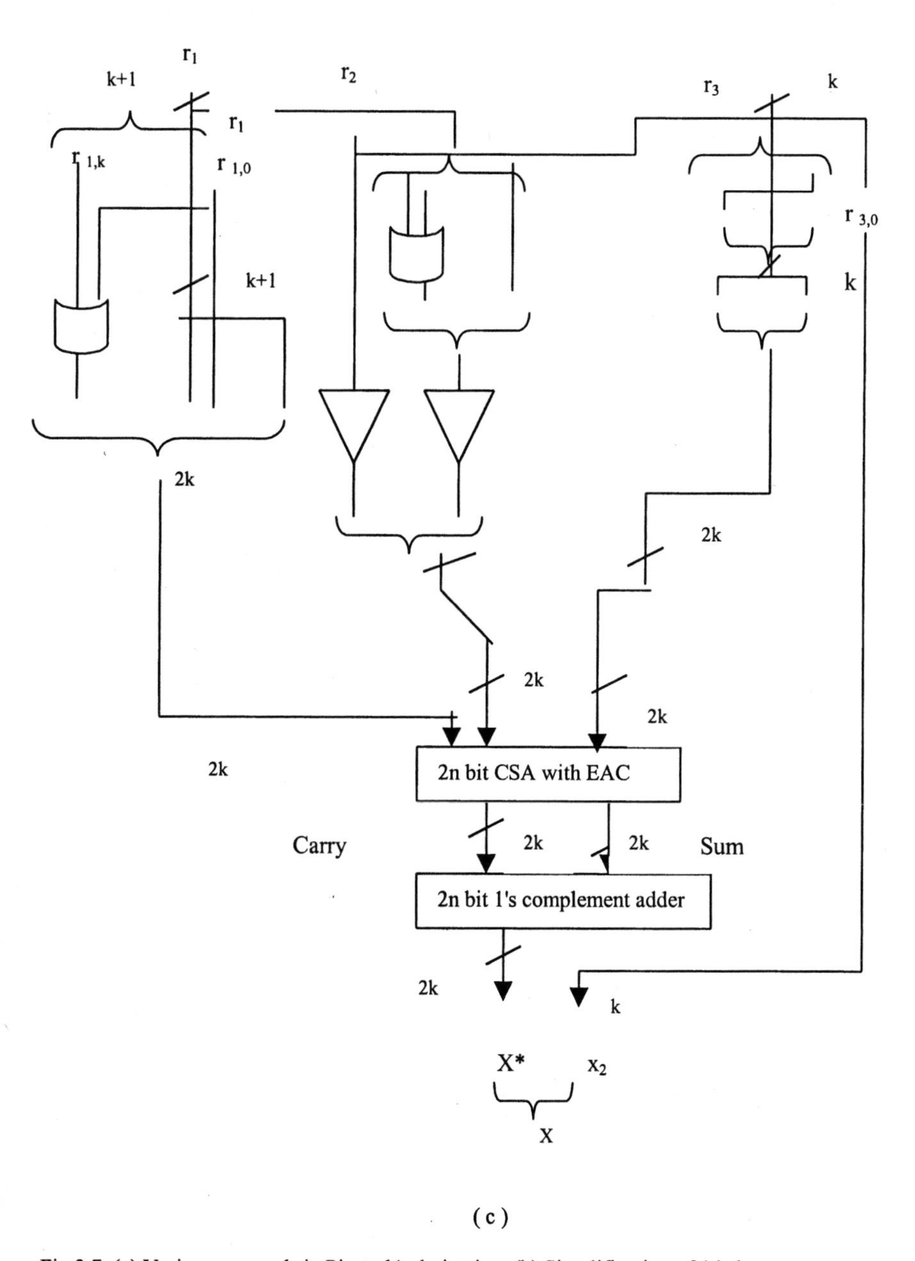

(c)

Fig 3.7. (a) Various operands in Piestrak's derivation, (b) Simplification of (a) due to Dhurkadas (c) Hardware implementation of Dhurkadas's converter (adapted from [Dhur98] 1998©IEEE)

As an illustration for n=16 RNS (i.e.48bit dynamic range), we obtain the following results:
Piestrak Cost Effective design:

$$Area_{PCE} = 256.A_{FA} \quad (3.27a)$$

$$Delay_{PCE} = 14.\Delta_{FA} \quad (3.27b)$$

Piestrak High Speed (PHS) design:

$$Area_{PHS} = 448.A_{FA} + 2 A_{MUX} \quad (3.28a)$$

$$Delay_{PHS} = 8.\Delta_{FA} + \Delta_{MUX} \quad (3.28b)$$

It can be seen that the Vinnakota and Rao converters of Fig 3.3 using cascade adders exhibit more delay than the Andraros and Ahmad design. Note that in the Andraros and Ahmad design the full adders needed can be reduced by noting that some bits in the input B in (3.22b) are constant 1. The high-speed version of the Vinnakota and Rao converter exhibits similar delay and area requirements as Andraros and Ahmad technique. However, both the Piestrak's cost effective and high-speed designs are superior to Vinnakota and Rao's high-speed as well as cost-effective designs.

3.3.8. Dhurkadas's modification of Piestrak converter

Dhurkadas [Dhur98] simplified matters further by having a close look at the operands A, B, C and $-r_1$ mod $(2^{2k}-1)$. This is considered next. It is useful to recollect the structure of A,B,C and $-r_1$ redrawn in Fig 3.7 (a). It can be seen that in two operands viz., B and $-X_1$, some bits are logic ones albeit in mutually exclusive locations. Dhurkadas recommends writing the new operands as shown in Fig 3.7 (b). Since only three operands are involved to be added, second level of carry save adders in Fig 3.6 (b) need not be employed, thus enhancing the speed as well as reducing the area.

This is explained as follows. When $x_{1,k} = 1$, the 2^{k-1} column in Piestrak's matrix of Fig 3.7 (a) becomes $p=x_{3,0}$, q=1, r=1 and s=1 corresponding to A, B, C and $-r_1$. In the 2^k column, $x_{1,k}$ is zero. Thus by adding the q and r bits in the 2^{k-1} column gives a zero in the 2^{k-1} column of B and the carry generated makes the 2^k column in $-X_1$ as one. Thus all elements in all the bit positions in the Piestak matrix becomes 1 thus facilitating elimination of one row since residue of that word with respect to $2^{2k}-1$ is zero. Thus the zeroing of q conditionally, when only $x_{1,n} = 1$ is needed which is accomplished as shown in Fig 3.7 (b). With these improvements, Andraros-Ahmad technique with modifications by Piestrak and Dhurkadas is the fastest possible RNS to binary converter in this moduli set. The hardware implementation is as shown in Fig 3.7 (c).

The area and time requirements of this approach using cascade adders and using parallel adders are as follows:
Dhurkadas Modified Piestrak High Speed (PHS) design:
$\text{Area}_{DMP} = 4n(\log_2 2n+1)A_{FA} + 2n.A_{FA} + 2\,A_{MUX}$ (3.29a)
$\text{Delay}_{DMP} = (\log_2 2n+1)\Delta_{FA} + \Delta_{FA} + \Delta_{MUX}$ (3.29b)
As an illustration for n=16RNS (i.e.48bit dynamic range), we obtain the following results:
Dhurkadas Modified Piestrak High Speed (PHS) design:
$\text{Area}_{DMP} = 416.A_{FA} + 2\,A_{MUX}$ (3.30a)
$\text{Delay}_{DMP} = 7.\Delta_{FA} + \Delta_{MUX}$ (3.30b)

3.3.9. Wang, Jullien and Miller modification of Piestrak's technique

Wang, Jullien and Miller [Wan00] suggested a technique similar to Dhurkadas's simplification of Piestrak's method which will be described next. In this technique, all the four operands to be added in Andraros and Ahmad and Piestrak's technique in (3.17) will be reduced to three thus saving one level of carry-save-adders. Wang et al [Wan00] suggest rewriting $B + C - r_1$ as two words instead of three needed in Andraros and Ahmad method. The result will be two words one exclusively dependent on bits of r_1 and another on bits of r_2. It is easy to see that
$B = [(2^{2k}-2^k-1).r_2] \bmod (2^{2k}-1) = [-r_2] \bmod (2^{2k}-1)$ (3.31a)
Denoting the bits of r_2 as $b_{k-1}\, b_{k-2} \ldots b_1 b_0$, we can obtain b as before by ones complementing the bits of r_2 appended by k zeroes. Wang et al split this word as two words to yield
$B = 0\ldots\ldots01\ldots.1 + b'_{k-1}\, b'_{k-2} \ldots b'0\ 000\ldots0$ (3.31b)
Note that k number of zeroes and ones are present in the two 2k bit words. Next,
$(C-r_1) \bmod (2^{2k}-1) = [(2^{2k-1} + 2^{k-1} - 1).r_1] \bmod (2^{2k}-1) = [(-2^{2k-1} + 2^{k-1}).r_1] \bmod (2^{2k}-1)$ (3.32)
Noting that r_1 is a k+1 bit number denoted by $c_k\, c_{k-1} \ldots c_1 c_0$, $(C-r_1+B) \bmod (2^{2k}-1)$ is the sum of four words:
$(C-r_1+B) \bmod (2^{2k}-1) = [c'_0 c'_{k-1} \ldots c0c'_{k-1} \ldots c'_1 + c_k\, 1 \ldots 1c'_k 0 \ldots.0$
$+ 0\ldots\ldots01\ldots.1 + b'_{k-1}\, b'_{k-2} \ldots b'_0\ 000\ldots0] \bmod (2^{2k}-1)$ (3.33a)
Here, they observe that the (3.33) can be simplified as
$(C-r_1+B) \bmod (2^{2k}-1) = c'_0 c'_{k-1} \ldots c'_0 c'_{k-1} \ldots. c'_1 + b_{k-1}\, b'_{k-2} \ldots b'_0\, b_{k-1}\, b_{k-1} \ldots b_{k-1}$ (3.33b)

We interpret this step as adding $(2^k-1).2^{k-1}$ to B and subtracting same numbe from $C-r_l$. An example will illustrate the simplification. Consider k=3, i which case the original $(C-r_l+B)$ mod 63 will be
$(C-r_l+B) \bmod 63 = c'_0c_2c_1c_0c'_2c'_1 + c_311c'_300 + 000111 + b'_2b'_1b'_0000$ =
$c'_0c_2{}^*c_1{}^*c_0{}^*c'_2c'_1 + b_2b'_1b'_0b_2b_2b_2$ (3.33c)
Note that c^*_i means that c_i OR c_k. Note that the last two terms in (3.33c) firs equality, when added with 35 =(-28) mod 63 yield the single term in (3.33c second equality. Note that 28 corresponds to $(2^k-1).\ 2^{k-1}$ for k=3. Similarly the first two terms when added with 28 yield the first term. Since when c_3 =1 all other bits will be zero in r_l and when c_3=0, for all other values of r_l, th addition is replaced by OR-ing as indicated by asterisks. Evidently, the are and time requirements of Wang et al are same as in Dhurkadas technique.

3.3.10. Bharadwaj, Premkumar and Srikanthan design o RNS to binary converter.

Bharadwaj et al [Bhar98] have observed that in the expressions for A,B,C an $-r_l$ in Piestrak's technique (see Fig 3.7 (a)), the left and right fields in A and C are same. Similarly in B and $-X_1$, the bits with logic one value are symmetri in the left and right fields except for one bit x_{1k}' in $-X_1$. Thus, they observe that this property can be used to reduce half the number of adders in the firs level of carry save adders in Piestrak's implementation. Note that the presence of x_{1k}' bit however needs one extra full adder cell for computation of the lef field. This architecture is shown in Fig 3.8 (a).

Bharadwaj et al suggest also an additional modification so as to increase the conversion speed while using a CPA stage in the second stage of Piestrak's converter in the high-speed version. The original Piestrak converter i redrawn in Fig 3.8 (b) in which it can be seen that instead of 2k bit CPA, we have shown two k bit CPAs cascaded. It is interesting to note that the inputs for n bit CPAs 1 and 3 are same except for the carry input. Similarly, the inputs for the n bit CPAs 0 and 2 are same except that the carry input C_1 and C_3 can be 0 or 1. Thus instead of waiting for the carry C_1 and C_3 to determine C_0 and C_2, Bharadwaj et al suggest that the CPA2 can be fed with 0 as input carry and CPA0 with input carry as 1, thus making all the options available. Then, the carries C_1, C_2, C_3 and C_4 can be used to select the appropriate output of the lower field from CPA1 and CPA3 and the upper field from one of CPA2 and CPA0 as shown in Fig 3.8 (b). The logic signals g and f are related to C_1, C_2, C_3 and C_4 as follows:

$f(C_0, C_1, C_2, C_3) = C_1.C_3 + C_0.C_3 + C_1\,C_2'$ (3.34a)

$g(C_0, C_1, C_2, C_3) = C_1C_2 + C_0C_1'$ (3.34b)

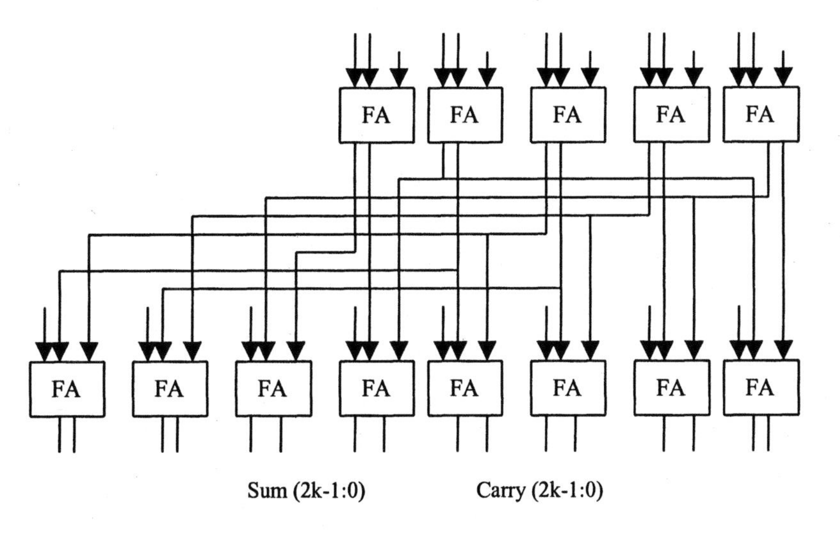

(a)

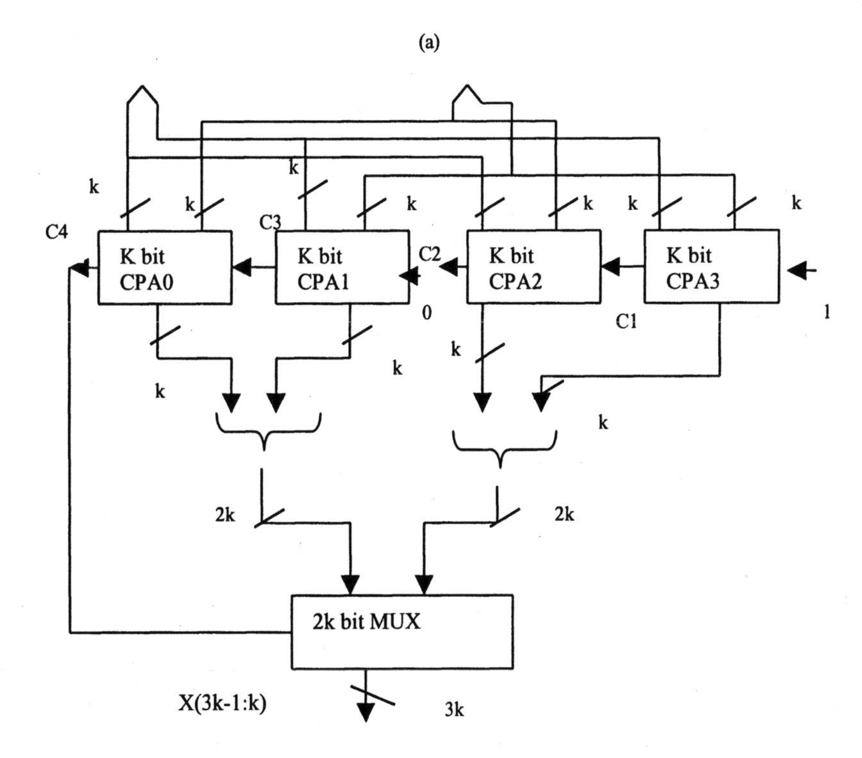

(b)

Fig 3.8 (contd)

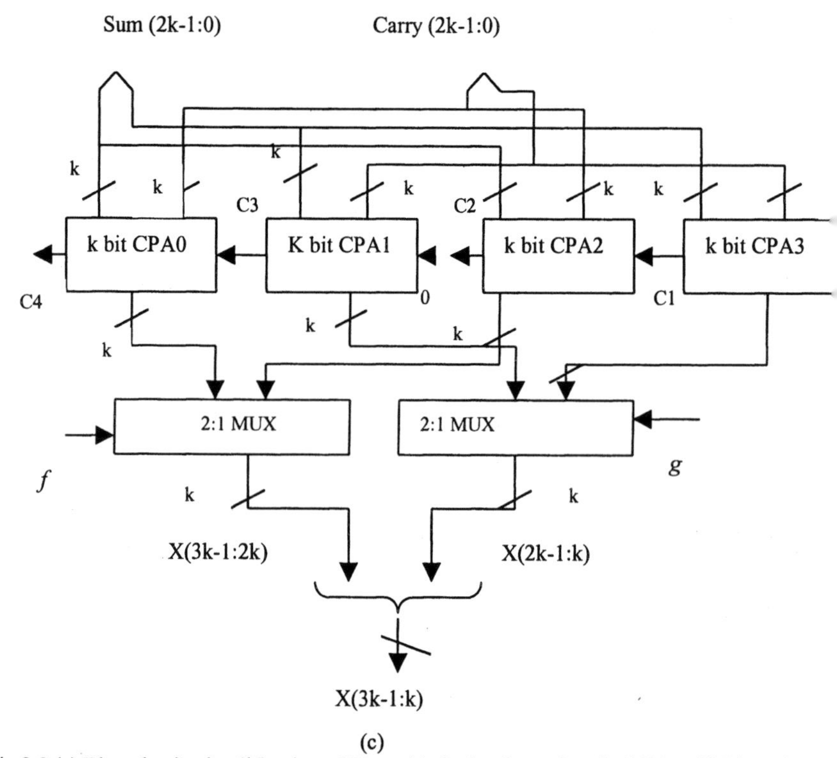

Fig 3.8 (a) Bharadwaj's simplification of Piestrak's design for saving (k-1) FAs, (b) Piestrak's final CPA architecture (redrawn) and (c) Modification to reduce the adder delay (adapted from [Bhar98] 1998©IEEE)

With this modification, it can be seen that the delay for RNS to binary conversion reduces to that based on a k bit CPA rather than a 2k bit CPA. For small bit length RNS, this may be attractive in that the area is not increased due to the change suggested. However, Ananda Mohan [Anan01] observed that if e.g., Brent-Kung adder is employed, the area and computation time evaluation shows that the area of Piestrak's approach is only 12.5% greater in the second level than that of Bharadwaj et al approach and further, the decrease in delay in Bharadwaj et al approach is marginal –just a full adder delay. This is explained by the area expression: $4k(\log_2 2k+1).A_{FA}$ vis a vis $4.k.(\log_2 k+1).A_{FA}$ and delay expressions $(\log_2 2k+1)\Delta_{FA}$ and $(\log_2 k+1)\Delta_{FA}$. Typically, for k =64, i.e. a 192 bit dynamic range RNS, areas are $2048.A_{FA}$ and $1792A_{FA}$ whereas the delays are $8\Delta_{FA}$ and $7\Delta_{FA}$ for Piestrak's and Bharadwaj et al techniques respectively.

Bharadwaj et al also observe that the Piestrak technique has a problem for RNS to binary conversion for residues of the form {x, x, x}. These get mapped to 111111...1x, where x is the given k bit binary word preceded by 2k logic one bits. While this conversion maps to a unique range between $2^k(2^{2k}-1)$ and 2^{3k}, it needs a correction that MSBs shall be detected for all ones condition and zeroed. Bharadwaj et al suggest that for this anomalous case, $C_0 = C_1 = 0$ and $C_2 = C_3 = 1$. Hence, modification of f and g given in (3.17a) and (3.17b) is needed as follows:

$$f'(C_0,C_1,C_2,C_3) = f(C_0,C_1,C_2,C_3) + C_0\, C_1'\, C_2C_3 \quad (3.35a)$$

$$g'(C_0,C_1,C_2,C_3) = g(C_0,C_1,C_2,C_3) + C_0'C_1'C_2C_3 \quad (3.35b)$$

Note, however, that Piestrak has suggested the use of 2-input AND gate in the EAC (end around carry) path to alleviate this problem, following Wakerley's suggestion.

3.3.11. Gallaher, Petry and Padmini RNS to binary converter

An interesting approach to RNS to binary conversion for the moduli set $\{2^k+1, 2^k, 2^k-1\}$ has been considered by Gallaher, Petry and Srinivasan [Gall97], which though not superior to DMP design, is considered for the sake of completeness. In this method, called Digit Parallel Method, the three k bit fields of the desired binary word are parallelly generated.

Already, we have observed that if D_2, D_1, D_o are the three k bit fields in a 3k bit binary number to be determined, the residues mod 2^k, mod 2^k-1, and mod 2^k+1 can be expressed as

$$r_2 = D_o \quad (3.36a)$$

$$r_3 = (D_2 - D_1 + D_o) \bmod (2^k+1) \quad (3.36b)$$

and

$$r_1 = (D_2 + D_1 + D_o) \bmod (2^k-1) \quad (3.36c)$$

Gallaher et al express r_1 and r_3 as

$$r_3 = (D_2 - D_1 + D_o) - n(2^k+1) \quad (3.37a)$$

$$r_1 = (D_2 + D_1 + D_o) - m.(2^k-1) \quad (3.37b)$$

where m and n need to be determined uniquely. Note that D_o is already known from (3.36a). They rewrite (3.37a) and (3.37b) as

$$D_1 = X_1 + L_1 = \{(r_1 - r_3)/2\} + \{m.(2^k-1)/2 - n(2^k+1)/2\} \quad (3.38a)$$

$$D_2 = X_2 + L_2 = \{(r_1 + r_3)/2 - r_2\} + \{m.(2^k-1)/2 + n(2^k+1)/2\} \quad (3.38b)$$

Gallaher et al observe that among the choices of m and n possible, it turns out that m can be {0, 1, 2} and n can be {0, -1, 1}. Among the pairings possible for m and n, some are ruled out and the remaining are available in two classes depending on whether (m + n) is even or (m+n) is odd. These are as follows:

	even (m+n)			odd (m+n)	
	(m , n)	m+n		(m,n)	m + n
Case 1	(0, 0)	0	Case 5	(0,-1)	-1

Case 2	(1, -1)	0	Case 6	(1,0)	1
Case 3	(1, 1)	2	Case 7	(2,1)	3
Case 4	(2, 0)	2			

The implementation is next as shown in Fig 3.9. There are five blocks in this implementation. The top two blocks evaluate X_1, X_2 and C_1 which are defined in the Figure and can be identified as the first terms in (3.38a) and (3.38b). C_1 denotes whether $r_1 + r_3$ is odd or even. Based on the values of X_1 and X_2, the (m +n) pair can be selected using the logic described in Fig 3.10. Next, based on the chosen (m, n), the second terms in (3.35a) and (3.35b) viz., L_1 and L_2 are evaluated which when added to the first terms already available yield the desired D_1 and D_2 values. The authors suggest that L_1 and L_2 can be stored in a PROM for high-speed realizations. Only n bit adders are needed and two adder stages in cascade are needed.

Ananda Mohan [Anan00a] observed that the algorithm of Fig 3.11 needs some change to take care of the case (m, n) = (1, 1) and (1, -1) when X_1 =0 and X_2 = 0 respectively. As an illustration, the following two examples will be considered:
Example 1. r_1 =1, r_3=1 and r_2 =7.
For this case, X_1 = 0 and X_2 =-6. According to Gallaher et al, (1, 1) needs to be selected for (m, n) yielding L_1 =-1 and L_2 =8. Thus, D_1 = -1 and D_2 =2 needing additional operations to obtain the result.
Example 2. r_1 = 3, r_3=5 and r_2 = 4.
For this case, X_1 = -1 and X_2 = 0. According to Gallaher et al, (1, -1) needs to be selected for (m, n) yielding L_1 = 8 and L_2 = -1. Thus, D_1 = 7 and D_2 = -1 needing additional operations to obtain the result.

We note that correct results can be obtained in both the cases by choosing (m, n) as (2, 0) yielding L_1 = 7 and L_2=7. Thus, the selection function in Fig 3.11 for (2, 0) needs to be modified as (2, 0) = $C_1 \wedge (C_7.C_4 \vee C_8.C_3)$ where C_7= X_1 is zero and C_8= X_2 is zero. The detection of C_7 and C_8 needs two (k+1) input AND gates arranged as $\log_2(k+1)$ level 2-input AND gate array.

Ananda Mohan also pointed out the equivalence of Andraros-Ahmad technique and Gallaher et al technique. In Andraros-Ahmad method described in section 3.2.6, CRT is used to determine W, a 2k bit word $D_2.2^k + D_1$. From (3.38a) and (3.38b), the evaluation of W yields

$$W = 2^{k-1}(r_1 + r_3)-2^k.r_2 + ((r_1-r_3)/2)+((m+n)/2).(2^{2k}-1) \quad (3.39a)$$

We note that W mod $(2^{2k}-1)$ eliminates the need for knowing (m +n) i.e. the last term in (3.39a). Next (1/2) mod $(2^{2k}-1)$ can be seen to be 2^{2k-1} yielding W as

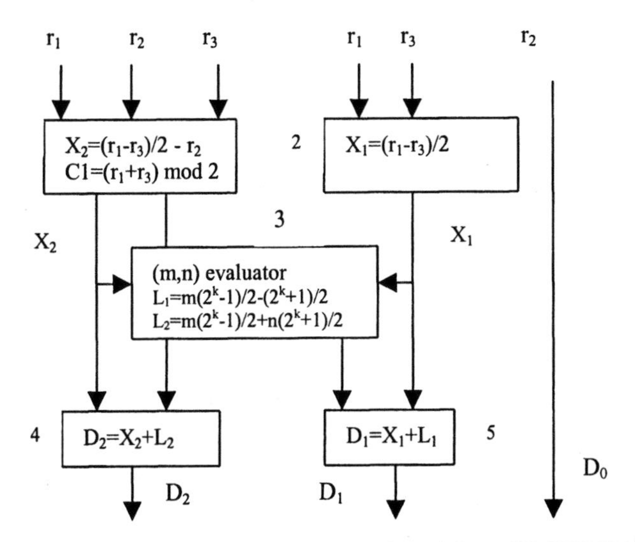

Fig 3.9. Gallaher et al Reverse Converter architecture (adapted from [Gall97] 1997©IEEE)

Function	Meaning
$C_1= (a_{10} \oplus a_{30})'$	LSBs of a_1 and a_3 equal?
$C_2=x_{1\ k+1} \vee x_{2\ k+1}$	X_1 or X_2 is negative?
$C_3 = x_{1\ k+1}$	X_1 is negative?
$C_4=x_{2\ k+1}$	X_2 negative?
$C_5 = x_{2\ k+1}{}' \wedge x_{2k}$	$X_2 \geq 2^k$?
$C_6 = x_{2\ k+1} \wedge x_{2\ k}'$	$X_2 \leq -2^k+1$?

C_2 will be expressed as $C_3 v C_4$ in the (b) part of this figure.

(a)

$(0,0) = C_1 \wedge C_3' \wedge C_4'$
$(1,-1)= C_1 \wedge C_3 \wedge C_4'$
$(1,1)= C_1 \wedge C_3' \wedge C_4$
$(2,0) = C_1 \wedge C_3 \wedge C_4$
$(0,-1)= C_1' \wedge C_5$
$(2,1) = C_1' \wedge C_6$
$(1,0) = C_1' \wedge C_5' \wedge C_6'$

(b)

Fig 3.10 (a) Selection functions and (b) selection logic for (m, n) pairs (adapted from [Gall97] 1997©IEEE)

$$W = r_1(2^{k-1} + 2^{2k-1}) + r_3(2^{k-1} - 2^{2k-1}) - 2^k.r_2 \qquad (3.39b)$$

which is same as (3.21a) of Andraros and Ahmad technique.

The Gallaher et al technique needs a 3-input k bit adder and a 2-input k bit adder working in parallel. The (m, n) evaluation block yielding L_1 and L_2 comprises of two 7:1 k bit multiplexers. This is followed by two 2-input k bit adders operating in parallel. The area and time requirements of Gallaher et al (denoted by the suffix GPS) architecture assuming regular VLSI layout fo adders can be shown to be as follows:

$$A_{GPS} = kA_{FA} + 4k(\log_2 k+1)A_{FA} + 4kA_{MUX} + 2k\log_2 k.A_{AND}+2kA_{INV} \quad (3.40a)$$

$$\Delta_{GPS} = \Delta_{FA} + 2(\log_2 k+1)\Delta_{FA} \quad (3.40b)$$

Note that in (3.40 a), the first term corresponds to a 3-input carry-save-adder's first level of adders and the second term to two parallel adders computing X and X_2. The third term is for selecting the summand based on (m, n) logic and the last term corresponds to the zero detection logic for both X_1 and X_2. Note that the MUXes can be replaced by a ROM of 14k bit size, if so desired. In (3.40), A_{FA}, A_{MUX}, A_{INV} and A_{AND} denote the areas of a full-adder, a k bit 2:1 multiplexer, an inverter and a two-input AND gate respectively whereas Δ_{FA}, Δ_{MUX}, Δ_{INV} and Δ_{AND} denote the delays.

We next consider an alternative proof for the derivation of L_1 and L_2 without needing the analysis about (m, n) pairs possible and the choice based on the range of X_1 and X_2. The starting point of this alternative derivation is the availability of X_1 and X_2 i.e. the first terms in (3.38a) and (3.38b). Evidently $r_1 \le 2^k-2$, $r_3 \le 2^k$ hold, whence we obtain the bounds on X_1 and X_2 as

$$-2^{k-1} \le X_1 \le 2^{k-1} - 1 \quad (3.41a)$$

and

$$-(2^k-1) \le X_2 \le (2^k-1) \quad (3.41b)$$

Note that X_1 and X_2 assume integer values for $(r_1 \pm r_3)$ even and are odd multiples of ½ when $(r_1 \pm r_3)$ is odd. In the even case, it may be seen from (3.41) that four cases arise viz., (a) X_1 and X_2 positive, (b) X_1 positive and X_2 negative, (c) X_1 negative and X_2 positive and (d) X_1 and X_2 negative.

The aim of the digit parallel method is to add quantities L_1 and L_2 to X_1 and X_2 to yield the desired positive D_1 and D_2 values such that $0 \le D_1 \le 2^k-1$ and $0 \le D_2 \le 2^k-1$. Further, the quantity added is such that $(2^k.L_2+L_1) \bmod (2^{2k}-1) = 0$ so that the addition of L_1 and L_2 does not change the original word $(2^kX_2+X_1)$. The even and odd cases are considered separately.

(a) $(r_1 \pm r_3)$ even:

In case (a), $L_1=L_2=0$, since both X_1 and X_2 are already positive. In the case (b), $L_1= -1$ and $L_2=2^k$ whereas in the case (c), $L_1 =2^k$ and $L_2 = -1$. In the case (d), we have $L_1 = 2^k-1$ and $L_2 = 2^k-1$.

In the case (b), note that the negative X_2 is made positive by adding 2^k which increases the value of $(2^k.X_2+X_1)$ by 2^{2k}. This increase is compensated by subtracting 1 from X_1 since $2^{2k} \bmod (2^{2k}-1) =1$. In a similar manner for case (c), negative X_1 is made positive by adding 2^k and this addition is compensated by subtracting from X_2 unity since weight of X_2 is already 2^k. In case (d), follows that the addition of 2^{2k}-1 yields a positive answer. Note that the cases (a) –(d) correspond to (m, n) pairs viz., (0, 0), (1, 1), (1, -1) and (2,0).

(b) $(r_1 \pm r_3)$ odd :

The case for $(r_1 \pm r_3)$ odd needs to be considered next. The presence of ½ term in both X_1 and X_2 needs a different approach. First, the ½ term is eliminated (rounded) by adding $(2^k-1)/2$ to both X_1 and X_2. It may be verified that this addition amounts to the addition of $(2^{2k}-1)/2$ which mod $(2^{2k}-1)$ means that the value of $(2^k.D_2+D_1)$ remains unchanged. The compulsory addition of this value $(2^k-1)/2$ to both X_1 and X_2 alters them to $X_1'$and X_2' where $X_1' = X_1+(2^k-1)/2$ and $X_2' = X_2+(2^k-1)/2$. The range of X_1' and X_2' is such that X_1' is always positive and less than 2^k-1 whereas $(-2^k + 2^{k-1} + 1) < X_2' < (2^k + 2^{k-1} - 2)$. Thus based on the value of X_2', three cases arise viz., $X_2' < 2^k-1$, $X_2' > 2^k-1$ and $X_2' < 0$.

These cases correspond respectively to the cases where $|X_2| < (2^k-1)/2$, $X_2 > (2^k-1)/2$ and $X_2 < -(2^k-1)/2$. The L_1 and L_2 values for these cases are (e) $L_1 = (2^k-1)/2$ and $L_2 = (2^k-1)/2$, (f) $L_1 =(2^k+1)/2$ and $L_2= -(2^k+1)/2$ and (g) $L_1=(2^{k-1}+1)/2$ and $L_2 =2^k+(2^k-1)/2$ respectively. Note that in addition to the $(2^k-1)/2$ value being added due to the ½ term, in these cases, additionally L_1' and L_2' values are added such that (e) $L_1'=0$, $L_2'=0$, (f) $L_1'=1$, $L_2'=-2^k$ and (g) $L_1'=-1$ and $L_2'=2^k$. Evidently, in case (f), since L_2' is greater than 2^k-1, -2^k is added to make $D_2 < 2^k-1$ and this addition is compensated by subtracting 1 from D_1. The case (g) needs addition of 2^k to L_2' and compensation of this addition by subtracting 1 from L_1'. Note that these cases correspond to (m, n) pairs (1, 0), (0, -1) and (2, 1) respectively.

3.3.12. Conway and Nelson technique of RNS to Binary Conversion

Conway and Nelson [Con99] have suggested an alternative technique for RNS to binary conversion in the moduli set $\{2^n-1, 2^n, 2^n+1\}$. This method starts with CRT and uses a decomposition of the terms in CRT so as to avoid the modulo M reduction. From the CRT, we have

$$X = [\ 2^n.(2^n+1).((r_1/M_1) \bmod (2^n-1)) + (2^{2n}-1).\ ((r_2/M_2) \bmod (2^n)) + \ 2^n.(2^n-1).((r_3/M_3) \bmod (2^n+1))\ ] \bmod M \quad (3.42a)$$

where $M_1=M/m_1$,$M_2=M/m_2$ and $M_3 = M/m_3$. We wish to express next X as an integer multiple of $(2^{2n}-1)$ and the fractional part as

$$X=\{(2^{2n}-1).[((r_1/M_1) \bmod (2^n-1)) + ((r_2/M_2) \bmod (2^n)) + \ ((r_3/M_3) \bmod (2^n+1))\ ] \bmod (2^n) + (2^n+1).((r_1/M_1) \bmod (2^n-1)) - (2^n-1).\ ((r_3/M_3) \bmod (2^n+1))\} \bmod M \quad (3.42b)$$

or

$X = 2^{2n}[((r_1/M_1) \bmod (2^n-1)) + ((r_2/M_2) \bmod (2^n))+ ((r_3/M_3) \bmod (2^n+1)) - p.2^n]+ 2^n. [((r_1/M_1) \bmod (2^n-1)) - ((r_3/M_3) \bmod (2^n+1))+p] - ((r_2/M_2) \bmod (2^n))$ (3.42c)

Note that the term $-p.2^n$ is added to take note of the modulo 2^n reduction needed for the coefficient of $(2^{2n}-1)$ term in (3.42b). This is corrected by adding the $p.2^n$ term. In essence $-p. (2^{3n}-2^n)$ is added to (3.42b) which evidently does not change the value of X.

We next observe that $((r_1/M_1) \bmod (2^n-1))$ can be seen to be $(r_1.2^{n-1}) \bmod (2^n-1)$. Conway and Nelson suggest writing r_1 as $(2r_1'+\alpha)$ where α can be zero or one depending on x_1 being even or odd. Then, we have

$(r_1.2^{n-1}) \bmod (2^n-1) = [2^n.r_1'+\alpha.2^{n-1}] \bmod (2^n-1) = r_1'+\alpha.2^{n-1}$. (3.43a)

Similarly for $((r_3/M_3) \bmod (2^n+1))$ written as $((r_3.(2^{n-1}+1)) \bmod (2^n+1))$, expressing r_3 as $(2r_3'+\beta)$ where β can be zero or one depending on r_3 being even or odd, we have

$(r_3.(2^{n-1}+1)) \bmod (2^n+1) = [2^n.r_3'+\beta.(2^{n-1}+1)] \bmod (2^n+1) = r_3'+\beta.(2^{n-1}+1)$ (3.43b)

The term $((r_2/M_2) \bmod (2^n))$ is $(-r_2) \bmod 2^n$.

Using these relationships in (3.42c), we obtain

$X = 2^{2n}.[r_1'+\alpha.2^{n-1}-r_2+r_3'+\beta(2^{n-1}+1) - p.2^n] + 2^n.[r_1'+\alpha.2^{n-1} - r_3'-\beta(2^{n-1}+1)+p] + r_2$ (3.44a)

Conway and Nelson add one more $(2^{3n}-2^n)$ term to X in (3.44) yielding

$X = 2^{2n}.[r_1'+\alpha.2^{n-1} -r_2 +r_3'+\beta(2^{n-1}+1) + (2^n-1) -r.2^n] + 2^n.[r_1'+\alpha.2^{n-1}-r_3'-\beta(2^{n-1}+1)+r+ (2^n-1)]+r_2$ (3.44b)

or

$X = 2^{2n}. [X_1'+L_1'] + 2^n.[X_2'+L_2'] + x_2$ (3.44c)

They suggest the evaluation of r first in the coefficient of 2^{2n} and then using r in the coefficient of 2^n to yield the correct result in the next step.

Conway and Nelson define various terms T, S and U to define the α and β dependent terms in (3.44) which is quite involved unnecessarily. Conway and Nelson suggested using two different equations in place of (3.44b), for the case with $x_3 =0$ and $x_3 \neq 0$. In order to avoid two different types of hardware being used, they suggest the use of only the case $x_3 \neq 0$, which leads to the limitation that that the decoding can not be done for certain range of numbers in the available dynamic range. Note that in their approach, the term (2^n-1) added for the coefficients of both the 2^{2n} term and 2^n term instead of $2(2^n-1)$. As a result the value of 2^n term can be -1 in some cases. In order to remove this problem and to make the terms positive, a {0, 0, 1} is recommended to be added to the original number. In other words, the method is not valid for $(1/(2^n+1))$} of the total dynamic range. This is an unnecessary restriction. This problem arises because of the definition of T_3 (x_{m3}). Changing the definition given in (16) in [Con99] as $T_3 (x_{m3}) = 1$ if $x_{m3} = 0$, will make (19) of [Con99] applicable in all the cases.

Some simplifications can be done by noting that $x_1' + \alpha.2^{n-1}$ can be easily obtained by suitable rotation of the bits of x_1. Hence, there is no need for defining α and x_1'. The architecture of Conway and Nelson uses two (n+1) bit carry- save adders followed by three (n+1) bit conditional sum adders corresponding to r=0, r=1 and r=3 respectively and several 2:1 and 3:1 n bit multiplexers. The implementation of (3.44b), which is more general than in Conway and Nelson [Con99], can proceed on the same lines as in [Con99]. Note that the terms $2^n\text{-}1 - x_2$ and $2^n\text{-}1 - x_3'$ in (3.44b) can be realized by one's complementation of x_2 and x_3'.

It is interesting to compare (3.44b) with (3.38a) and (3.38b). In both the techniques, the x_1, x_2 and x_3 values decide the first term and x_1, x_3 values decide the second term. Correction terms are added to both these terms to obtain the result. This addition is fully decoupled facilitating separate realization of blocks computing upper and middle bits of X in Gallaher et al approach whereas this is not so in the case of Conway and Nelson approach. It is interesting to see that the range of the operands in Conway and Nelson approach is quite large compared to Gallaher et al approach. As an illustration, for the moduli set {7, 8, 9}, it may be noted that the range of the coefficients of 2^{2n} and 2^n terms before subtracting $r.2^n$ term can be seen from (5a) to be between 21 and 0, 13 and -1 respectively in Conway and Nelson technique. In the Gallaher et al approach, the bounds on the coefficients of 2^{2n} term and 2^n term are respectively between 7/2 and -4, 15/2 and -7 respectively.

The architecture of Conway and Nelson [Con99] is presented in Fig 3.11. They compute X as

$$X = r_2 + 2^n.(s_0+c_0+r) + 2^{2n}.(s_1+c_1-r.2^n) \qquad (3.45)$$

where

$$s_0+c_0 = r'_1 + \alpha.2^{n-1} - r_3' - \beta(2^{n-1}+1) - 1 \qquad (3.46a)$$

and

$$s_1 + c_1 = r_1' + \alpha.2^{n-1} - r_2 + r_3' + \beta(2^{n-1}+1) + 2^n \qquad (3.46b)$$

Equations (3.46) are computed using architectures of Fig 3.11 (a) and (b) which are n+1 bit carry save adders each with outputs s_0, c_0 and s_1, c_1

respectively. Next s_1 and c_1 are added as shown in Fig 3.11 (c) in a CLA2 whose carry bits representing r shall be subtracted from CLA1 adding s_0 and c_0. Note, however, that here conditional sum adders can be used with a MUX to select the correct output based on the carry bits of the CLA2. Note that the carry bit of the adder obtaining the middle bits, feeds the adder obtaining the upper n bits of X.

The results for Conway and Nelson approach are as follows:

$$A_{CN} = (2k+1)A_{FA} + 5k(\log_2 k+1).A_{FA} + kA_{MUX\ 2:1} + kA_{MUX\ 3:1} + 2k.A_{INV} \quad (3.47a)$$

$$\Delta_{CN} = 2\Delta_{FA} + (\log_2 k+1)\Delta_{FA} + \Delta_{MUX} + \Delta_{INV} \quad (3.47b)$$

In the case of Dhurkadas's [Dhur98] modified Piestrak converter and Wang et al [Wan00] modified Piestrak converter, since only three operands need to be added as against four in Piestrak's technique, the hardware and delay requirements are slightly less than those in Piestrak's converter:

$$A_{DMP,WANGet\ al} = 2kA_{FA} + 4k(\log_2 2k+1).A_{FA} + 2k.A_{MUX} \quad (3.47c)$$

$$\Delta_{DMP,WANGet\ al} = \Delta_{FA} + (\log_2 2k+1)\Delta_{FA} + \Delta_{MUX} \quad (3.47d)$$

As an illustration for a 96 bit RNS i.e. k=32, (3.45a) - (3.45d) yield the following results:

$$A_{CN} = 1025A_{FA} + 32A_{2:1MUX} + 32A_{3:1MUX} + 64A_{INV} \quad (3.47e)$$

$$\Delta_{CN} = 8\Delta_{FA} + \Delta_{MUX} + \Delta_{INV} \quad (3.47f)$$

$$A_{DMP,WANGet\ al} = 960A_{FA} + 64A_{2:1MUX} \quad (3.47g)$$

$$\Delta_{DMP,WANGet\ al} = 8\Delta_{FA} + \Delta_{MUX} \quad (3.47h)$$

3.4. FORWARD AND REVERSE CONVERTERS FOR THE MODULI SET $\{2^K, 2^K-1, 2^{K-1}-1\}$

Szabo and Tanaka have suggested the use of other moduli sets related to powers of two. However, they have not pursued these rigorously. Recently, Ahmad and Hoda [Hias98], have considered a moduli set which, however, may not have any advantage over the previous moduli set considered in section 3.2. This moduli set has a dynamic range of $2^k(2^k-1)(2^{k-1}-1)$ which is close to $2^{3k}-1$ than the dynamic range of the previous moduli set viz., $2^k(2^{2k}-1)$ which is much less than $2^{3k}-1$. The reason is that the residue corresponding to 2^k+1 needs one extra bit. As an illustration for k = 4, the moduli set {15, 16, 17} has a dynamic range 4080 as against that of a 13 bit dynamic range (=8192) i.e. less than 50%. Ahmad and Hoda moduli set {16, 15, 7} has a dynamic range of 1680 which may be compared with a 11bit dynamic range (=2048) greater than 80%.

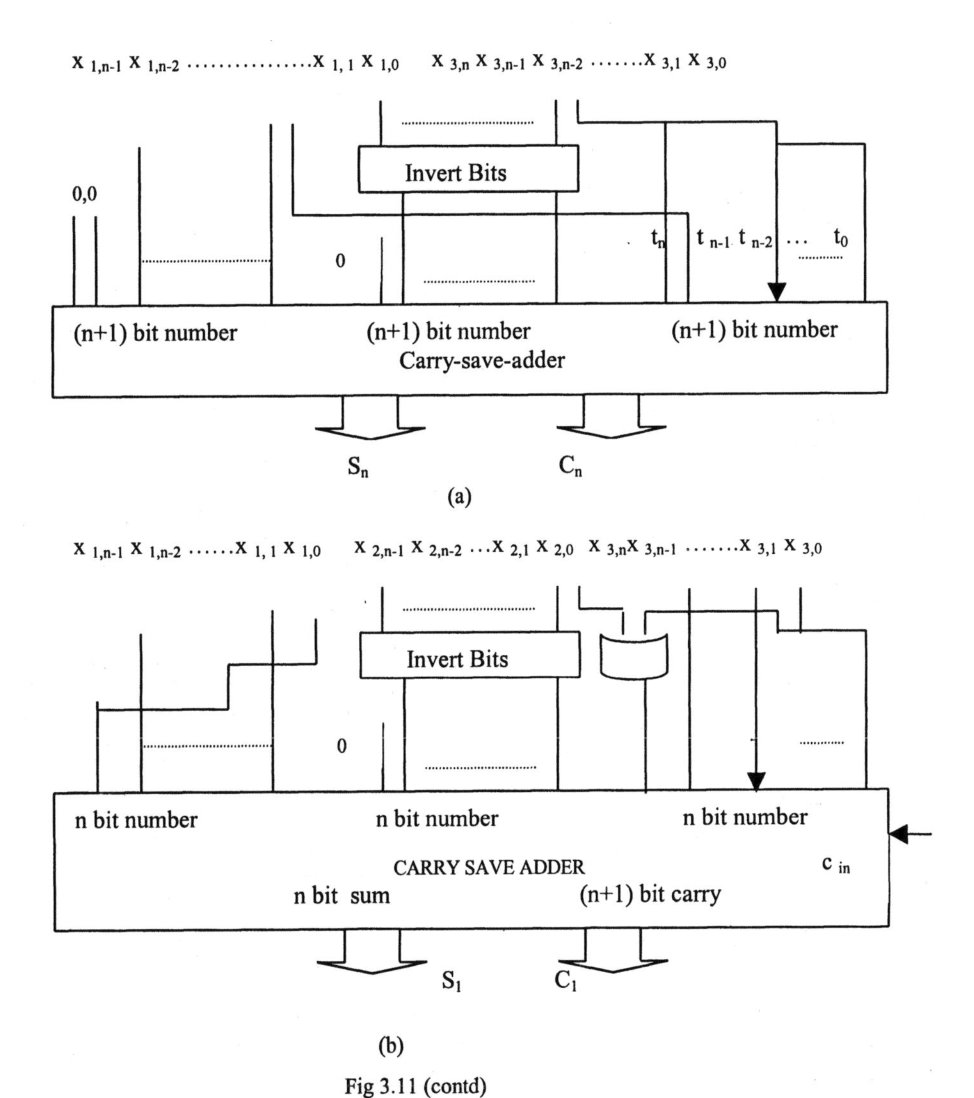

Fig 3.11 (contd)

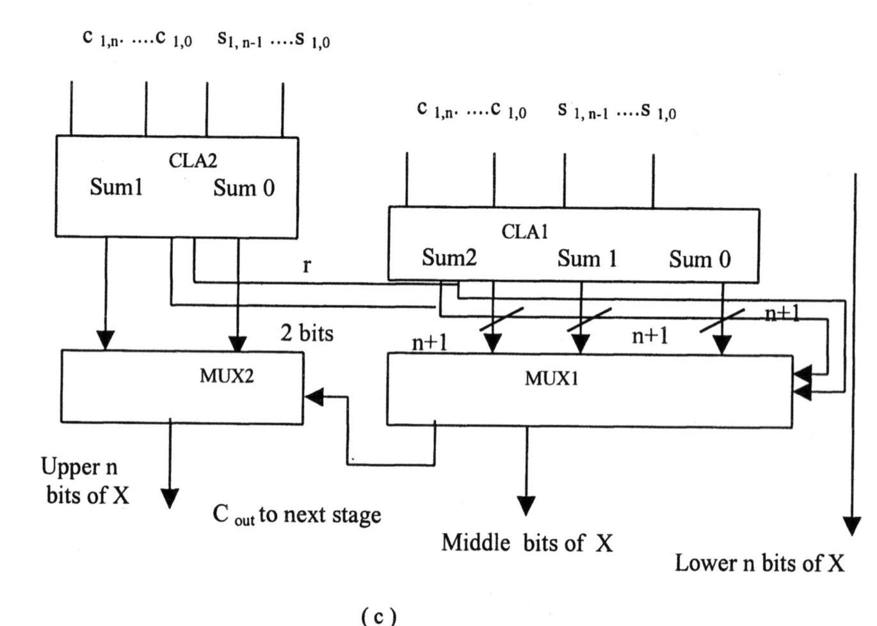

(c)

Fig 3.11. Conway and Nelson architecture for RNS to Binary conversion: (a) CSA to compute s_0 and c_0, (b) CSA to compute s_1 and c_1 and (c) Final CPA using CLA stages (adapted from [Con99] 1999©IEEE)

However, the other advantages of the moduli set also need to be compared especially that of RNS to binary conversion. The binary to RNS conversion is same as that described in section 3.1 since the modulus $2^{k-1}-1$ is similar to 2^k-1. Note, however, that we need to take a k-1 bit fields for computing the residue thus needing four fields to be added rather than three in the case of other moduli set. The advantage of the moduli set may be that unlike in the case of modulus (2^k+1), the operations for all moduli in this moduli set are simpler. Ahmad and Hoda [Hias98] described a RNS to binary converter which follows a similar approach as Andraros and Ahmad technique described at length in Section. 3.2.6. Using CRT for the present moduli set, we have

$$W = [(2^{2k-2}-1)(2^k-1)r_1 + 2^k(2^{k-1}-1)(2^k-3)r_2 + 2^k(2^k-1)(2^{k-2})r_3] \bmod (2^k.(2^k-1).(2^{k-1}-1)) \quad (3.48a)$$

Ahmad and Hoda suggest first the evaluation of integer part of $W/(2^k(2^k-1))$ which can be written as

$$\lfloor (W/(2^k(2^k-1)) \rfloor = \{2^{k-2}-(1/2^k))r_1-(1-(1/(2^k-1)))r_2 + 2^{k-2}r_3\} \bmod ((2^{k-1}-1) \quad (3.48b)$$

This can be simplified as the two equations:

$\lfloor(W/(2^k(2^k-1))\rfloor = (2^{k-2}r_1 - r_2 + 2^{k-2}r_3) \bmod (2^{k-1}-1)$ (3.48c)
$\lfloor(W/(2^k(2^k-1))\rfloor = (2^{k-2}r_1 - r_2 + 2^{k-2}r_3 - 1) \bmod (2^{k-1}-1)$ (3.48d)
Note that the sum of the fractional terms is rounded to the nearest integer and can be zero or -1. When $r_2 \geq r_1$, this term is zero otherwise it is 1. Next, using the basic definition of RNS, we have
$W = \lfloor(W/(2^k(2^k-1))\rfloor . 2^k.(2^k-1) + W \bmod (2^k(2^k-1))$ (3.49)
where the second term is evaluated using CRT with the moduli set $\{2^k, 2^k-1\}$. This term is evidently the residue of W with respect to $2^k.(2^k-1)$. We note using CRT that
$W \bmod (2^k.(2^k-1)) = (2^k.(r_2-r_1) + r_1) \bmod (2^k.(2^k-1))$ (3.50)
The desired value of W is obtained by appending $\lfloor W/2^k \rfloor$ with r_1 since r_1 forms the LSBs already available as data. Hence, we find the MSBs from (3.48) as
$\lfloor W/2^k \rfloor = [W/(2^k(2^k-1))].(2^k-1) + \{[W \bmod(2^k(2^k-1))]/2^k\}$ (3.51)
Substituting for the second term from (3.50), we obtain
$\lfloor W/2^k \rfloor = [W/(2^k(2^k-1))].2^k + (r_2 - r_1) \bmod (2^k-1) - [W/(2^k(2^k-1))]$ (3.52a)

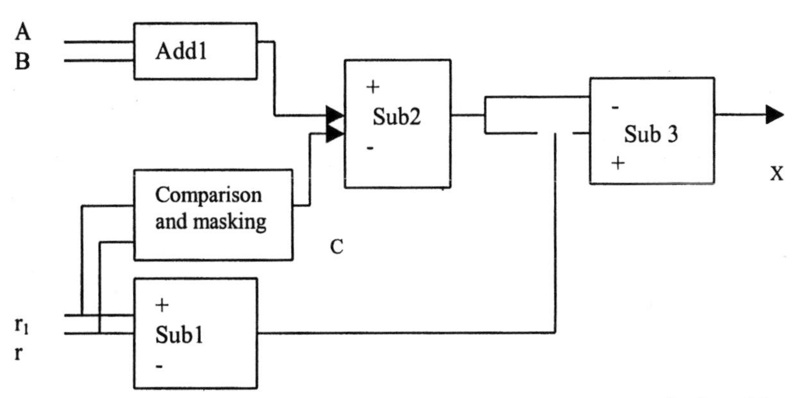

Fig 3.12. Ahmad and Hoda Reverse converter architecture for the moduli set $\{2^k, 2^k-1, 2^{k-1}-1\}$ (adapted from [Hias98] 1998©IEEE)

The hardware implementation is as shown in Fig 3.12. Note that similar to Andraros and Ahmad technique for moduli set $\{2^k-1, 2^k, 2^k+1\}$, the implementation of (3.48c) and (3.48d) needs only circular shifts followed by additions. We define
$A = (2^{k-2}.r_3) \bmod (2^{k-1} - 1)$ (3.52b)
$B = (2^{k-2}.r_1) \bmod (2^{k-1}-1)$ (3.52c)
$C = (r_2) \bmod (2^{k-1} - 1)$ or $(r_2+1) \bmod (2^{k-1}-1)$ (3.52d)

i.e. (3.48) is written as
$\lfloor (W/(2^k(2^k-1)) \rfloor = (A + B - C) \bmod (2^{k-1}-1)$
Note that (3.52d) is implemented by first comparing r_1 and r_2 and then adding 1 in case $r_2 \leq r_1$. This can be done using a Programmable logic array in the block shown as comparison and masking. The adder ADD1, SUB2 realize (3.48d). The second term in (3.52a) is implemented by SUB1 and we next concatenate it as LSBs with MSB bits of output of block SUB2 to result in sum of the first two terms in (3.52a). Finally, SUB3 subtracts the third term from this result to yield the final result. Note that totally five adders are required of sizes one of k bits, three of (k-1) bits and one of (2k-1) bits. The reader is referred to [3.19] for a detailed implementation of this converter which is not speed and area efficient when compared to Piestrak's modified converter.

3.5. FORWARD AND REVERSE CONVERTERS FOR THE {2n+1, 2n, 2n-1} MODULI SET

Premkumar [Prem92] has proposed the use of the moduli set {2n+1, 2n, 2n-1} which he considers to be an extension of the $\{2^n+1, 2^n, 2^n-1\}$ moduli set extensively discussed in this Chapter. This will be considered in some detail in this section. This moduli set firstly does not lend itself to faster binary to RNS conversion unlike the latter. The RNS to binary conversion is also time consuming as will be demonstrated next.

The use of CRT for obtaining the binary number W corresponding to the residues $\{r_1, r_2, r_3\}$ in the moduli set $\{m_1, m_2, m_3\}$ where $m_1 = 2n+1$, $m_2 = 2n$ and $m_3 = 2n-1$ yields

$$W = \{((1 + m_1)/2).m_2.m_3.r_1 + m_1m_3(m_2-1)r_2 + r_3m_1m_2((m_3+1)/2)\} \bmod M \quad (3.53a)$$

where $M = m_1.m_2.m_3$. Note that in (3.53a), the multiplicative inverses of $M_i = M/m_i$ where $M = m_1.m_2.m_3$ are in brackets and can be verified by the reader. Equation (3.53a) can be rewritten to eliminate the integer multiples of $m_1m_2m_3$ to give W as

$$W=\{(m_1m_2m_3/2)(r_1+r_3)+(m_2m_3/2).r_1+(m_1.m_2/2).r_3-m_1m_3r_2\} \bmod M \quad (3.53b)$$

Evidently, when $r_1 + r_3$ is even, the first term can be deleted since it is an integer multiple of M. In case $(r_1 + r_3)$ is odd however, the first term can be treated as M/2 which needs to be added to the remaining terms.

It can be seen that in case (r_1+r_3) is even, the value of W can be negative e.g. when $r_1=r_3=0$) and then M needs to be added once to get the correct result. In other cases of r_1, r_2 and r_3 values, the maximum positive value of X is still less than 2M in spite of addition of M/2 and thus only one subtraction is needed to get the correct result X. The architecture of Fig 3.13 (a) is a direct implementation of (3.53b). The complexity of the architecture can be appreciated since three multipliers (2bxb bits) where $b = \log_2 (2n+1)$ are needed together with a four-input 3b bit adder. Note also that just observing the LSBs of x_1 and x_3 is sufficient to enable M/2 thus needing no adder as shown in Premkumar's architecture of Fig 3.13 (a).

Premkumar, Bharadwaj and Srikanthan [Prem98] suggested a simplified version of their RNS to Binary converter which needs one 2bxb bit multiplier and one bxb bit multiplier. In this technique, we start with (3.53a) and then determine the integer part of
$(W/m_2) \bmod (m_1.m_3)$: $\lfloor W/m_2 \rfloor = \{((1+m_1)/2)m_3r_1 - m_1m_3r_2/m_2\ r_3m_1((m_3+1)/2)\}$ mod $(m_1.m_3)$ (3.54a)
Substituting the values of m_1 and m_3, (3.54a) yields,

$$\lfloor W/m_2 \rfloor = \{n(r_1+r_3-2r_2) + ((r_1 - r_3+m_1m_3(r_1+r_3))/2)\} \bmod(m_1m_3) \quad (3.54b)$$

Similar to the previous design, depending on the even or odd nature of r_3 and r_1, $(m_1.m_3)/2$ can be selectively added:

$$\lfloor W/m_2 \rfloor = n(r_1-2r_2+r_3) + ((r_1-r_3)/2) \quad (3.55a)$$

$$\lfloor W/m_2 \rfloor = n(r_1-2r_2+r_3) + ((r_1-r_3 + m_1m_3)/2) \quad (3.55b)$$

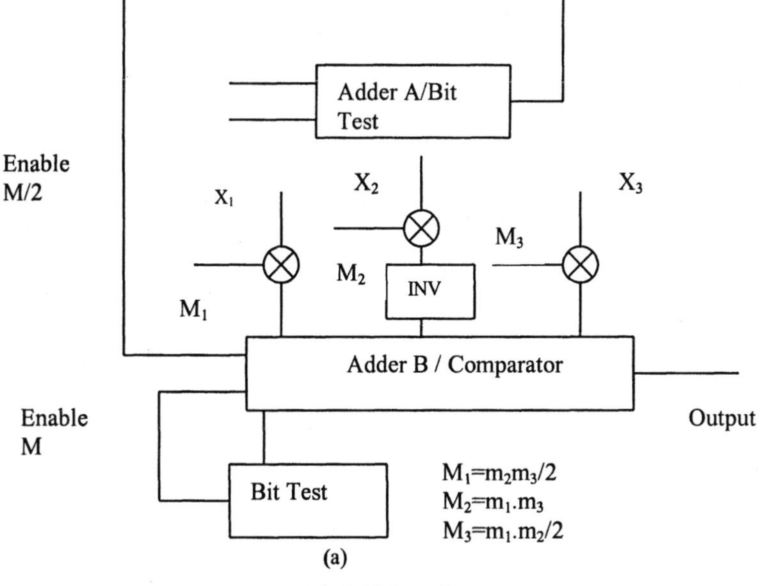

(a)

Fig 3.13 (contd)

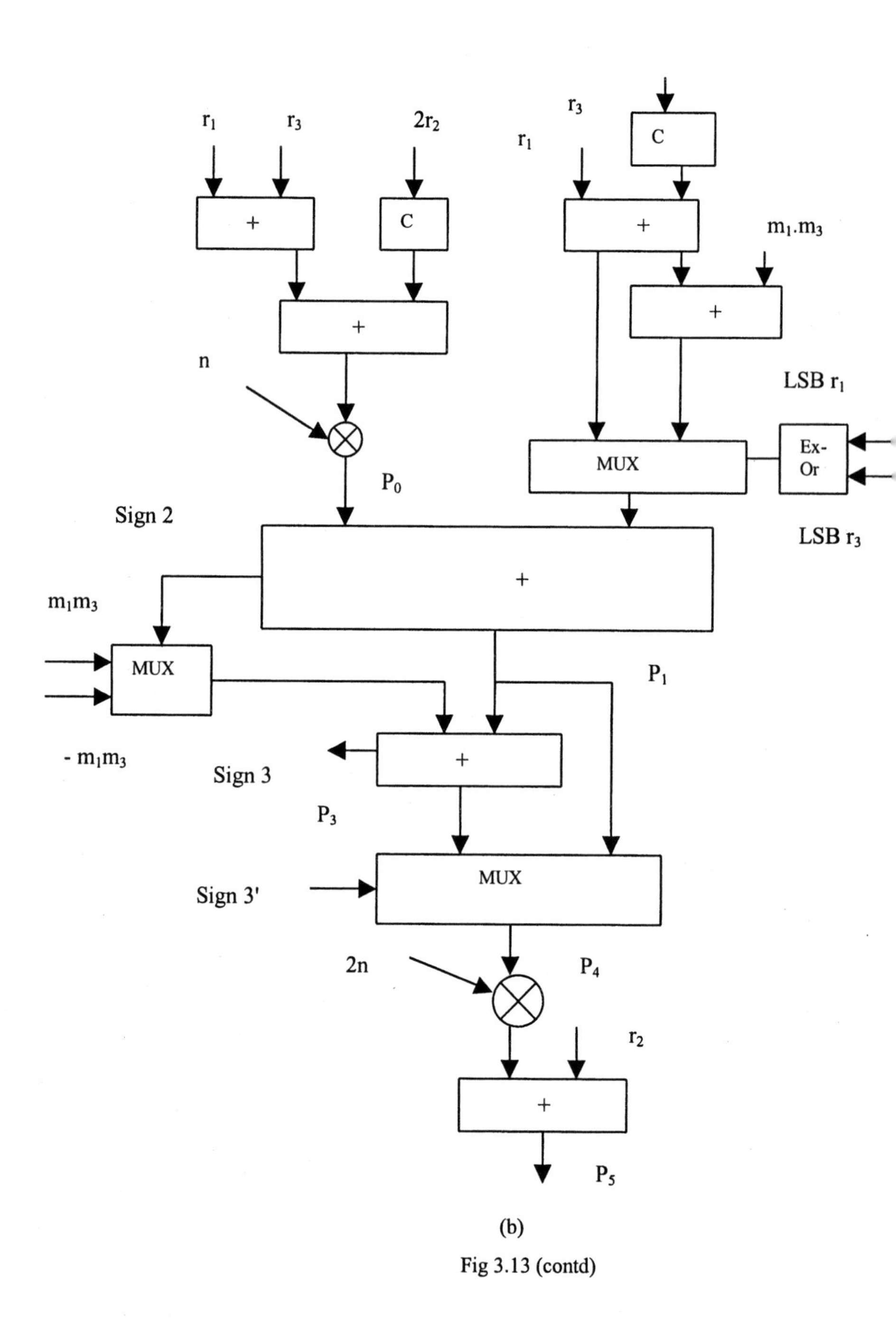

(b)

Fig 3.13 (contd)

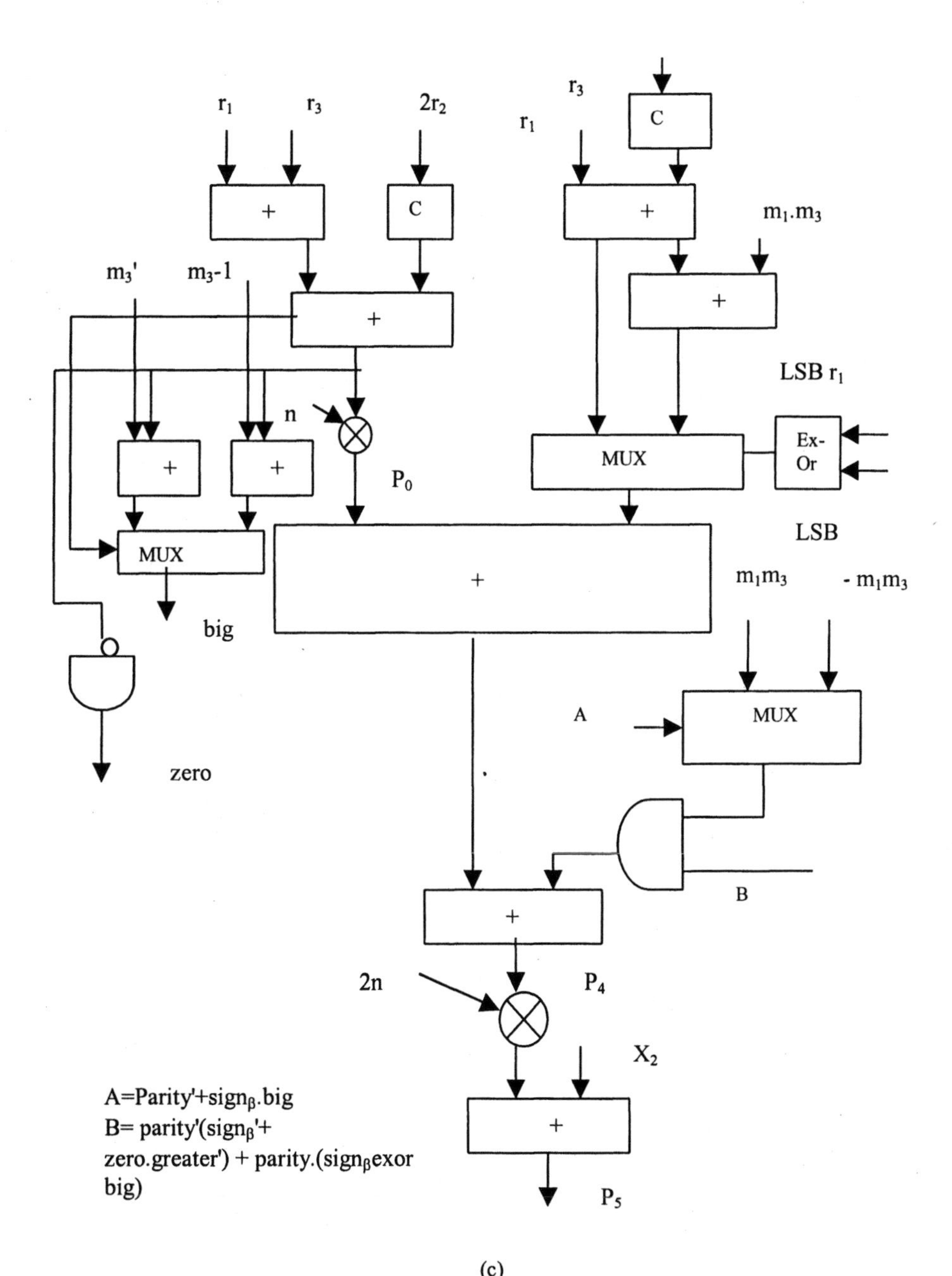

(c)

Fig 3.13. Reverse converter architectures for the moduli set {2n, 2n-1, 2n+1} (a), (b) Low-cost version of (a) and (c) High-speed version. ((a) adapted from [Prem95] 1995©IEEE) and (b), (c) adapted from [Prem98] 1998©IEEE)

The architecture for implementation is shown in Fig 3.13 (b) wherein the first and second terms in (3.55) are implemented in parallel and then added after reduction mod (m_1m_3). The last stage implements the hardware to realize x as

$$W = \lfloor W/m_2 \rfloor .m_2 + r_2 \quad (3.56)$$

Premkumar et al suggested a faster implementation of the scheme as shown in Fig 3.14(c) wherein fast modulo addition and subtraction operations have been implemented using additional hardware. The reader is referred to their work for more details.

Premkumar [Prem95] has also suggested the use of another moduli set {2n, 2n+1, 2n+2} which has 2 as common factor. Evidently, the set reduces to either {n, 2n+1, 2n+2} or {2n, 2n+1, n+1} where unique representation in the available dynamic range is possible. The derivation of RNS to binary conversion is similar to that described in the case of {2n, 2n+1, 2n-1} and hence not considered here. This set also does not appear to be attractive over the moduli set { 2^n+1, 2^n, 2^n-1}. It may also be mentioned that Wang, Swamy and Ahmad [Wan99] have considered Premkumar et al moduli set. Interestingly, however, they put a restriction that the representable numbers in this moduli set must have the property that r_1 and r_3 must be either both odd or both even. Subject to this constraint, they derive the following expression for RNS to binary conversion:

$$W = r_2 + (2n+1).\ \{(r_2 - r_3) + (r_1 - 2r_2 + r_3)(n+1)\} \bmod (2n(n+1)) \quad (3.57)$$

Note that Premkumar's suggestion of replacing residues mod (2n+2) with residue mod (n+1) yields a formula same as (3.57) which has $(n+1)^2$ in place of (n+1) in the second term .

In other words, even though the full dynamic range of moduli set shall be available, only half is available. Such restriction also will map the residues uniquely to the lower half of the dynamic range. However, as can be seen in the conversion formula (3.57), the range of operands is smaller thus simplifying the hardware in Wang et al [Wan99] technique. The reader is referred to their work for more details.

3.6. CONCLUSION

In this Chapter, a comprehensive description of the RNS to binary and Binary to RNS conversion techniques available for the powers of two related moduli sets have been described. These also have been evaluated regarding their area and time requirements. The next chapter deals with the topic of Modulo multiplication in detail.

4

MULTIPLIERS FOR RNS

4.1. INTRODUCTION

The design of RNS based signal processors needs invariably multipliers. Much attention has been paid to the topic of design of modulo multipliers as they have application in other areas such as Cryptography as well. The realization of multipliers could be using ROMs or could be without ROMs. Both these approaches will be studied in detail in this Chapter.

4.2. MULTIPLIERS BASED ON INDEX CALCULUS

4.2.1. Soderstrand and Vernia realization

Szabo and Tanaka [Sza67] and Soderstrand and Vernia [Sode80] suggested multiplier implementation based on index calculus. In this technique shown in Fig 4.1, the multiplier and multiplicand are first converted to indexes by ROM look-up. The index is defined with respect to a base e.g. 2 and the modulus under consideration. As an illustration for the modulus 5, the various numbers 1, 2, 3, 4 can be represented as indices viz., 4, 1, 3, 2. In other words, $2^4 \text{mod } 5 = 1$, $2^1 \text{mod } 5 = 2$, $2^3 \text{ mod } 5 = 3$ and $2^2 \text{mod } 5 = 4$. Evidently, multiplication implies adding these indices and reading back the original number using a PROM. As an example, 2.3 mod 5 is (1+3)=4 which corresponds to 1. Note that zero detection logic is required to handle the case separately, if the input is zero, since it does not have an index.

4.2.2. Dugdale's multiplier implementations

Dugdale [Dug94] has suggested a solution for the use of index calculus to realize multipliers with composite moduli without needing the zero detection logic. Extra code is used to indicate instead this state so that the result is still accurate. His architecture for this purpose is shown in Fig 4.2 (a). The composite modulus can be written evidently as the product of prime moduli. Hence, multiplication can be performed by taking the residues of each

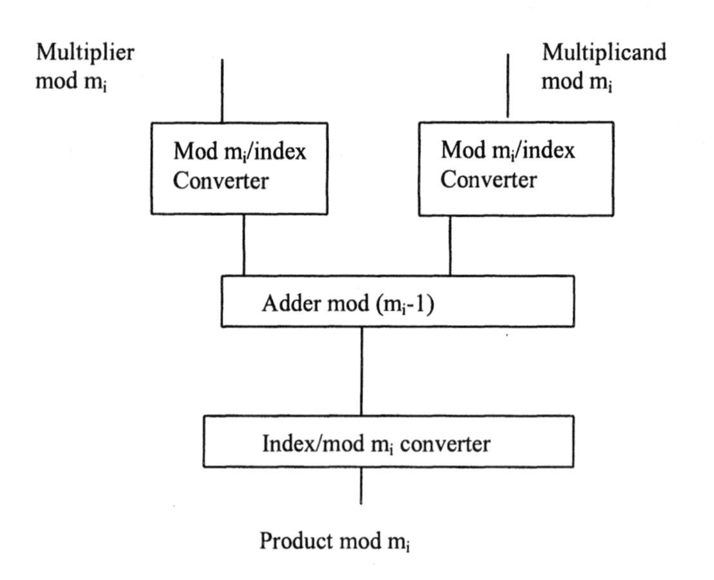

Fig 4.1. RNS multiplier based on index calculus (adapted from [Sode80] 1980©IEEE)

multiplicand and multiplier with respect to all moduli, multiplying these residues and then applying CRT to get the result. If the factor is non-prime, direct multiplication of the residues with respect to that modulus can be done. In the case of prime moduli, index calculus can be used. In Fig 4.2 (a), at left we have shown a case of non-prime modulus and at right the case of a prime modulus.

A specific example will be considered next in order to outline Dugdale's approach. Consider the composite modulus 28, which is realized as the RNS {7, 4}. The numbers between 0 and 27 shall be uniquely represented in this system as normal residues but however using index calculus, for 7, the residue state 0 can not be represented by any index. As a solution, Dugdale first suggests evaluating residue mod 7 which is represented next in another RNS {2, 3}. Thus, the table shown in Fig 4.2 (b) is employed where the base used is 3. Note that for residue of 0, an unused state in {2, 3} viz., {2, 0} has been employed. Thus, as shown in Fig 4.2 (b), the top level converter converts the given X to I (X mod 7) mod 2 and I (X mod 7) mod 3 and X mod 4. Next, the second stage which is fed with similar residues corresponding to Y also adds these indices for the moduli 2 and 3 whereas for the last modulus a modulo multiplier mod 4 is used. The last stage considers these indices and result mod 4 to obtain the final result. A specific example is considered next.

X = 14 and Y =9 for which the results at the first stage output are {2, 0, 2} and {0, 2, 1}. The second stage yields {2, 2, 2} which when fed to third stage yields the result as 14. Note that in the last table, the states {2, 1, 0} and {2, 2, 0} are superfluous which however are interpreted as zeroes.

Dugdale has studied exhaustively the index calculus technique as well as non-index calculus based technique regarding the ROM needed to implement multipliers. He denotes these as factored decomposition multipliers. He observes that smallest memory was needed when index calculus was used for prime factors. Their results for moduli 217-256 are presented in Table. 4.1. Note that index calculus was used. The ROM requirements of the three stages are clearly identifiable.

As an illustration, consider the decomposition of modulus 221 viz., (13.17). The first ROM level needs 3978 bits since the two operands X and Y need to look into ROMs to get their indices:
3978 = 2.221.(4+5)
Similarly, in the layer 2, the indices need to be added. The total memory requirement is 2121 bits:
2121 = 13.13.4 + 17.17.5
The third ROM has 221 locations each 8 bit wide to yield the result with a total requirement of 1768 bits. Note that the exact memory requirement is indicated whereas in practice while using standard cell approach, the memory requirements may be much larger as they are usually powers of two.

4.2.3. Ramnarayan Multipliers for mod (2^n-2^k-1)

Ramnarayan [Ram80] has described a modulo multiplier for special moduli of the type 2^n-2^k-1 as shown in Fig 4.3 (a). This uses index calculus. The indices read from PROMs are added in a mod (p-1) adder which can be easily implemented using combinational logic. Due to the special nature of the modulus chosen, (p-1) is (2^n-2^k) i.e. of the form 111111..1 00000 (n-k ones and k zeroes). Hence, if (x+y) <(2^n-2^k), the result is (x+y) itself. If x+y is between 2^n and (2^n-2^k), the result is x+y- (2^n-2^k). If the value of x+y is greater than 2^n, then the result is {(x+y) mod 2^n}+2^k. These three conditions

modulus	factorisation	ROM1 bits	ROM2 bits	ROM3 bits	Total bits
217	7(2x3)x31(2,3,5)	4774	147	3240	8161
218	2x109(4x27)	3924	3724	2160	9808
219	3x73(8x9)	4380	666	1944	6990
220	4x5x11	3960	591	1760	6311
221	13x17	3978	2121	1768	7867
222	2x3x37(4x9)	4440	421	2160	7021
223	223(2,3,37)	4460	8250	2664	15374
224	32x7	3584	5267	1792	10643
225	9x25	4050	3449	1800	9299
226	2x113(16,7)	3616	1220	2048	6884
227	227(2,113)	4086	89401	2712	96199
22	4x3x19(2,9)	4560	392	2592	7544
229	229(4,3,19)	4122	1869	2432	8423
230	2x5x23(2,11)	4600	581	2640	7821
231	3x7x11	4158	649	1848	6655
232	8x29(4,7)	4176	414	2240	6830
233	233(8,29)	4194	4529	2088	10811
234	2x9x13(4,3)	4212	392	2304	6908
235	5x47(2,23)	4700	2738	2760	10198
236	4x59(2,29)	4248	4255	2784	11287
237	3x79(2,3,13)	4740	730	2808	8278
238	2x7x17	4284	1596	1904	7784
239	239(2,7,17)	4780	1610	2856	9246
240	16x3x5	4320	1117	1920	7357
241	241(16,3,5)	4338	1131	2560	8029
242	2x121	3872	102491	1936	108299
243	243	3888	472392	1944	478224
244	4x61(4,3,5)	4392	171	2560	7123
245	5x49	4410	14481	1960	20851
246	2x3x41(8,5)	4428	322	2304	7054
247	13(4,3)x19(2,9)	4940	406	3458	8802
248	8x31(2,3,5)	4960	303	2880	8143
249	3x83(2,41)	4980	10122	2952	18054
250	2x125	4000	109379	2000	115379
251	251(2,125)	4518	109393	3000	116911
252	4x9x7	4536	503	2016	7055
253	11x23	4554	3129	2024	9707
254	2x127(2,9,7)	5080	493	3024	8597
255	3x5x17	5100	1538	2040	8678
256	256	4096	524288	2048	530432

Table. 4.1. Hardware requirements for moduli 217-256 (adapted from [Dug94] 1994©IEEE)

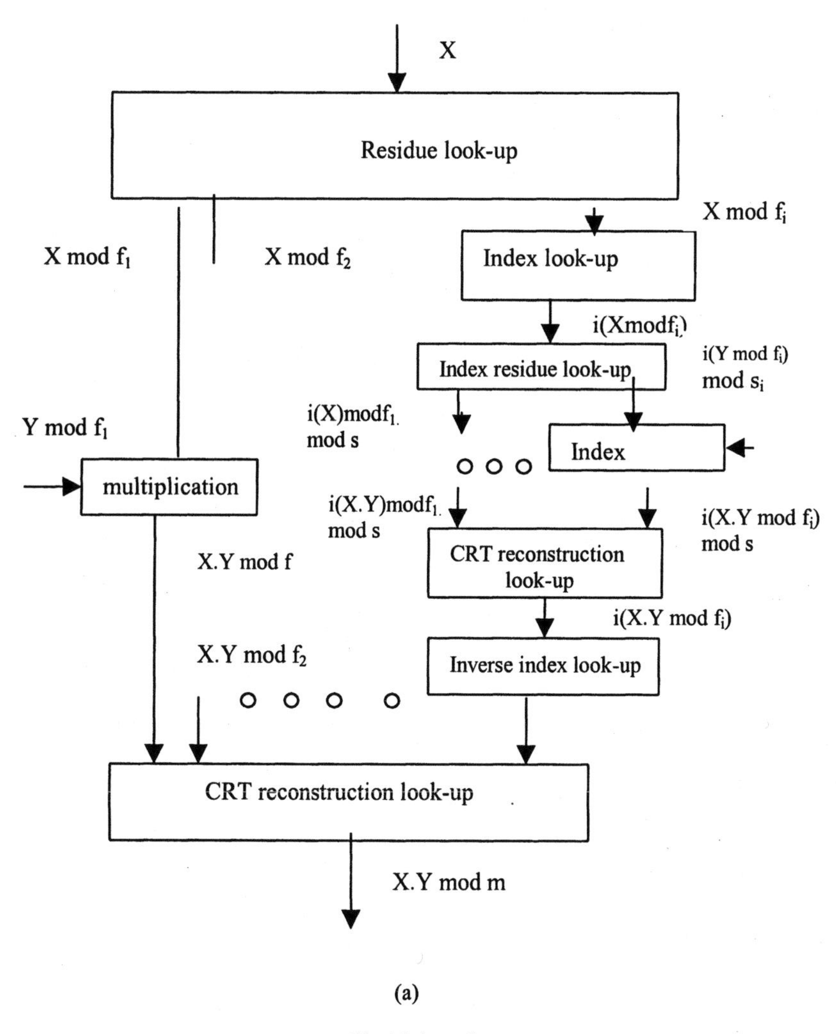

(a)

Fig 4.2 (contd)

can be checked easily using combinational logic. The architecture of the mod (p-1) adder is as shown in Fig 4.3 (b). The result of the mod (p-1) adder addresses a ROM to obtain the actual number. Note that logic needed to detect the input condition zero is also shown in Fig 4.3 (a).

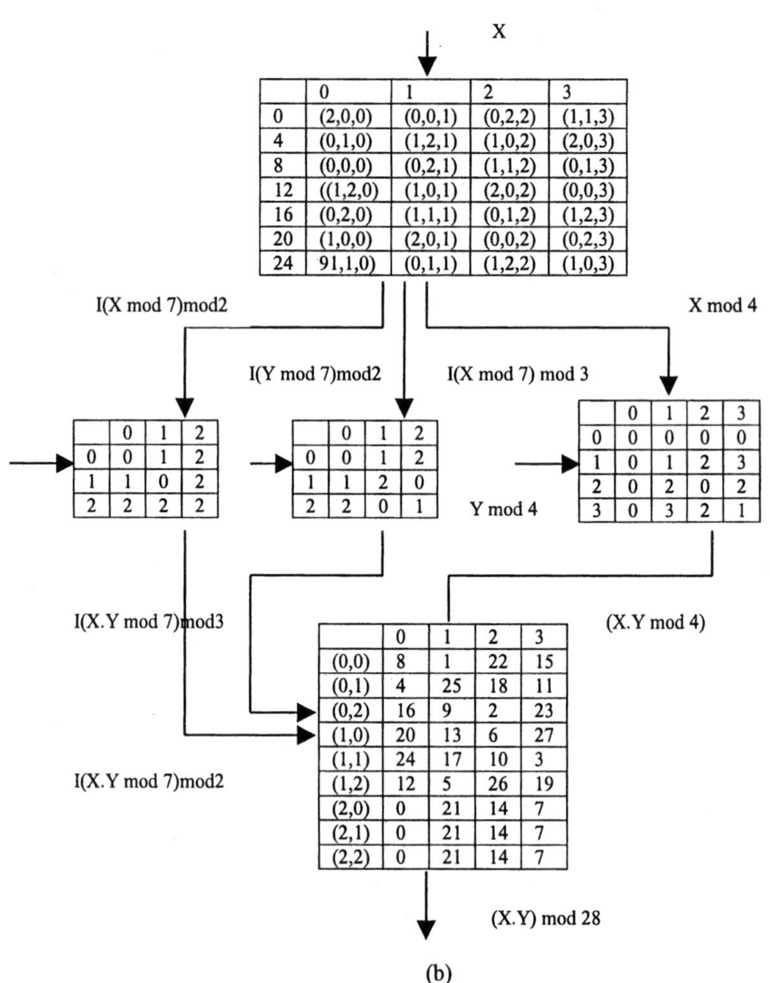

(b)

Fig 4.2 (a) Factored decomposition multiplier architecture due to Dugdale and (b) application to modulo 28 multiplier (adapted from [Dug94] 1994©IEEE)

4.2.4. Radhakrishnan and Yuan RNS multipliers based on index calculus

Radhakrishnan and Yuan [Radh92] described an index calculus based multiplier using sub-modular decomposition. As an illustration consider the modulus 29. We wish to evaluate x.y with x=5 and y=12. The indices corresponding these can be found to be 22 and 7 respectively considering a primitive element of 2. In other words, 2^{22} mod 29 =5 and 2^7 mod 29 = 12. Now, using the sub-moduli 7 and 4 we can express these indices as (1, 2) and (0, 3), Thus, adding these yields (1, 5) which corresponds to 2. The procedure is graphically illustrated in Fig 4.4 (a).

Note that the method is applicable only in cases where factorization of (p-1) is possible. As an example 17 does not permit such evaluation. The hardware requirements in this method are less than those in Jullien's technique especially the ROMs and adders needed. As an illustration for the example 29 considered above, the ROM size is 280 bits and adders are 2 bit and 3 bit. The inverse index ROM is 1540 bits. Radhakrishnan and Yuan also observe that for a given modulus, optimum factorization can be chosen so as to reduce the hardware requirements. As an illustration, for modulus 31, p-1 = 30 yielding several factorizations: 2.3.5, 3.10, 5.6 or 2.15. However, the moduli set {2, 15} gives the minimum $\Sigma_{j=1}^{r} \log_2 m_j = 5$ whereas individual bit sizes if considered, will give the minimum $\log_2 m_j$ for {2, 3, 5} and {5,6}.

Radhakrishnan and Yuan suggest another architecture for SRNS (symmetric residue number system). In this, the sign bits are treated separately and magnitude bits are used to look into indices as illustrated for an example modulus 29. Choosing the root g=2, the dynamic range +14 to –14 is represented as in Table.4.2. The given residue is represented by a sign bit and an index. If the number corresponding to i_x is larger than 14, it is complemented with respect to 29. As an example, 2^4 = 16= -13. The negative sign is indicated by the v_x bit being zero. Contrast this with v_x=1 which corresponds to 13. Consider multiplication mod 29 of x=-9 and y=5. From the Table 4.2, we have $(v_x, i_x) = (1, 10)$ and $(v_y, i_y) = (1, 8)$. Hence, the sum is 18 mod 14 = 4. Since the sum before modulo reduction is 18, the number is negative. Interestingly, the sign can be found as $v_z = s_x \oplus v_x \oplus s_y \oplus v_y \oplus h$ where h =0 if $(i_x+i_y) < ((p-1)/2)$ and h=1 otherwise. The architecture is presented in Fig 4.4 (b). Since the magnitude is less than or equal to ((p-1)/2), the sizes of the ROMs are reduced. The sign detection logic is shown in the inset at top.

Residue Number x	Vx	Index i_x
1	0	0
2	0	1
3	0	5
4	0	2
5	1	8
6	0	6
7	0	12
8	0	3
9	0	10
10	1	9
11	1	11
12	0	7
13	1	4
14	0	13
-14	1	13
-13	0	4
-12	1	7
-11	0	11
-10	0	9
-9	1	10
-8	1	3
-7	1	12
-6	1	6
-5	0	8
-4	1	2
-3	1	5
-2	1	1
-1	1	0

Table. 4.2. A mod 29 SRNS index table with pseudo primitive root g=2
(adapted from [Radh92] 1992©IEEE)

4.2.5. Jullien's ROM based multipliers

Jullien [Jull80] has described modulo multipliers using ROMs which use three stages denoted as (a) sub-modular reconstruction, (b) modulo index addition and (c) reconstruction of desired result. As an illustration, a mod m_i (=19) multiplier is shown in Fig 4.5 (a), wherein the ROM contents are shown. In this example, sub moduli m_1= 6 and m_2= 7 are chosen. Jullien observes that the sub-moduli shall be chosen such that $m_1.m_2>2m_i$. For obtaining the indices a primitive root such as 2 can be selected. As an illustration, 2 is selected in this example. The three stages can be clearly seen. In fact, since we know that index for 0 does not exist, a value 7 is written which will never appear as a sub-modular result. As an example, consider X=12,Y=17 and X.Y = (12.17) mod 19 = 14. We see that 12 = 2^{15} mod 19 and 17=2^{10}mod 19 for which indices in (6, 7) are (3, 1) and (4, 3).

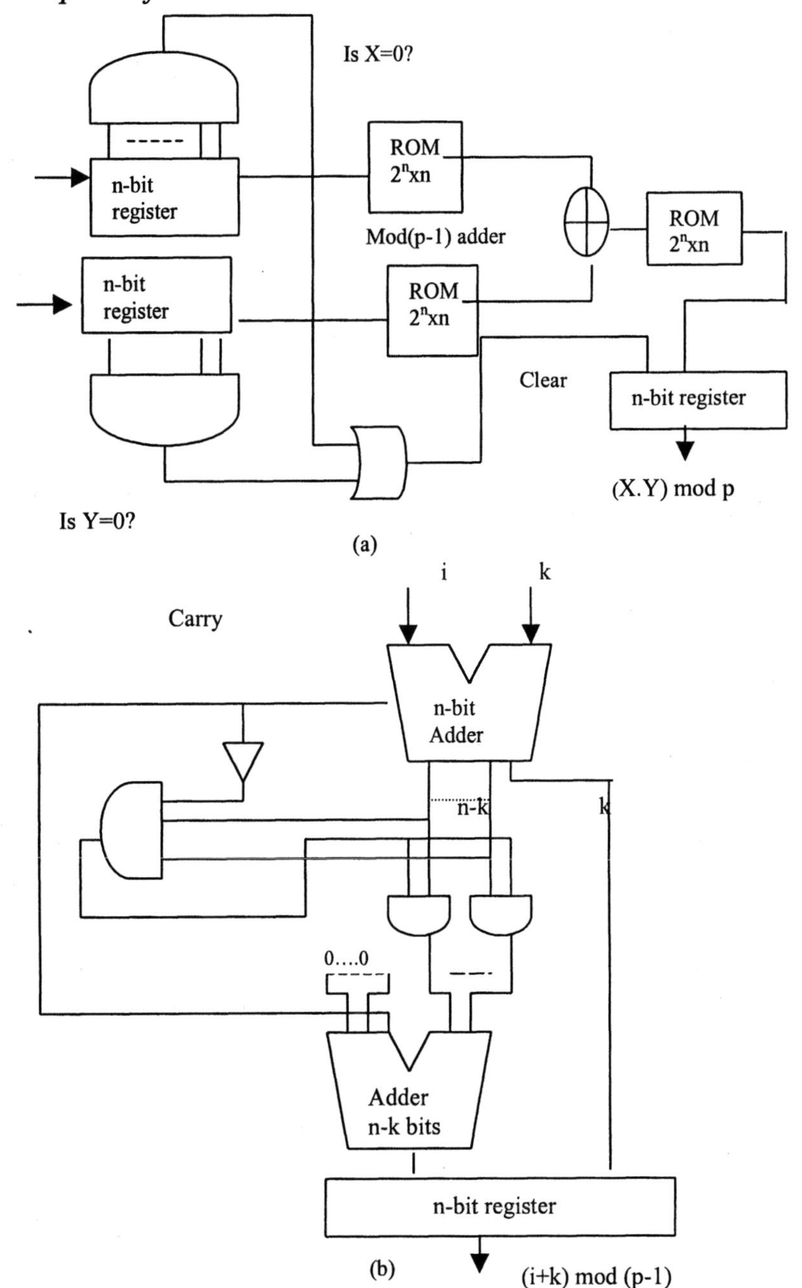

Fig 4.3. Ramnarayan's architecture for (a) mod p, p prime multiplier and (b) mod (p-1) adder. (adapted from [Ram80] 1980©IEE)

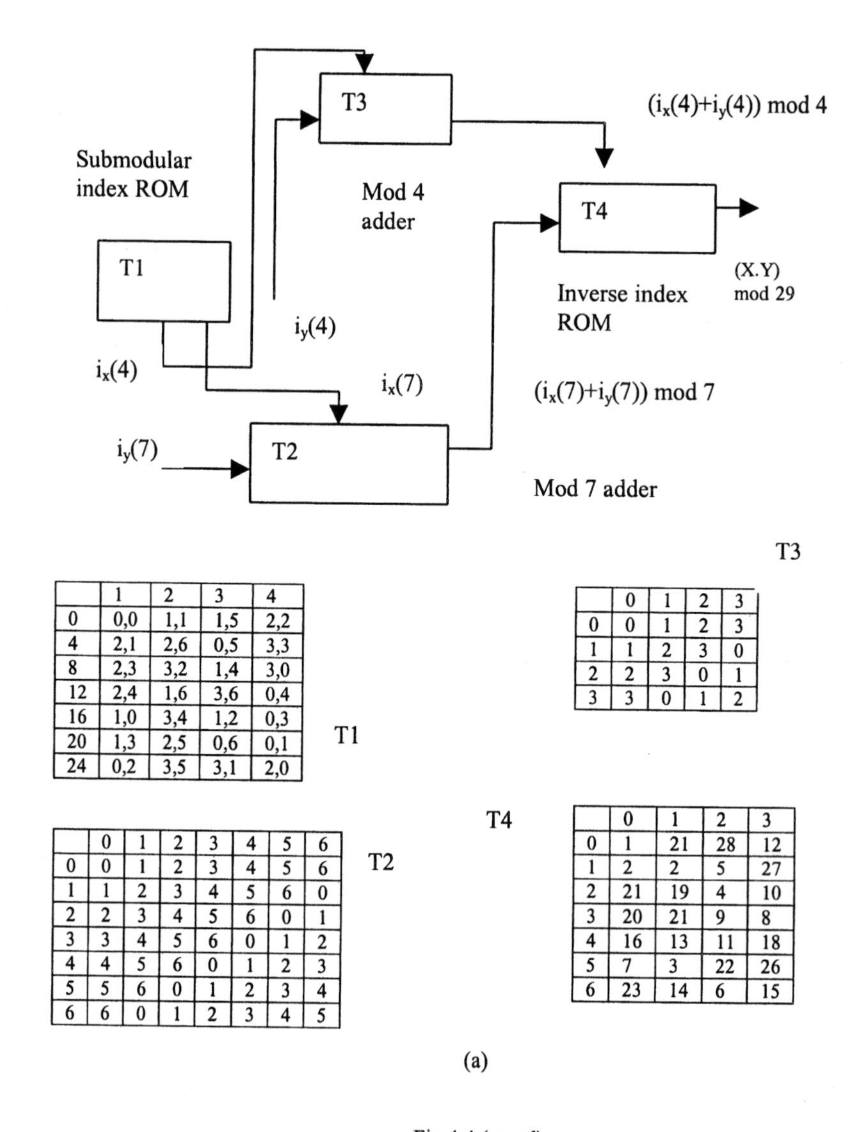

T1

	1	2	3	4
0	0,0	1,1	1,5	2,2
4	2,1	2,6	0,5	3,3
8	2,3	3,2	1,4	3,0
12	2,4	1,6	3,6	0,4
16	1,0	3,4	1,2	0,3
20	1,3	2,5	0,6	0,1
24	0,2	3,5	3,1	2,0

T3

	0	1	2	3
0	0	1	2	3
1	1	2	3	0
2	2	3	0	1
3	3	0	1	2

T2

	0	1	2	3	4	5	6
0	0	1	2	3	4	5	6
1	1	2	3	4	5	6	0
2	2	3	4	5	6	0	1
3	3	4	5	6	0	1	2
4	4	5	6	0	1	2	3
5	5	6	0	1	2	3	4
6	6	0	1	2	3	4	5

T4

	0	1	2	3
0	1	21	28	12
1	2	2	5	27
2	21	19	4	10
3	20	21	9	8
4	16	13	11	18
5	7	3	22	26
6	23	14	6	15

(a)

Fig 4.4 (contd)

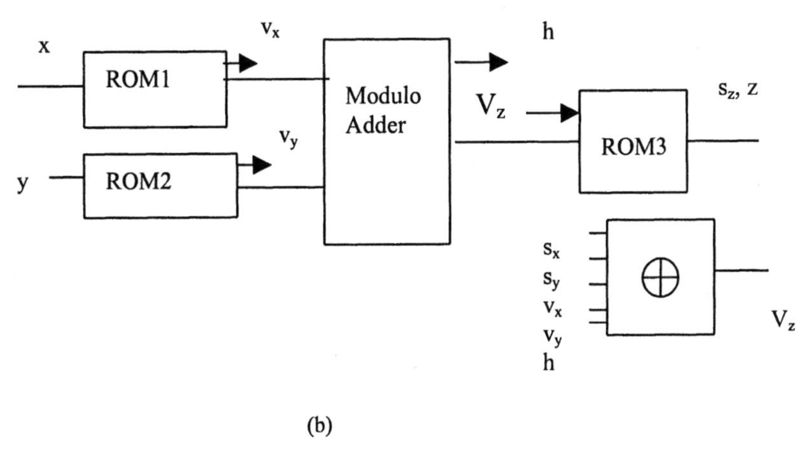

(b)

Fig 4.4 (a) Radhakrishnan's mod 29 multiplier using sub-modular index addition and (b) A SNRS Galois field multiplier (adapted from [Radh92] 1992©IEEE)

The sum of these indices is (1, 4) which corresponds to 14. Note that pipelining is easily possible in the structure of Fig 4.5 (a).

The realization of a radix 2 NTT butterfly is next considered. It is well-known that this needs the computation of $A=a+\alpha^i.b$, $B=a-\alpha^i.b$. The implementation mod 19 is shown in Fig 4.5 (c). Note that subtraction tables are also needed. Note also that $\alpha^i.b$ residues shall be available after the Sub-modular reconstruction. The schematic of the scheme is shown in (b).

4.3. QUARTER- SQUARE MULTIPLIERS

4.3.1. Soderstrand and Vernia implementations

Soderstrand and Vernia [Sode80] also suggested a square law mod m_i multiplier also known as Quarter Square multiplier, in which a.b is written as

$$a.b=[(a+b)^2-(a-b)^2]/4 \qquad (4.1)$$

The architecture for implementation is shown in Fig 4.6. Addition and subtraction are needed to evaluate (a+b) mod m_i and (a-b) mod m_i. Squaring is done in two ROMs and another subtractor is used to evaluate the numerator in (4.1). A final table look-up is needed to implement multiplication by (1/4) mod m_i. Note that (1/4) mod m_i does not exist for m_i which are multiples of 4. However, this is not an objectionable restriction,

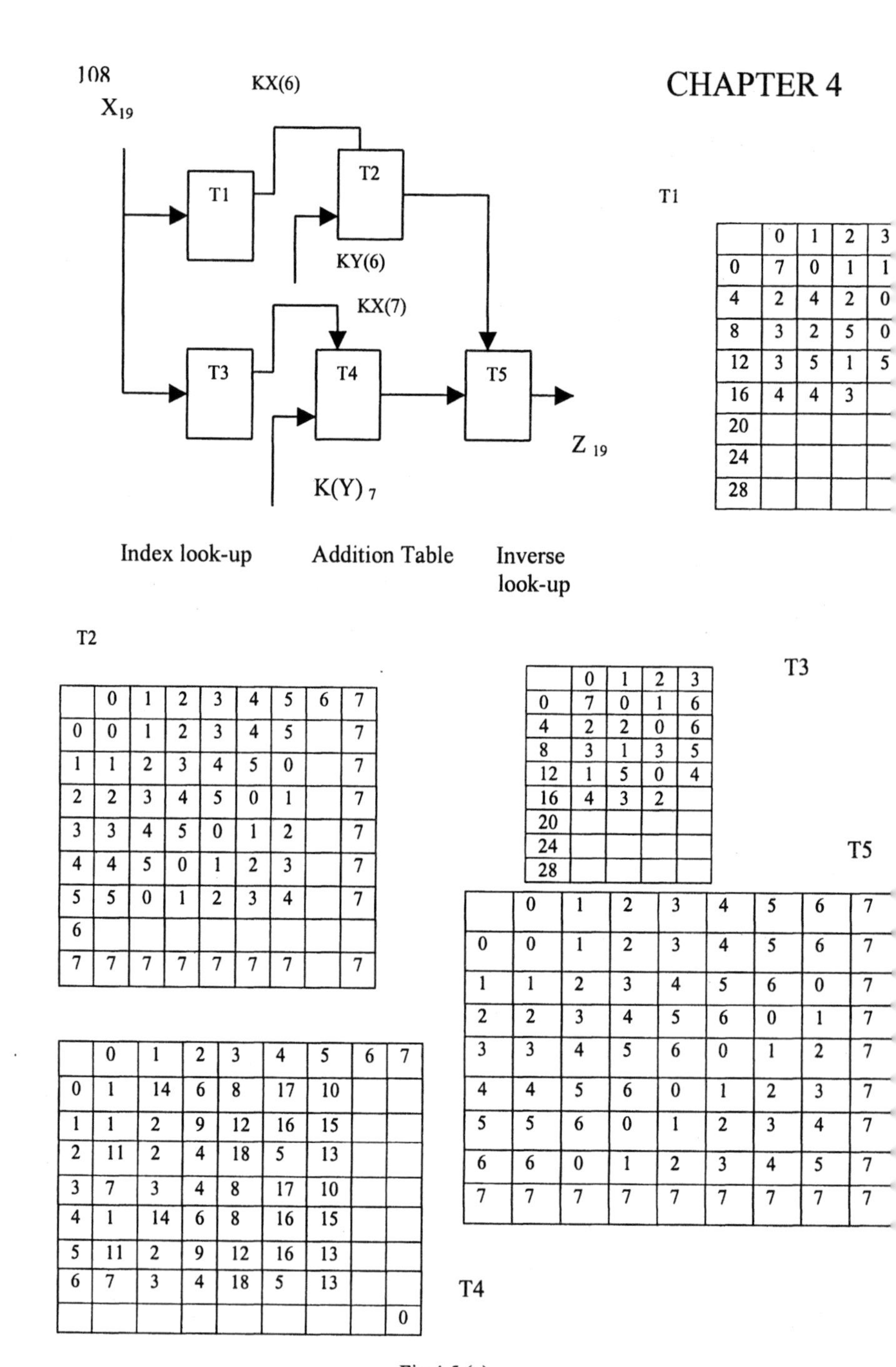

T1

	0	1	2	3
0	7	0	1	1
4	2	4	2	0
8	3	2	5	0
12	3	5	1	5
16	4	4	3	
20				
24				
28				

T2

	0	1	2	3	4	5	6	7
0	0	1	2	3	4	5		7
1	1	2	3	4	5	0		7
2	2	3	4	5	0	1		7
3	3	4	5	0	1	2		7
4	4	5	0	1	2	3		7
5	5	0	1	2	3	4		7
6								
7	7	7	7	7	7	7		7

T3

	0	1	2	3
0	7	0	1	6
4	2	2	0	6
8	3	1	3	5
12	1	5	0	4
16	4	3	2	
20				
24				
28				

T5

	0	1	2	3	4	5	6	7
0	0	1	2	3	4	5	6	7
1	1	2	3	4	5	6	0	7
2	2	3	4	5	6	0	1	7
3	3	4	5	6	0	1	2	7
4	4	5	6	0	1	2	3	7
5	5	6	0	1	2	3	4	7
6	6	0	1	2	3	4	5	7
7	7	7	7	7	7	7	7	7

T4

	0	1	2	3	4	5	6	7
0	1	14	6	8	17	10		
1	1	2	9	12	16	15		
2	11	2	4	18	5	13		
3	7	3	4	8	17	10		
4	1	14	6	8	16	15		
5	11	2	9	12	16	13		
6	7	3	4	18	5	13		
								0

Fig 4.5 (a)

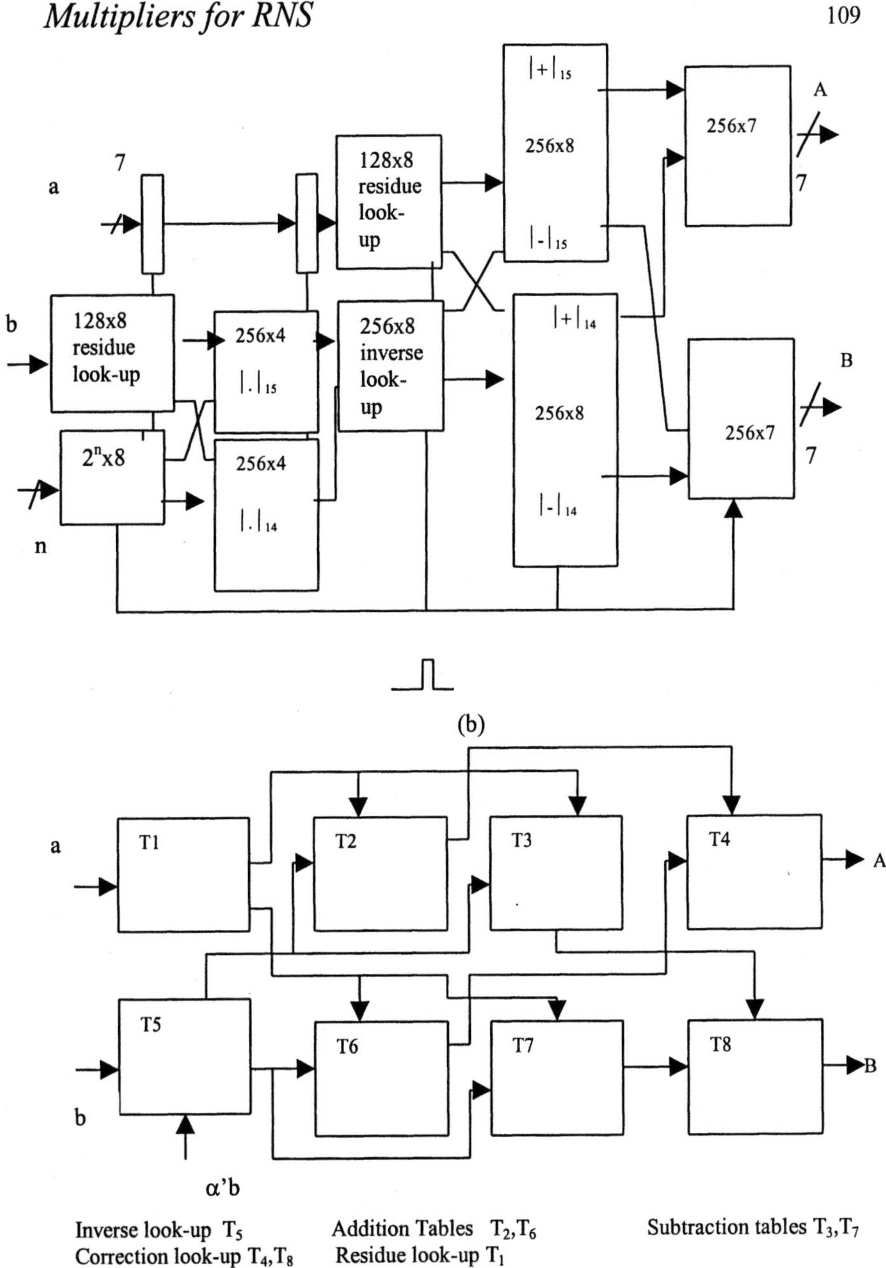

a
7
b
128x8
residue
look-up
2^nx8
n
256x4
|.|15
256x4
|.|14
128x8
residue
look-
up
256x8
inverse
look-
up
|+|15
256x8
|-|15
|+|14
256x8
|-|14
256x7
A
7
256x7
B
7
(b)
a
T1
T2
T3
T4
A
T5
T6
T7
T8
B
b
α'b
Inverse look-up T5
Correction look-up T4,T8
Addition Tables T2,T6
Residue look-up T1
Subtraction tables T3,T7

Fig 4.5 (contd)

	0	**1**	2	3
0	0	1	2	3
4	4	5	0	1
8	2	3	4	5
12	0	1	2	3
16	4	5	0	
0	0	1	2	3
4	4	5	6	0
8	1	2	3	4
12	5	6	0	1
16	2	3	4	

	0	1	**2**	3	4	5	6	7
0	0	1	2	3	4	5		
1	1	2	3	4	5	0		
2	2	3	4	5	0	1		
3	3	4	5	0	1	2		
4	4	5	0	1	2	3		
5	5	0	1	2	3	4		
6								
7								

	0	1	2	**3**	4	5	6	7
0	0	1	2	3	4	5		
1	5	0	1	2	3	4		
2	4	5	0	1	2	3		
3	3	4	5	0	1	2		
4	2	3	4	5	0	1		
5	1	2	3	4	5	0		
6								
7								

	0	1	2	3	**4**	5	6	7
0	0	7	14	2	9	16		
1	17	1	8	15	3	10		
2	11	18	2	9	16	4		
3	5	12	x	3	10	17		
4	18	6	13	x	4	11		
5	12	0	7	14	X	5		
6	6	13	1	8	15	X		
7								

	0	1	2	3	4	**5**	6	7	0	1	2	3	4	5	6	7
0	1	2	0	2	5	4			1	0	6	1	3	3		
1	1	2	3	0	4	3			1	2	2	5	2	1		
2	5	2	4	0	5	1			4	2	4	4	5	6		
3	1	2	0	2	4	3			1	0	6	1	2	1		
4	5	2	3	0	4	1			4	2	2	5	2	6		
5	1	3	4	0	5	1			0	3	4	4	5	6		
6																
7																

	0	1	2	3	4	5	**6**	7
0	0	1	2	3	4	5	6	
1	1	2	3	4	5	6	0	
2	2	3	4	5	6	0	1	
3	3	4	5	6	0	1	2	
4	4	5	6	0	1	2	3	
5	5	6	0	1	2	3	4	
6	6	0	1	2	3	4	5	
7								

	0	1	2	3	4	5	6	**7**
0	0	1	2	3	4	5	6	
1	6	0	1	2	3	4	5	
2	5	6	0	1	2	3	4	
3	4	5	6	0	1	2	3	
4	3	4	5	6	0	1	2	
5	2	3	4	5	6	0	1	
6	1	2	3	4	5	6	0	
7								

	0	1	2	3	4	5	6	7
0	0	7	14	17	5	12		
1	13	1	8	15	18	6		
2	7	14	2	9	16	0		
3	1	8	15	3	10	17		
4	18	2	9	16	4	11		
5	12	15	3	10	17	5		
6	6	13	16	4	11	8		
7								

(c)

Fig 4.5 (a) A modulo 19 ROM based multiplier due to Jullien (b) a radix-4 butterfly architecture and (c) a mod 19 ROM array based implementation. (adapted from [Jull80] 1980©IEEE)

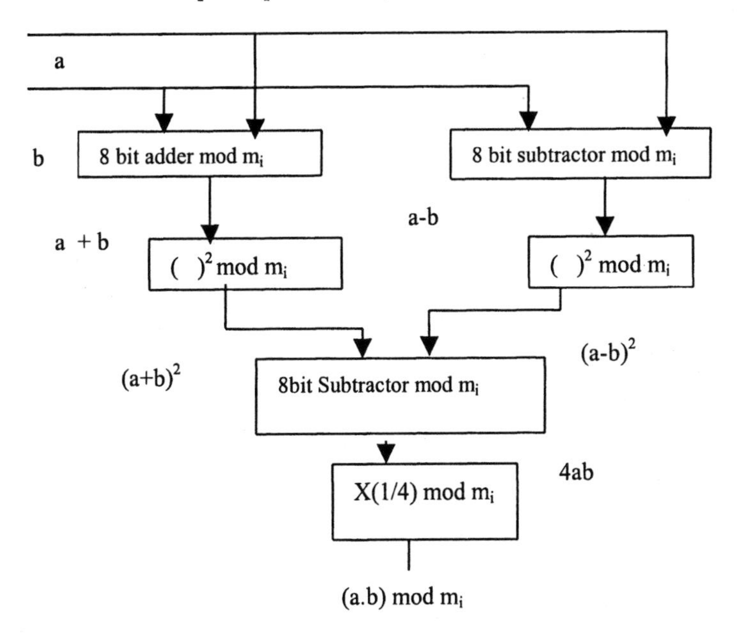

Fig 4.6. Quarter- Square multiplier architecture for RNS multiplier (adapted from [Sode80] 1980©IEEE)

Since only one modulus is allowed of this type and if it is also a power of 2, simple multiplier can be used.

Soderstrand and Vernia considered a multiplier mod a composite number as well. Architectures for realizing mod 249 multiplier are presented in Fig 4.7. Note that in (a), a direct implementation using a RNS technique in a three moduli RNS is presented which is ROM intensive. An alternative two-moduli RNS by decomposition of 249 into 3x83 is presented in (b). In this, the residue multiplication for modulus 3 is done directly, whereas, for modulus 83 index calculus is used. In Fig 4.7 (c), a Quarter-Square multiplier based on ROMs is shown which does not need that much memory as in (a). This needs an adder and subtractor at the front end and subtractor in the middle together with ROMs.

4.3.2. Taylor's implementations

Taylor [Tay81a, Tay84, Tay82a] has extensively investigated the design of multipliers for the moduli set $\{2^n - 1, 2^n, 2^n+1\}$ using ROMs. For the case of moduli 2^n and 2^n-1, the ROM size can be 2^{2n} words each of n bits. However, in the case of 2^n+1, the residue could be 2^n as well, a n+1 bit word. It therefore appears at first sight that the memory size needed is $2^{(2n+2)}$ words each of (n+1) bits. Interestingly, however, in the case of 2^n+1, Taylor [Tay81a] observes that if the residue is zero, the result need not be looked in the ROM but that location instead can be used to handle the case when the residue is 2^n. Thus, the memory size can be 2^{2n} words. This architecture is shown in Fig 4.8(a). It is also clear that memory requirements in Quarter-square multiplier are substantially less than that in the direct approach. Herein, look-up tables are addressed by s^+ and s^- where $s^+ = (x+y)/2$ and $s^- = (x-y)/2$. Note that since the maximum value for s^+ is 2^n for modulus 2^n+1, the output of the adder realizing s^+ and s^- is (n+2) bits. However, the ROM size can be reduced to 2^{n+1} words using the arrangement of Fig 4.8 (b) which is an adaptation of (a). Here, the overflow bit serves to differentiate $s^\pm=0$ from 2^{n+1}.

Taylor has made some interesting observations on the implementation of the Quarter-square multipliers (see Fig 4.9). He deals with the size of the memories $\phi(s)^+$ and $\phi(s)^-$. It may appear that the resolution needed is two bits more than the modulus e.g. p=32, $s^+=9$, $\phi(s)$ will be $(9^2/4) \bmod 32 = 20.5$. However, it is sufficient to store the integer value of $\phi(s)$ due to the following. Denoting $(x+y)/2 = v+(k/2)$ and $(x-y)/2 = q+(k/2)$, where k = 0 or 1, we have

$$\begin{aligned} z &= [(x+y)^2/4] \bmod p - [(x-y)^2/4] \bmod p \\ &= [v^2+kv] \bmod p - [q^2+kq] \bmod p + [k^2/4] \bmod p - [k^2/4] \bmod p \\ &= [\phi(s+) - \phi(s-)] \bmod p \end{aligned} \tag{4.2}$$

It is clear that the last two terms cancel thus making $\phi(s+)$ and $\phi(s-)$ integers only. Hence, the existence of the multiplicative inverse of 4 is not an issue.

Taylor [Tay81a] also suggests modulo adder implementations. Basically, mod (2^n-1) and mod (2^n+1) adders need n bit addition with the result being incremented or decremented by unity. This operation can be done using PLAs. As an illustration, consider s a 11 bit word (decimal 92). If s-1 is desired, it can be seen that the MSBs do not change and only LSBs need to be altered. By using few masks stored in a PLA, the operation of addition of only LSBs can be accomplished. As an illustration, an example of addition of A=5 and B=6 mod 9 is presented in Fig 4.10 (a). The overflow bits, the sign bits of $\phi(s+)$ and $\phi(s-)$ and zero condition of the sum of the adder can be used to select the mask together with the value of the result. The possible masks of a 11 bit case are shown in Fig 4.10 (b). Note that addition of 1 to LSBs xxxx xxxx x01 yields xxxx xxxx x10 without the need for changing the other MSBs.

4.4. TAYLOR'S MULTIPLIERS

Taylor [Tay82a] described a VLSI nxn multiplier modulo 2^n, 2^n-1 and 2^n+1 which is considered next. In this technique, the words X and Y to be multiplied are split into two halves each first:

$$X = X_H \cdot 2^{n/2} + X_L \quad (4.3a)$$

$$Y = Y_H \cdot 2^{n/2} + Y_L \quad (4.3b)$$

Then, evidently,

$$X.Y = X_H.Y_H.2^n + X_H.Y_L.2^{n/2} + X_L.Y_H.2^{n/2} + X_L.Y_L \quad (4.4a)$$

The value of X.Y mod 2^n is then found to be

$$(X.Y) \bmod 2^n = X_H.Y_L.2^{n/2} + X_L.Y_H.2^{n/2} + X_L.Y_L \quad (4.4b)$$

Since, X_H, X_L, Y_H are n/2 bit numbers, the modulo 2^n operation means retaining the LSBs of $X_H.Y_L$ and appending it with n/2 zeroes. Denoting the partial products $X_H.Y_H$, $X_L.Y_L$, $X_L.Y_H$ and $X_H.Y_L$ in (4.4a) as a, b, c and d respectively, we have

$$(X.Y) \bmod 2^n = b + (c_L + d_L).2^{n/2} \quad (4.5)$$

where c_L and d_L are the lower n/2 bit words of c and d. Thus, three words need to be added with carry ignored to obtain the final result. In the case of (2^n-1), the computation of $(c+d).2^{n/2}$ has been achieved in a quite complicated way by Taylor. Denoting (c+d) as V, we have

$$V.2^{n/2} = V.(2^n-1)/2^{n/2} + V/2^{n/2} \quad (4.6)$$

The terms in the right hand side are treated as mixed fractions. The first term is obtained by a table look-up. The second term is easily obtained from V

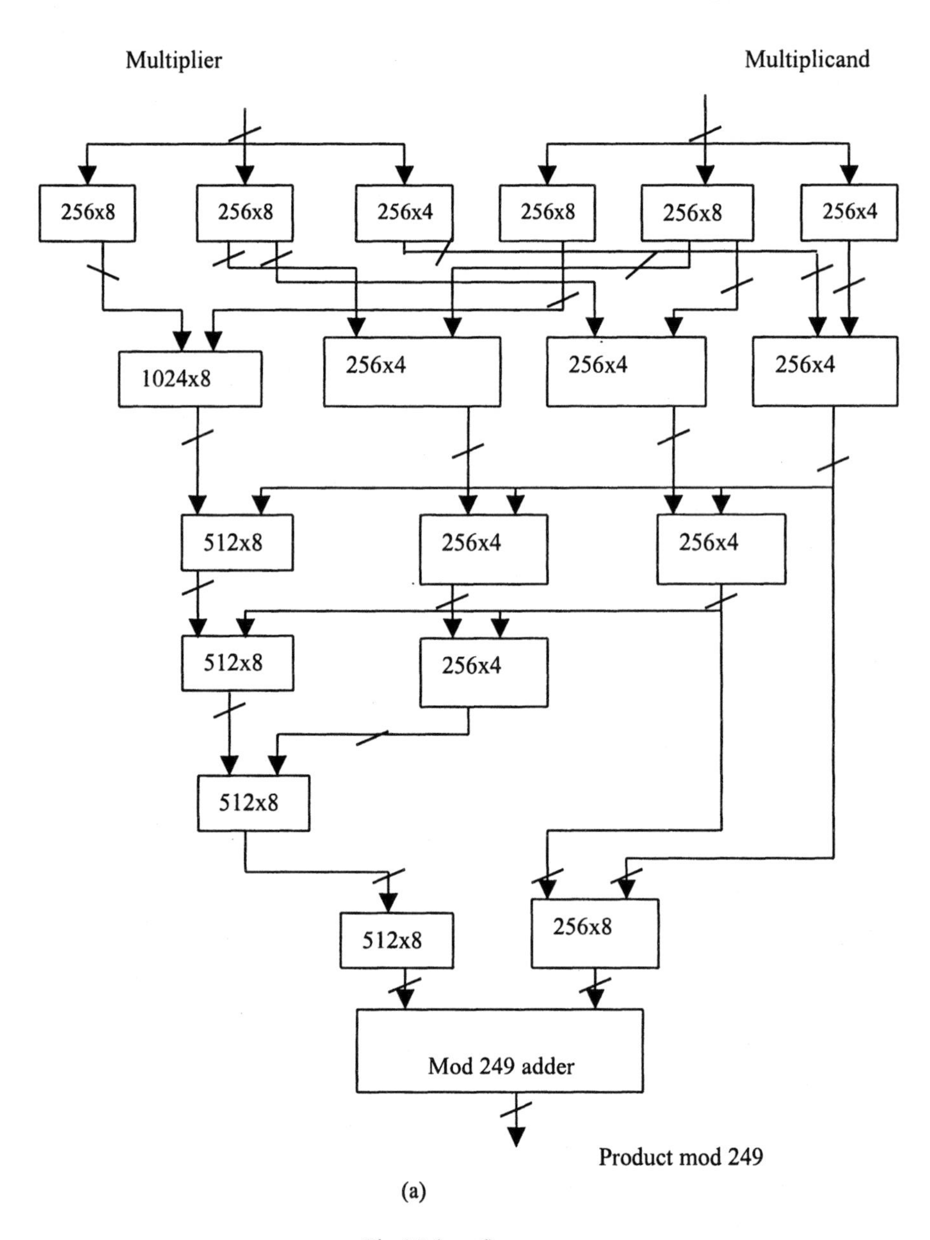

(a)

Fig 4.7 (contd)

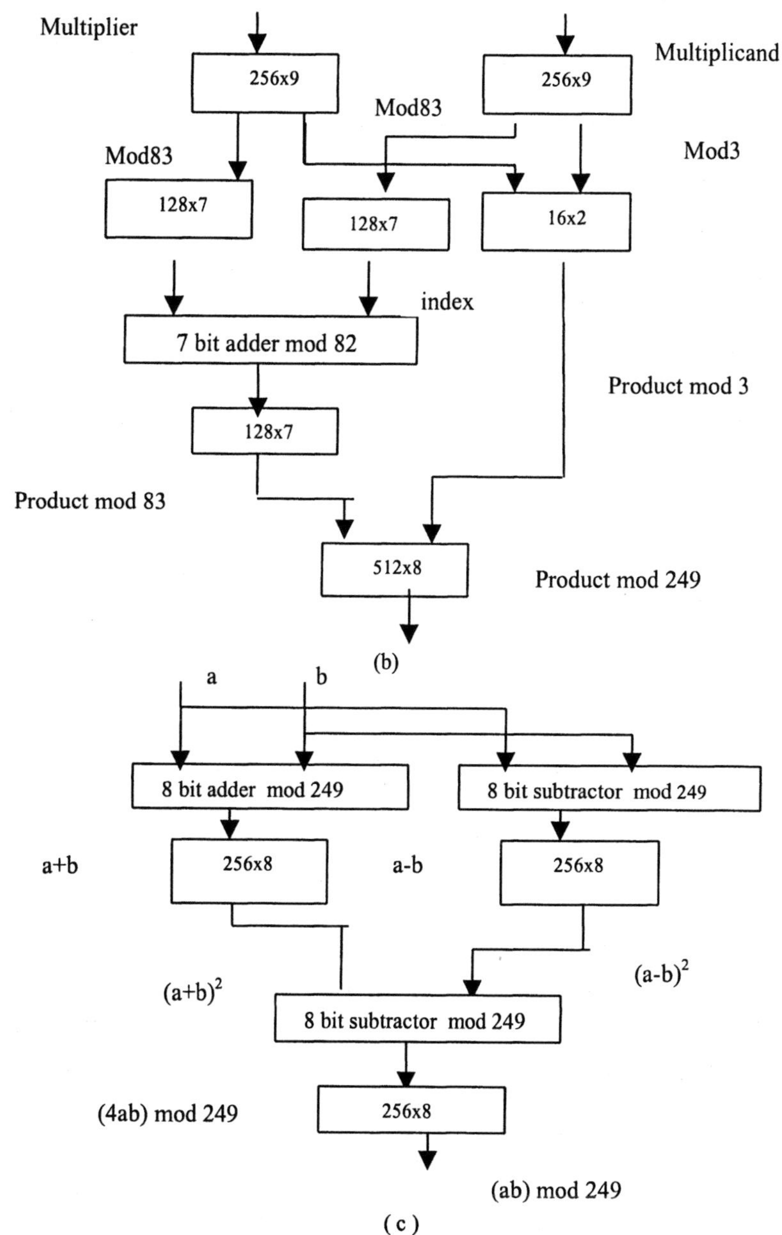

Fig 4.7. Multiplier Architectures for composite moduli (a) using RNS, (b) using index calculus and direct multiplication and (c) using Quarter Square multiplier (adapted from [Sode80] 1980©IEEE)

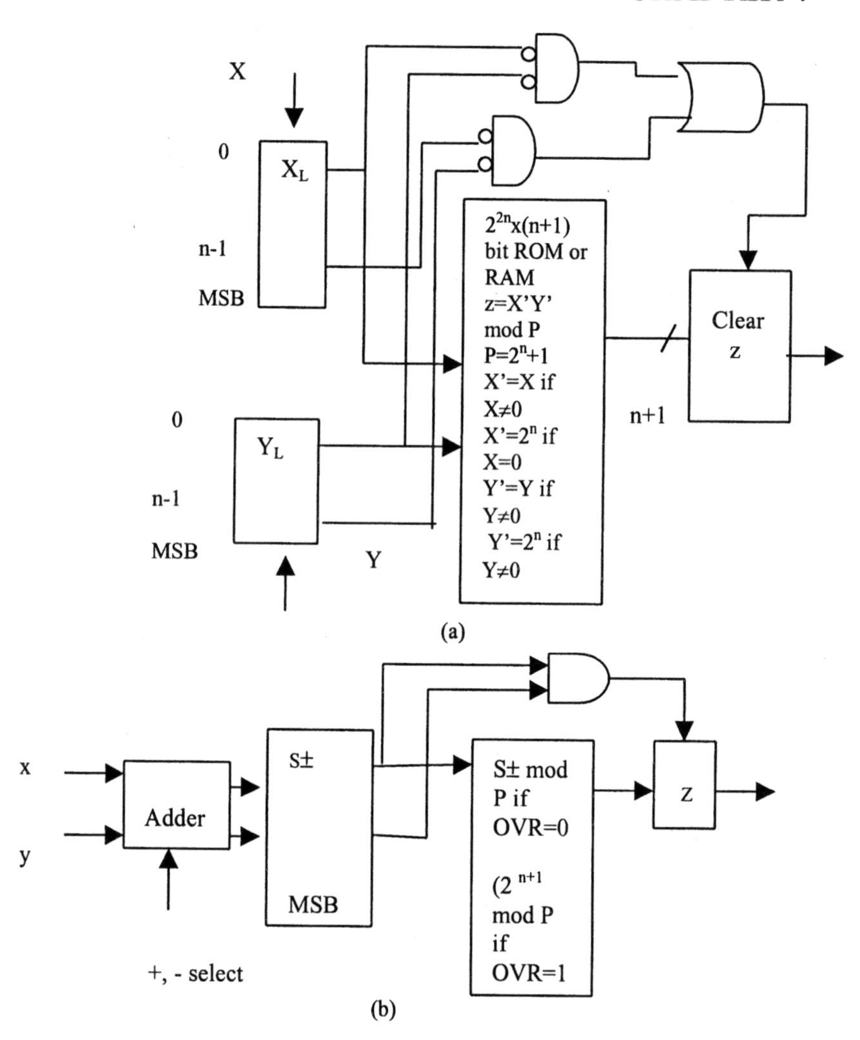

Fig 4.8 (a) Modulo 2^n+1 ALU due to Taylor and (b) memory compression technique for s+ (adapted from [Tay81a] 1981©IEEE)

with a decimal point assumed. Even though, both are fractions in (4.6), the sum is an integer. As an illustration for (c+d) = V = 20, we have $\langle V.8\rangle_{63} = \langle 20.63/8+20/8\rangle_{63} = \; = \langle 63(2+4/8) + 20/8\rangle_{63} = (31.5 + 2.5)_{63} = 34$. Note that V is expressed as IV+.XV. It may be recognized that writing V as $2^{n/2}.V_H+V_L$, then the result is $(V.2^{n/2}) \bmod (2^n-1)$ can easily be seen to be $V_L.2^{n/2} + V_H$, thus eliminating the ROM used in Taylor's approach. The case for modulus (2^n+1) follows similar lines with slight difference in the sign of certain terms. The general hardware architecture of Taylor is shown in

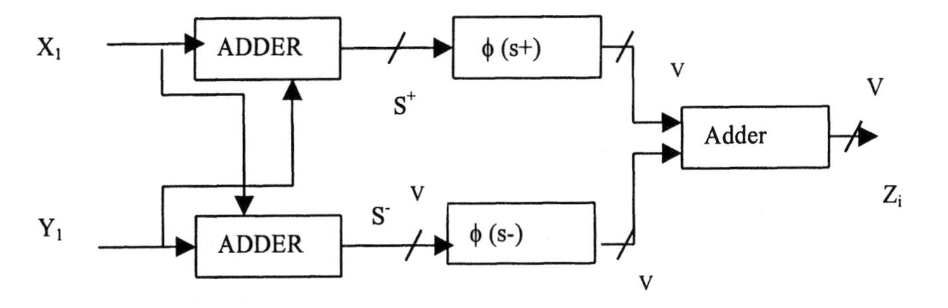

Fig 4.9. Basic Quarter square multiplier. (adapted from [Tay81a] 1981 ©IEEE)

Fig 4.11. There are four levels in the architecture. Note that W_1 and W_2 are unnecessarily long.

4.5. MULTIPLIERS WITH IN-BUILT SCALING

4.5.1. Soderstrand and Fields ROM based multipliers

Soderstrand and Fields [Sode77] suggested a fully ROM-based architecture for evaluation of y=X. k/N where division by constant N is achieved. As an illustration, consider the moduli {13, 7, 4} in the basic RNS and the N defined by two moduli 11.15=165. The architecture for base extension first of the given number X to the new moduli 11 and 15 is as shown in Fig 4.13 (a). Next, as shown in Fig 4.13 (b), RNS multiplication by k is performed and divided by 11.15. Note, however, that the memory requirements of this architecture are quite high.

4. 5 .2. Taylor and Huang Multipliers

Taylor and Huang [Tay82b] have described an over-flow free multiplier in the moduli set $\{2^n, 2^n-1, 2^n+1\}$. This integrates multiplication and dynamic range scaling in one step. We wish to calculate Z=c. X/V in this method. Z is represented by the symbol ϕ denoted as the auto scaled variable. In MRC, a smaller number is obtained after scaling which needs to be base-extended. The auto-scaled variable is

$$x_{RNS} \Rightarrow \{x_1, x_2, \ldots, x_L\}$$
$$x_{\phi} \Rightarrow \{z_1, z_2, \ldots z_L\} \; : Z=[c.X/V]_R$$

where R stands for rounding.

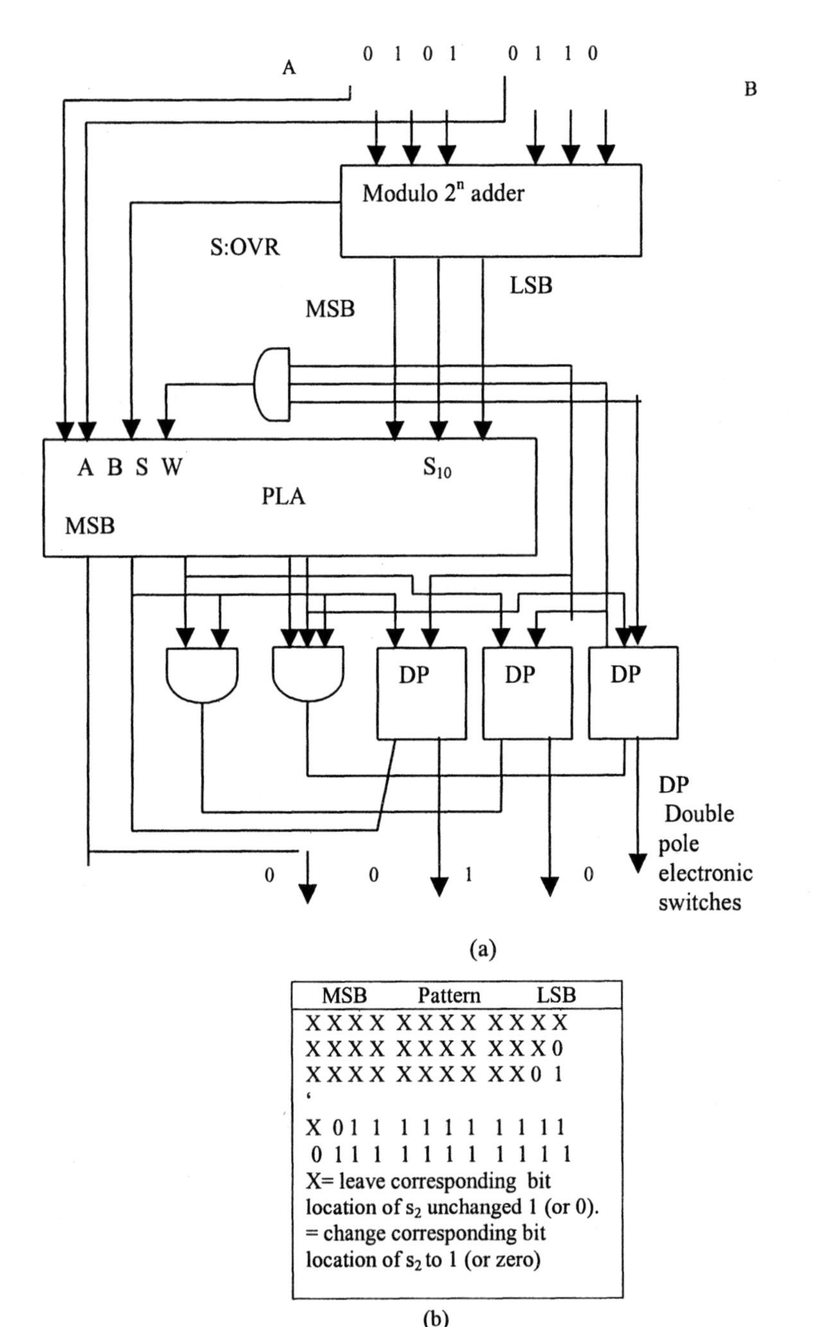

Fig 4.10 (a) Modulo 2^n+1 adder implementation using PLA and (b) the masks used in the PLA in (a) (adapted from [Tay81a] 1981 ©IEEE)

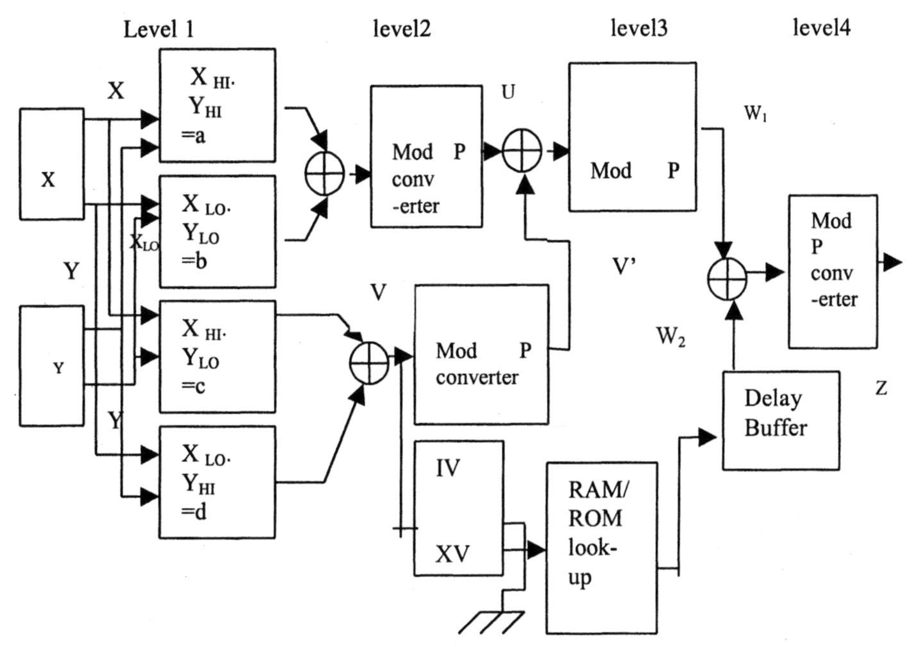

Fig 4.11. Taylor's general VLSI multiplier architecture for the moduli set $\{2^n, 2^n\text{-}1, 2^n\text{+}1\}$ (adapted from [Tay82a] 1982 ©IEEE)

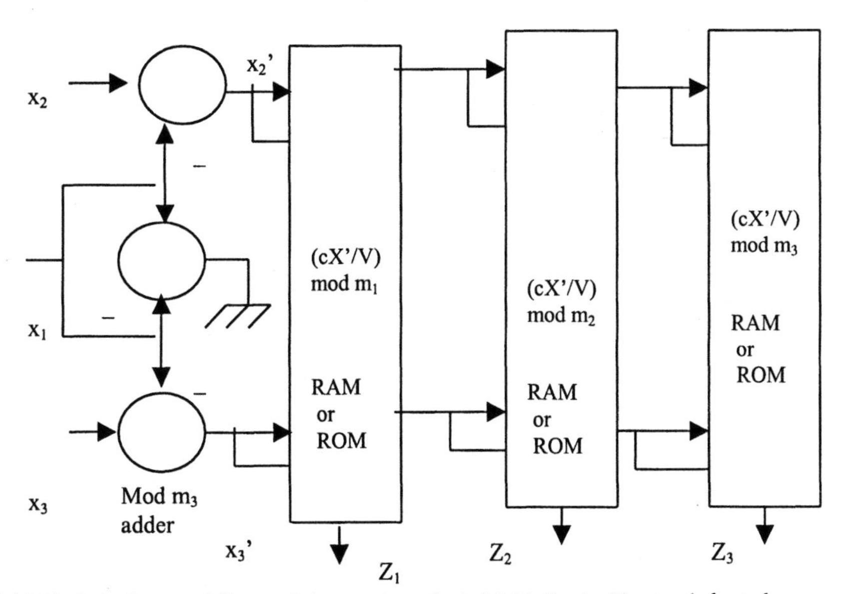

Fig 4.12. Taylor's three moduli extended range Auto-Scale Multiplier Architecture(adapted from [Tay82b] 1982 ©IEEE)

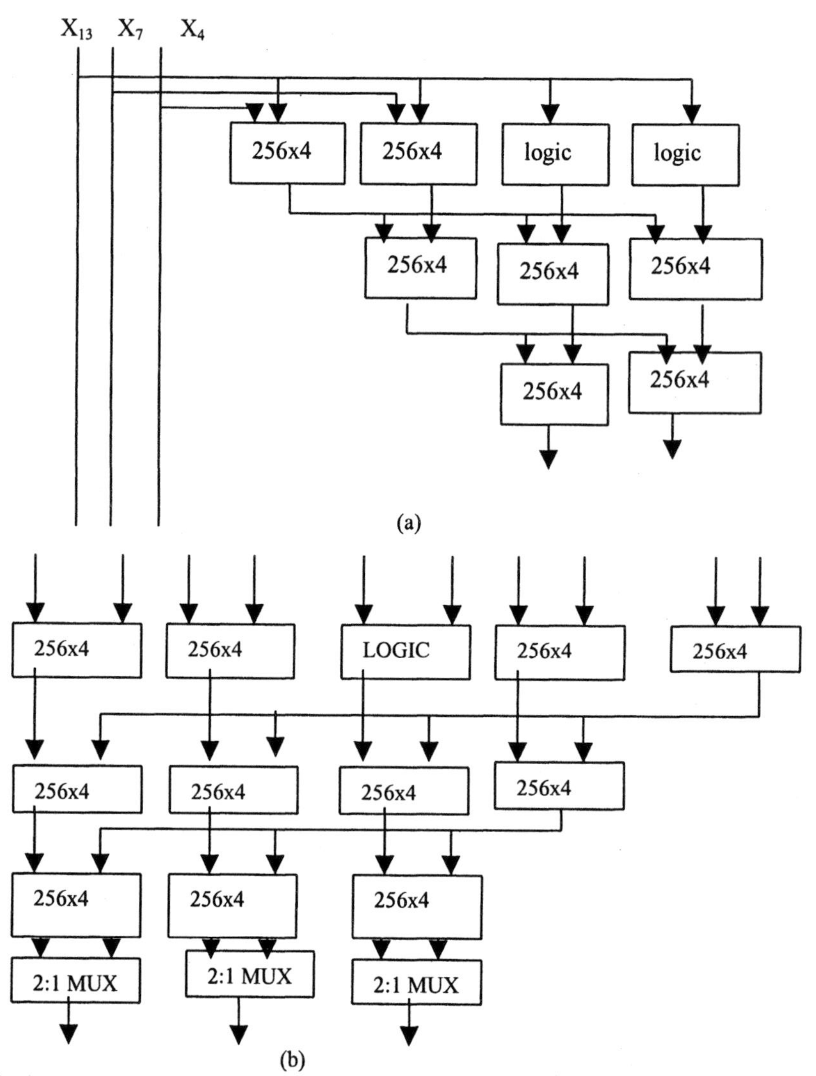

Fig 4.13. Soderstrand and Fields scaling multiplier architecture using ROMs (adapted from [Sode77] 1977 ©IEEE)

Taylor and Huang observe that the required ROM size for the above can be reduced by considering in stead of $\{x_1, x_2, x_3\}$ for a three moduli RNS, the residues $\{0, x_2\text{-}x_1, x_3\text{-}x_1\}$. Next, corresponding to these using a memory of only 2b bit addresses, the scaling as required can be performed as illustrated in Fig 4.12. As an illustration, consider the moduli set {8, 7, 9}. Let c=X=500.Thus Z yields Z=[X.c/M]R = [500.500/504]R=496. Using the architecture of Fig 4.12, however, we have the residues {0, 6, 1} in place of the original {4, 3, 5}. This corresponds to 496, whence Z=492. Thus, there will be an error. The reader is referred to Taylor and Huang for a detailed error analysis.

Huang and Taylor [Hua79] studied the ROM based modulo multiplier implementation. They observe that there are four regions in the multiplication table written as a pxp array where p is the number of bits needed to represent m. Since the entries in these areas have some similarity (see Fig 4.14 (a)), it is possible to reduce them to two types D_1 and D_2. Next, depending on the multiplicand and multiplier x and y, these memories are addressed to obtain the result. Note that D_1 contains all non-redundant diagonal modular products and D_2 contains all redundant diagonal products appended to the anti-diagonal. We define s as s =(p-1)/2 or p/2 based on whether p is odd or even. Then, the procedure is as follows:

Step 1: If x, y=0, the result is zero. We also define x', y' as follows:
x'=x if x<s, set x=0
x'=p-x if x>s, set X=1
y'=y if y<s, set y=0
y'=p-y if y>s set y=1.
Step 2: If x'=y', go to step 1. If x≠y, set

x	y	a	b
0	0	max(x', y')	min (x', y')
0	1	min(x', y')	max (x', y')
1	0	min(x', y')	max(x', y')
1	1	max (x', y')	min (x', y')

Step 1: Read $D_1(x)$ = (x.y) mod p; Terminate
Step 2: Read $D_2(x)$ = (x.y) mod p; Terminate.
The hardware implementation is as shown in Fig 4.14 (b). An example will be next presented. Consider p = 7 and x=4, y=5.
Step 1: x'=7-4=3
y'=7-5=2
Step 2: If x'≠y', a=3, b=2 Result is D_2 = (3, 2) =6.

	0	1	2	3	....	P3	p-2	p-1
0	**0**	0	0	0	...	0	0	0
1	0	**1**	2	3	A1...	p-3	p-2	**p-1**
2	0	2	**4**	6	...	p-6	**p-4**	p-2
3	0	3	6	**9**	....	**p-9**	p-4	p-3
. . .		**A2**					**A3**	
p-3	0	p-3	p-6	**p-9**	**A4**	**9**	6	3
p-2	0	p-2	**p-4**	p-6	....	6	**4**	2
p-1	0	**p-1**	p-2	p-3	...	3	2	**1**

(a)

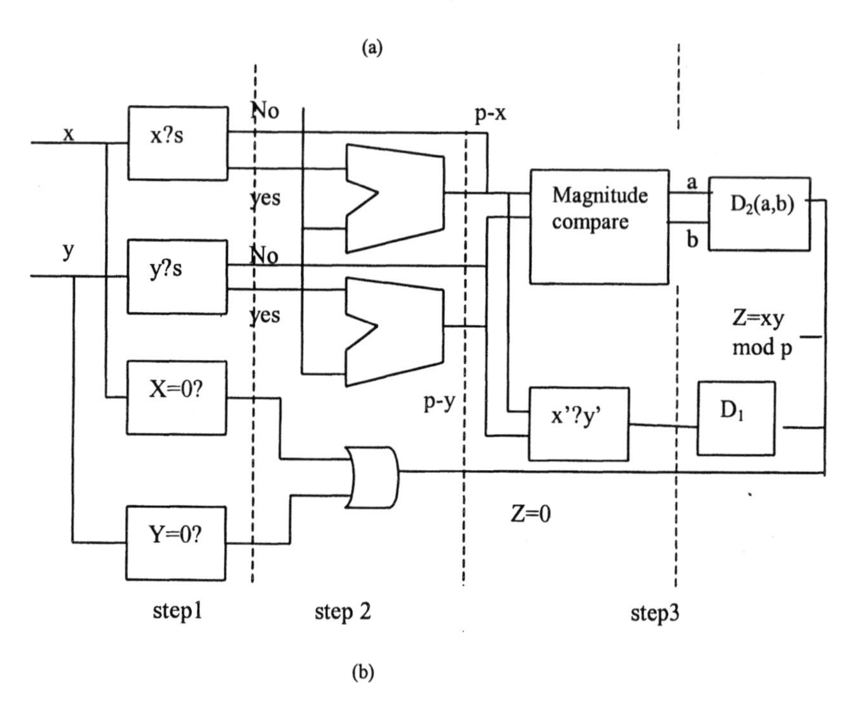

(b)

Fig 4.14. Huang and Taylor memory compression scheme based on the multiplier matrix (a) and hardware implementation (b) (adapted from [Hua79] 1979 ©IEEE)

4.6. RAZAVI AND BATTELINI ARCHITECTURES USING PERIODIC PROPERTIES OF RESIDUES

Razavi and Battelini [Raz92] sketched the architecture of a residue multiplier using small moduli. They realize multipliers using the periodic properties of $2^x \bmod m_i$. Next, the residues are used to obtain the final result using the

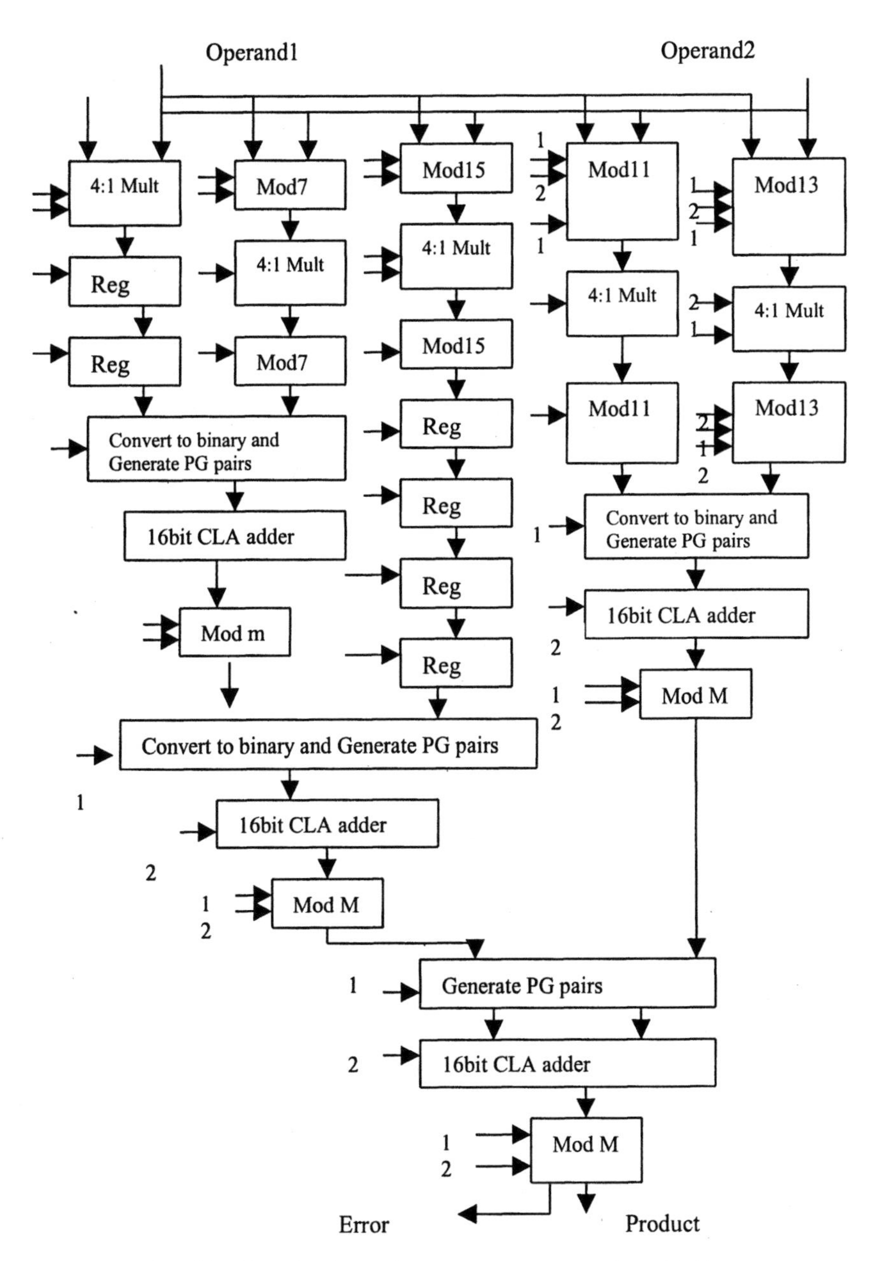

Fig 4.15. Razavi and Battelini architecture for modulo multiplier (adapted from [Raz92] 1992 ©IEE)

CRT using LUTs. This approach is straight- forward and can be pipelined easily. Their architecture for a moduli set {8, 7, 11, 13, 15} is shown in Fig 4.15. The input word is first converted into residues corresponding to these moduli and then multiplication of residues is carried out. The last stage converts the residues to the final binary number. The conversion for moduli 11 and 13 is done in two stages as illustrated next. For the modulus 11, first the given word is reduced mod 33 by considering five bit words. Then this word is reduced mod 11 as illustrated by an example.

Consider X = 86 = 0101 0110

Then N_1 = X mod 33 = $((010)_2 . f(33) + (10110)_2)$ mod 33 = $((10110)_2 - (010)_2)$ mod 33= 010100

$N_2 = N_1$ mod 11 = $((010100)_2)$ mod 11 = $((101)_2 . f(11) + (0100)_2)$ mod 11 = $(1001)_2$

Consider another example for m_i=13.

Then N_1 = X mod 65 = $((01)_2 . f(65) + (010110)_2)$ mod 65 = $((010110)_2 - (01)_2)$ mod 65 = $(0010101)_2$

$N_2 = N_1$ mod 13 = $((0010101)_2)$ mod 13 = $((001)_2 . f(13) + (0101)_2)$ mod 13 = $((0101)_2 + (011)_2)$ mod 13 = $(1000)_2$.

Note that we have used the fact that $f(33) = 2^5 - 33 = -1$ and $f(65) = 2^6 + 1$.

4.7. HIASAT'S MODULO MULTIPLIERS

Hiasat has described [Hias00a] a RNS multiplier for medium and large moduli. This is similar to Stouraitis et al architecture to be described in Section 4.10, with a small variation. The bit products corresponding to each power of two are added to yield a ($\log_2 n$+1) bit word. Then, each of these words V_j corresponding to each column in the product matrix, has a weight of 2^j. Each V_j is used to address a ROM to obtain $(V_j . 2^j)$ mod M which are next added using a multi-operand modulo adder. The architecture is sketched in Fig 4.16. The encoders are summers of one-bit partial products.

Hiasat has described [Hias00b] another multiplier (X.Y) mod m_i as will be described next. Herein, first the 2n bit product is determined which is divided into four fields A, B, C and D such that

$$Z= D.(2^{2n-k-1}) + C.2^n + B.2^{n-1} + A \quad (4.7)$$

Note that A, B, C and D are respectively n-1, 1, k and n-k bit words. Note that k corresponds to the number of bits needed to represent $a = (2^n - m_i)$. The next step is to evaluate Z mod m_i. By definition of a, the term $(C.2^n)$ mod m_i can be seen to be C.a. Note that A is already a (n-1) bit word

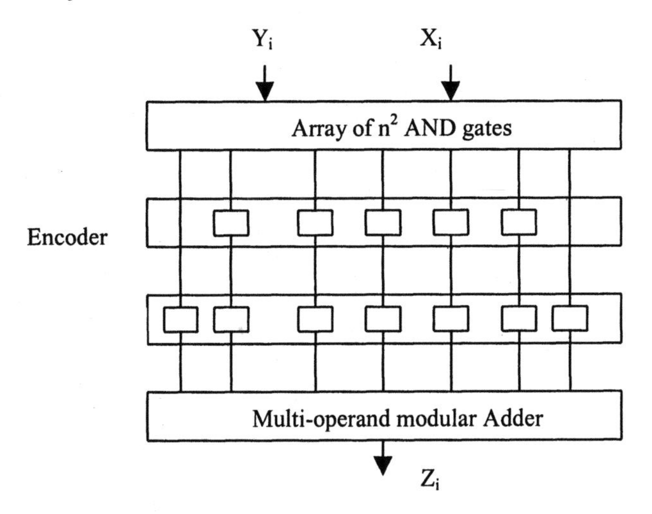

Fig 4.16. Hiasat's multiplier architecture for medium and large moduli (adapted from [Hias00b] 2000 ©IEEE)

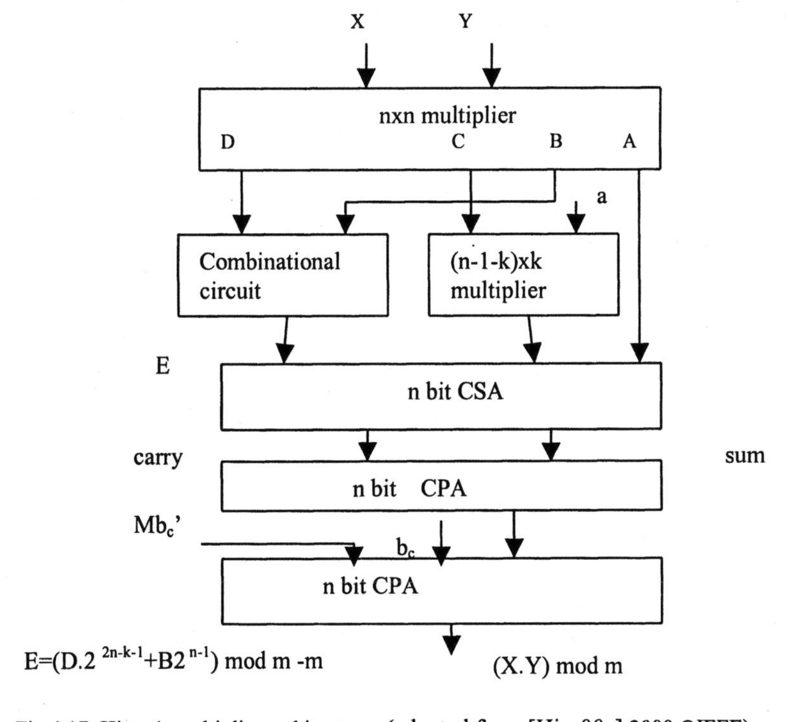

Fig 4.17. Hiasat's multiplier architecture (adapted from [Hias00a] 2000 ©IEEE)

needing no reduction. Hiasat suggests the evaluation of the remaining two terms by combinational logic which in our opinion is nothing but a binary to RNS converter mod m_i, whose output is negative. The next step is to add all these four operands in a carry-save adder followed by a carry-propagate adder. Based on the carry output of the n-bit CPA, the correction needs to be done by adding m_i to make the result positive. This method appears to be direct multiplication followed by modulo reduction. The architecture is presented in Fig 4.17.

4.8. ELLEITHY AND BAYOUMI MODULI MULTIPLICATION TECHNIQUE

Elleithy and Bayoumi [Ell95] described a modulo multiplier based on their modulo adder described in Section 2.3.4. They recommend three architectures two of which are shown in Fig 4.18 (a) and (b). In the architecture of Fig 4.18 (a), based on the sign of the evaluation $X_1.X_2$-k.m either $X_1.X_2$ or $X_1.X_2$-k.m is selected using a 2:1 multiplexer. In the architecture of Fig 4.18 (b), a ROM is used to reduce the 2n bit result of a normal multiplier mod m. Evidently, the ROM size is prohibitively large for for large moduli. The third method generates first the partial products as in a conventional multiplier. Then, these are added in a $\log_2 n$ level modulo adder tree. These modulo adders are constant speed adders using five levels as described in section.2.3.4. However, we note that the design is not correct since the partial products are all not of the same weight.

4.9. BRICKELL'S ALGORITHM BASED MULTIPLIERS AND EXTENSIONS

4.9.1. Modulo multipliers based on Brickell's algorithm

Brickell's [Blak83, Bric83] algorithm for modulo multiplication yields AB mod N as desired. This is illustrated below using an example. Consider A = 13 and B= 14 and N =15. The calculation starts with MSB of B. The result is expressed in each step as

$R_i = (2.R_{i-1} + b_i.A) \bmod N.$ (4.8)

The initial value of R i.e. R_{-1} is considered zero. Thus, we have the following four steps:

$R_0 = (2.0+1.13) \bmod 15 = 13$ $b_3=1$

$R_1 = (2.13 + 1.13) \bmod 15 = 9$ $b_2=1$
$R_2 = (2.9+ 1.13) \bmod 15 = 1$ $b_1=1$
$R_3 = (2.1+ 0.13) \bmod 15 =2$ $b_0=0$

The result can be verified to be true. It may be noted that the maximum value of the term Z in brackets can be atmost 3N. Hence, reduction mod N can be done using a parallel architecture to enhance the speed of modulo multiplication as shown in Fig 4.19. Note that the last four subtractions are

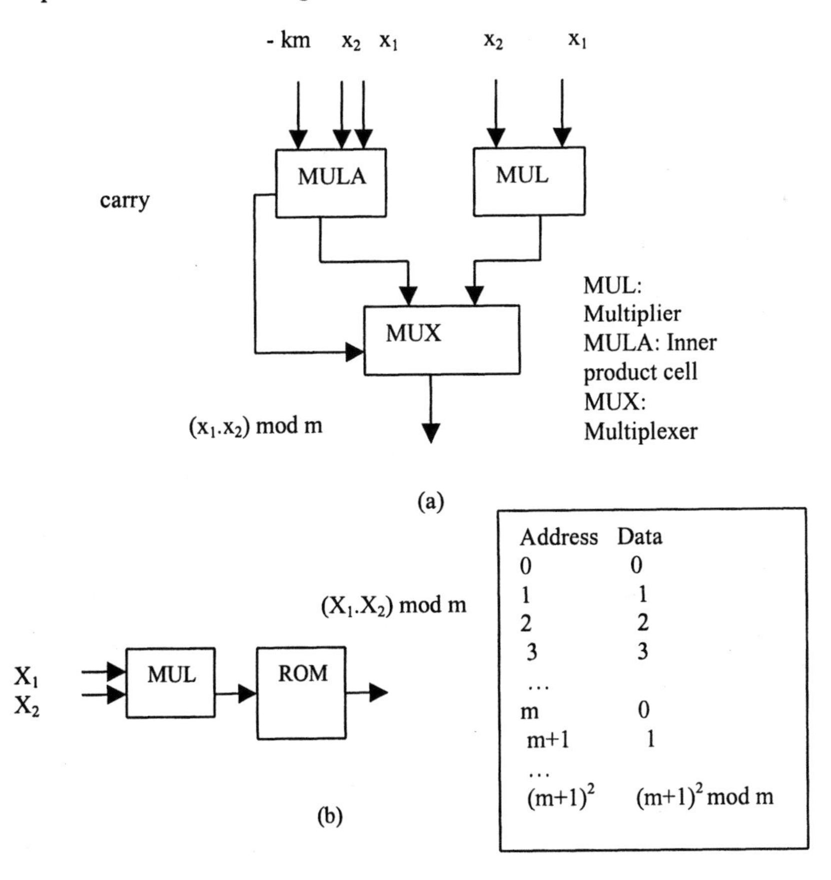

Fig 4.18. Elleithy and Bayoumi architectures for modulo multiplication: (a) using a multiplier and an inner product cell and (b) using look-up table (adapted from [Ell95] 1995 ©IEEE)

not needed. Other authors have considered the reduction of $2.R_i$ mod N first followed by another reduction mod N after adding b_i. A. However, this evidently is slower compared to the former.

Prasanna and Ananda Mohan [Pras94, Pras91] have considered extension of the method considering two bits at a time of B. Evidently, in this case, the number of steps are reduced by half. However, each step has the bracketed term with a magnitude at most 7N:
$R_0= (4.0+ b_1b_0.A)$ mod N
$R_2 = (4.R_0+b_3b_2.A)$ mod N
Evidently, four possibilities of multiples of A exist 0, 2A, 3A, 4A which need to be available at the inputs of a multiplexer and can be selected. The 2A can be obtained by left shifting whereas 3A is obtained by adding A and 2A. Note that the modulo reduction needs six parallel subtractors to subtract N, 2N, 3N, .. 6N and then based on the sign bits of the outputs of the subtractors, the correct value is selected as shown in Fig 4.19. This method is twice faster than Brickell's technique. The other techniques [Beth89] use scanning the bits, coding strings of logic one bits as signed operands etc for which the reader is referred to the literature.

4.9.2. Sloan's observations on Brickell's technique.

Sloan Jr. points out that [Sloa85] the quotient of the operation (A.B)/N is inherently present in the process of A.B mod N evaluation and not recognized by Brickell. This can be seen as follows:
Assume R=0 and Q=0 in the beginning where Q and R are the quotient and remainder of the operation (A.B)/N. The modulo multiplication is same as that of Brickell starting from the MSB of B and contains n steps where $n = \lceil \log_2 N \rceil$. Each step evaluates a new R value and new Q value where Q is the desired quotient. The basic step is as follows:
$R = 2R_{old}$, $Q = 2Q_{old}$.
If $b_i=1$, R = R+A.
If R>M, R=R-M, Q = Q+1

An example will illustrate the procedure for evaluating 13.14 mod 15, where A=13, B=14 and N=15.
Step1: R=0, Q=0, $b_3 = 1$, R=0+13.R<15.Hence, R=13, Q=0.
Step2: R=26, Q=0,R=11, Q=1,b_2=1, R=11+13=24.R>15.Hence, R=9, Q=2.

Step3: R=18,Q=4,R=3, Q=5,b_1=1, R=3+13=16. R>15.R=1, Q=6.
Step4: R=2, Q=12,b_0=0, R=2+0=2.R<15.R=2, Q=12.
The answer can be verified to be true.

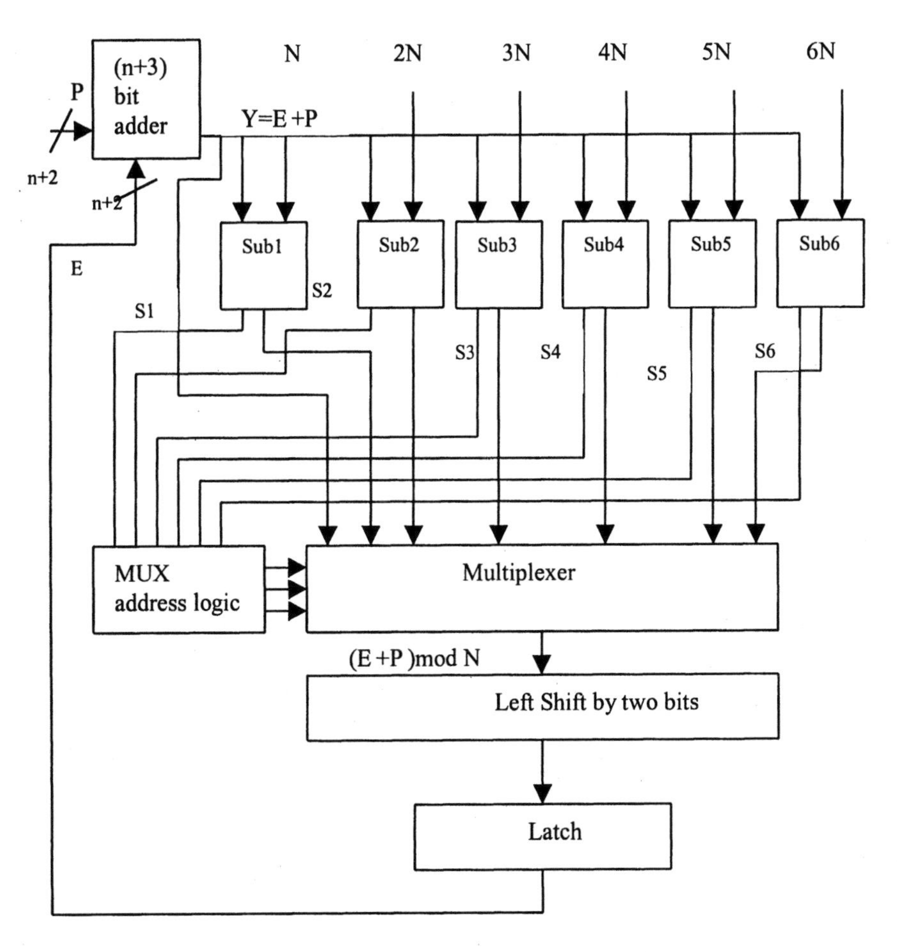

Fig 4.19. Architecture for reducing Z mod m_i.
(adapted from [Pras94] 1994 ©IEE)

4.9.3. Montgomery's modulo multiplication technique and extensions

In the Montgomery multiplication technique [Mont85], which is similar to Meehan et al binary to RNS converter studied in Chapter 3, only scaled result is obtained. In other words $(A.B\ .2^{-n})$ mod M is computed in stead of A.B mod M. Moreover, it is mandatory that the modulus M is odd. There are several ways of performing Montgomery multiplication which can be found in [Koc96].

In one method known as separated operand scanning, first normal multiplication is performed to find A.B a 2n-bit word assuming A and B are n bit words. Next a technique due to Montgomery which is similar to Meehan's technique described in Chapter II for binary to RNS conversion is applied to (A.B) to yield $A.B.2^{-2n}$ mod M. This computation needs 2n steps. Alternatively, $(A.B.2^{-n}$ mod M) can be evaluated [Yang98] by considering that

$$A.B = C_1.2^n + C_0. \qquad (4.9a)$$

Multiplying both sides by 2^{-n}, we have

$$A.B.2^{-n} = C_1 + C_0.2^{-n}. \qquad (4.9b)$$

Thus, only n steps are needed to evaluate the second term, which result is added modulo M to C_1 to yield the scaled result A.B. 2^{-n} mod M.

In another method, denoted as coarsely integrated operand scanning, the multiplication and subsequent modulo reduction with scaling are integrated into one step. Two examples are illustrated to illustrate the procedure. In the integrated multiplication and modulo reduction method, in each step starting from LSB, (b_i. A) is added and then the result is divided by two. This is similar to Brickell's technique. An example will be illustrative. Consider 11.11 mod 13 where A =11, B=11.

$(0/2 + 11/2)$ mod $13 = 12 = P_0$
$(12/2 + 11/2)$ mod $13 = 5 = P_1$
$(5/2 + 0/2)$ mod $13 = 9 = P_2$
$(9/2 + 11/2)$ mod $13 = 10 = P_3$.

Thus, based on the bit values of B starting from LSB, A/2 is added to the old value divided by two. Note that in the case of odd old values and A, the even part is added and ½ mod m_i is $(m_i+1)/2$. The division by two accomplishes the scaling of the result by 2. If the scaling by two is avoided, the algorithm reduces to Brickell's algorithm.

Alternatively [Su99], we can first multiply A and B to yield $121_{(10)} = 111\ 1001_{(2)}$. Then, using similar technique for the four-bit LSB word $1001_{(2)}$, we have the following:
$(0/2+1/2) \bmod 13 = 7 = P_0$
$(7/2+0) \bmod 13 = 10 = P_1$
$(10/2 + 0) \bmod 13 = 5 = P_2$
$(5/2 + 1/2) \bmod 13 = 3 = P_3$.

This corresponds to C_0 and $C_1 = 7$ i.e. the MSBs of A.B. Thus, the result is (3+7) mod 13 = 10,which is same as that obtained before. Note that application of this technique to complete word A.B yields $A.B.2^{-8} \bmod 13$, whereas in the above, we obtained $A.B.2^{-4} \bmod 13$.

It may be noted that often it is required to compute only A.B mod M, in which case a pre-multiplication of A or B by $2^n \bmod M$ will yield the correct result. Evidently, this needs additional computation time whereas in Brickell's method, the result is available directly in n steps.

It is important to note that the actual mechanization of the above algorithms is slightly different due to the presence of 1 /2 terms for division by two in all the steps. As an illustration consider the step (12/2 + 11/2) mod 13. The actual implementation of this tests the LSB of previous result (in this case LSB of 12 is zero) and present bit of B (in this case 1), to decide a quotient bit Q, a modulo 2 addition of these two bits (in this case 1). If Q is 1, an additional 13 is added and else zero is added. In short, we have (12+11+13)/2 = 18 mod 13 = 5.
Stated mathematically, new P_i is defined as follows:

$$P_i = (P_{i-1} + C_i + Q_i.M) \text{ div } 2 \qquad (4.10)$$

with $Q_i = (P_{i-1} + C_i) \bmod 2$
in the case of separate modulo reduction technique. In the case of integrated reduction technique, an additional term $b_i.A$ also will need to be added.

The reader is referred to other variations of Montgomery algorithm implementations to [Koc96]. It may be mentioned that extensions to the evaluation of X.Y mod M are of interest in cryptography which can be found in [Yang98].

4.9.4. Di Claudio et al Parallel Pseudo-RNS Multipliers

Di Claudio et al [Di95] have described a modulo multiplier which actually incorporates scaling also as in Montgomery's modular multiplication algorithm described before. However, as against the serial-parallel architectures used in earlier implementations of Montgomery's algorithm, this architecture is fully parallel.

The multiplier is presented in Fig 4.20. which can evaluate $(Z.Y.2^{-b})$ mod m_i. This follows similar lines as that described in section. First, Z.Y is evaluated as a 2b bit word. Denoting it as $X = X_H.2^b+X_L$, we have

$$2^{-b}(X_H.2^b+X_L) \bmod m_i = (X_H + X_L.2^{-b}) \bmod m_i \qquad (4.11)$$

The architecture in Fig 4.20 shows the two paths separately wherein the leftmost path evaluates X_H. The computation of $X_L.2^{-b}$ needs additions of multiples of m_i to X_L to make the result an integer multiple of 2^b. This helps to directly obtain the second term since 2^b and 2^{-b} terms when multiplied together yield unity. The operation performed in the right path is given by

$$\{(X_L.(2^b-\alpha_i) \bmod 2^b) \bmod 2^b.\ m_i\} \text{ Div } 2^b + \delta \qquad (4.12)$$

where $\delta=1$ if $(Z.Y)>0$ and $\delta=0$ when$(Z.Y)=0$. The derivation is as follows: We need to find K_i such that

$$[X_L+K_i.m_i] \bmod 2^b = 0 \qquad (4.13)$$

Thus,

$$K_i= 2^b-(\alpha_i X_L) \bmod 2^b \qquad (4.14)$$

where $\alpha_i = (1/m_i) \bmod 2^b$. Thus, X_L will be added $K_i.m_i$ to yield a multiple of 2^b:

$$p.2^b = X + [(2^b-\alpha_I).\ X_L) \bmod 2^b].\ m_i$$

$$=X_H.2^b + X_L +[((2^b- \alpha_i).X_L) \bmod 2^b].\ m_i \qquad (4.15a)$$

Evidently, the right hand side of the equation (4.15a):

$$(p-X_H)2^b=X_L+[((2^b-\alpha_i).X_L) \bmod 2^b].\ m_i \qquad (4.15b)$$

shall be a multiple of 2^b. Dividing both sides by 2^b yields

$$p = X_H + [m_i.[2^b - \alpha_i).X_L] \text{ Div } 2^b + (X_L/2^b) \qquad (4.15c)$$

Di Claudio et al [Di98] suggest the last term being replaced by δ as defined above.

As an illustration, consider Z=11,Y=11 with m_i =13 with $2^b = 16$. The result should be $(Z.Y.2^{-b})$ mod m_i = (121/16) mod 13 =10. We have α = (1/13) mod 16 =5,X_H=7 and X_L=9. Thus,

p= 7+ [13.[16-5).9) mod 16]] Div 16 + (9/16)
= 7+2+δ =10.
Di Claudio et al [Di98] observe that the hardware needed is about 2.5 the area of a multiplier since only LSBS are needed for one of the multipliers.

4.10. STOURAITIS ET AL ARCHITECTURES FOR (A.X+B) MOD m_i REALIZATION

Stouraitis et al [Soud97, Stou93] have proposed full – adder based implementations of (A.X+B) mod m_i. They observe that the higher order terms in A.X with bit positions larger than $\log_2 m_i$ will be expressed mod m_i so that the bit-length of the product A.X will be reduced from the original 2n bits where A and B each are of n bit length. Then in similar iterations denoted as functions g, the resulting value can be reduced mod m_i. They observed that generally three iterations are sufficient. They also suggested the use of certain moduli which will reduce the complexity of A.X + B mod m_i implementation. As an illustration, modulus 29 has the property that 2^5

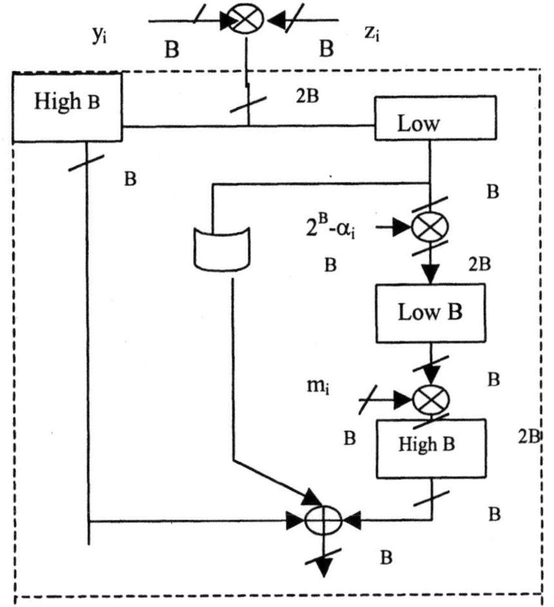

Fig 4.20. Di Claudio Direct implementation of Pseudo RNS multiplier (adapted from [Di95] 1995 ©IEEE)

mod m_i = 3, 2^6 mod m_i = 6, 2^7 mod m_i = 12 and 2^8 mod m_i = 2 all with at most two ones in any word. Thus, the partial product terms larger than 2^5 can be expressed as two bits in some positions yielding few bits to be added to obtain the final result. An example will be illustrated next.
Consider (A.B+X) mod m_i where A = $a_4a_3a_2a_1a_0$, B=$b_4b_3b_2b_1b_0$ and X = $x_4x_3x_2x_1x_0$. The bit products and bits to be added can be easily written as follows:

2^8	2^7	2^6	2^5	2^4	2^3	2^2	2^1	2^0
				a_4b_0	a_3b_0	a_2b_0	a_1b_0	a_0b_0
			$\underline{a_4b_1}$	a_3b_1	a_2b_1	a_1b_1	a_0b_1	
		$\underline{a_4b_2}$	$\underline{a_3b_2}$	a_2b_2	a_1b_2	a_0b_2		
	$\underline{a_4b_3}$	$\underline{a_3b_3}$	$\underline{a_2b_3}$	a_1b_3	a_0b_3			
$\underline{a_4b_4}$	$\underline{a_3b_4}$	$\underline{a_2b_4}$	$\underline{a_1b_4}$	a_0b_4				

There are thus ten terms - 4 of weight 2^5, 3 of weight 2^6, 2 of weight 2^7 and one of weight 2^8. These when reduced mod 29 yield the new matrix i.e. lower five column matrix in the above augmented by additional bit terms as follows:

2^4	2^3	2^2	2^1	2^0
a_4b_0	a_3b_0	a_2b_0	a_1b_0	a_0b_0
a_3b_1	a_2b_1	a_1b_1	a_0b_1	
a_2b_2	a_2b_1	a_2b_0	$\underline{a_4b_1}$	$\underline{a_4b_1}$
a_1b_3	a_0b_3	$\underline{a_4b_2}$	$\underline{a_3b_2}$	$\underline{a_3b_2}$
a_0b_4	$\underline{a_4b_3}$	$\underline{a_3b_3}$	$\underline{a_2b_3}$	$\underline{a_2b_3}$
$\underline{a_4b_4}$	$\underline{a_3b_4}$	$\underline{a_2b_4}$	$\underline{a_1b_4}$	$\underline{a_1b_4}$
	$\underline{a_4b_4}$	$\underline{a_4b_3}$	$\underline{a_4b_2}$	
			$\underline{a_3b_4}$	$\underline{a_3b_3}$
				$\underline{a_2b_4}$

Evidently, the sum of these bits yields at most 207, i.e. a 8 bit word. A second iteration on this word yields by rewriting the bits 2^7, 2^6, 2^5 as 12, 6 and 3 respectively (due to reduction modulo 29). Thus the resulting word is 52, a 6 bit word. This 6 bit word will yield at most 34 in the next iteration. Note, however, that in the last stage, if the result is larger than m_i, a ROM or modulo reduction hardware needs to be used. Stouraitis et al presented the hardware requirements for several moduli and suggested a pipeline implementation e.g. shown in Fig 4.21, for modulus 29 wherein the various latches are needed as illustrated.

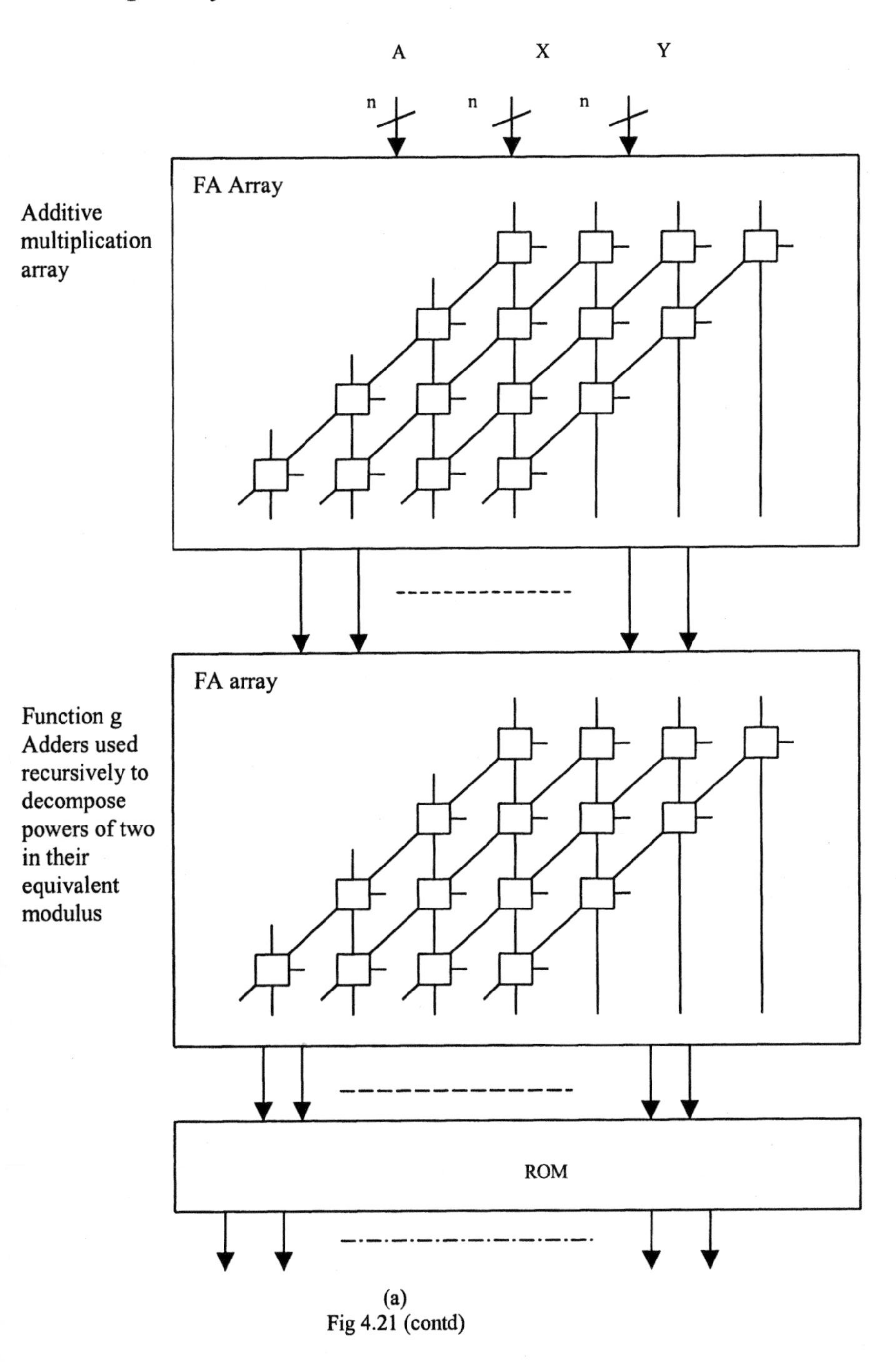

(a)
Fig 4.21 (contd)

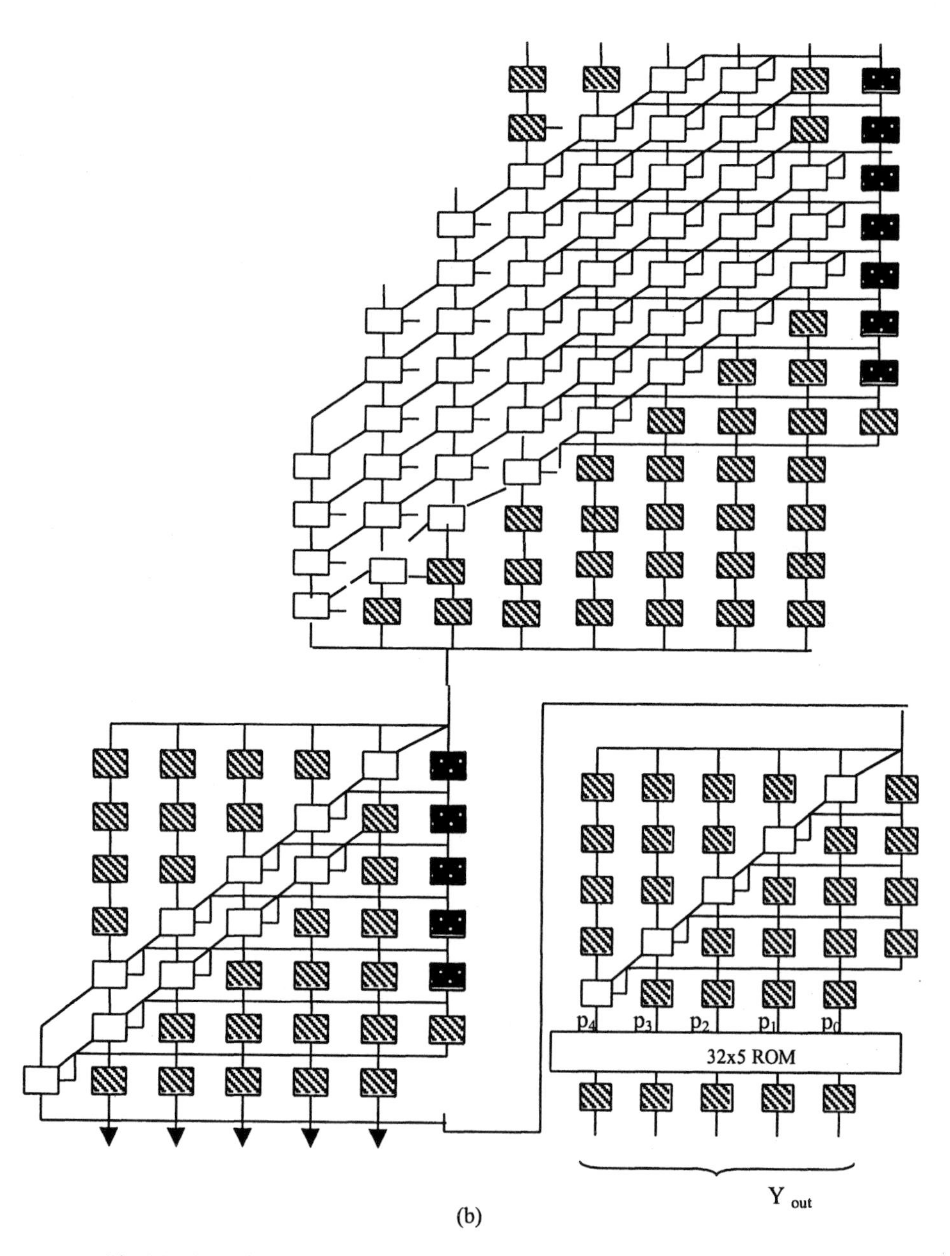

(b)

Fig 4.21. Stouraitis et al architecture for (A.X +B) mod m_i computation (a) Block diagram and (b) complete implementation mod 29 (adapted from [Stou93] 1993 ©IEEE)

4.11. MULTIPLICATION USING REDUNDANT NUMBER SYSTEM

The use of redundant binary representation where each digit can be 0,1 or −1 can yield a regular cellular array structure as well as improve the speed of computation. The use of this representation needs of course a binary to redundant binary converter and vice versa at the front and back-ends respectively. The remaining part of the hardware works in redundant binary form. In this technique, a given number can be represented in several ways. For instance, 0101, 0111', 11'01, 11'11', 101'1' all represent +5. However, 0 is uniquely represented. Owing to the redundancy, parallel addition of two numbers can be carry-free as will be illustrated next for the augend (87) and addend (101).

Augend	1 0 1'0 1'0 0 1'	(87)
Addend	1 1'1 0 0 1 1 1'	(101)
Intermediate sum	0 1 0 0 1'1'1 0	
Intermediate carry	1 1'0 0 0 1 0 1'	
Sum	1 1'1 0 0 0 1' 0 0	

Note that the sum of two bits x_i,y_i is represented as $x_i + y_i = 2c_i + s_i$ is computed in the first step. Herein, we ensure that both s_i, c_{i-1} are never 1. When one of the bits x_i or y_i is one and the other zero, we follow the following procedure:

If there is a possibility of a 1 carry from the previous position, we code c_i, s_i as 1, 1'. If there is a possibility of 1' carry, we code c_i, s_i as 0, 1. If there is no possibility of a carry from the previous position, we may code c_i, s_i as 1, 1' or 0, 1. Similarly, if one of x_i or y_i is 1', we let c_i, s_i as 0, 1' if there is a possibility of a carry from the previous bit position or as 1', 1 if there is no possibility of a negative carry. The possibility of a carry from the previous position can be found by examining the bits x_i, y_i, x_{i-1} and y_{i-1}. The sum is thus dependent on the six bits x_i, y_i, x_{i-1}, y_{i-1}, x_{i-2} and y_{i-2} only thus facilitating high-speed computation. The various options are listed in Table. 4.3. In the second step, the sum and carry bit from the previous position are summed.

We next consider a multiplication example using redundant binary representation. Note that the partial products are in RB form which are added two at a time to yield fewer partial products. These are in turn added to yield a RB result. This is split into A and B such that A-B is next evaluated using a two's complement carry-Look ahead adder. The result is a bit redundant binary integer and is converted into a 10 bit two's complement binary integer.

A 1101 (-3) 1' 1 0 1 $_{SD2}$
B 1011 (-5) 1' 0 1 1 $_{SD2}$

1' 10 1 $_{SD2}$
1'1 0 1 $_{SD2}$
0 0 0 0 $_{SD2}$
1 1' 0 1' $_{SD2}$

1' 1 1 0 0 1' $_{SD2}$
0 1 1' 0 1' 0 $_{SD2}$

0 0 0 0 1 0 0 0 1'$_{SD2}$

= 0 0 0 0 1 0 0 0 0
- 0 0 0 0 0 0 0 0 1
0 0 0 0 0 0 1 1 1 1

Therefore, P=11.

Takagi and Yajima [Taka92a] have described a modular multiplication algorithm using redundant arithmetic. This is based on the basic operation of modulo addition and its repeated application. This is briefly described next. Consider the evaluation of (A+B) mod Q and with A and B such that $-Q<A<Q$ and $-Q<B<Q$. The (A+B) mod Q evaluation is performed in two steps. In the first step, (A+B) is found using the carry propagation free addition procedure described above to obtain a (n+2) bit number. Next, by observing the three MSBs of the result, -Q or 0 or Q is added to obtain a (n+1) bit result.

Type	Augend digit (x_i)	Addend digit (y_i)	Digits at the next lower order position (x_{i-1}, y_{i-1})	Intermediate carry (c_i)	Intermediate sum digit (s_i)
1	1	1	------------------------	1	0
2	1	0	Both are nonnegative	1	1'
2	0	1	Otherwise	0	1
3	0	0	------------------------	0	0
4	1	1'	------------------------	0	0
4	1'	1	------------------------	0	0
5	0	1'	Both are nonnegative	0	1'
5	1'	0	Otherwise	1'	1
6	1'	1'	------------------------	1'	0

Table.4.3.Computation rules for first sep in carry propagation free first addition (adapted from [Taka85] 1985 ©IEEE)

Evidently, $-2Q<(A+B)<2Q$. The value t_v of the three most significant digits is $4t_{n+1}+2t_n+t_{n-1}$. If t_v is negative, Q can be added. If t_v is zero, nothing is added and if t_v is positive, -Q is added. The operation needed follows the tables presented in Fig 4.22. Note that here instead of finding $-Q$ exactly, ($-Q-1$) is determined by inverting all the LSBs except for first two MSBs. This can be seen by first noting that if Q is $1q_{n-2} \ldots q_1 q_0$, ($-Q-1$) is given by $1'0\, q'_{n-2} \ldots q'_1\, q'_0$. The value $-Q$ or 0 or Q is represented by u_i in Fig 4.22. Note that u_i can be 0 or 1 facilitating carry-free-addition. The intermediate sum and intermediate carry satisfy $2v_i+w_i=t_i+u_i$. Next, w_i and v_{i-1} is added to obtain s_i. Note that $v_{-1}$is defined as 1 when t_v is positive and otherwise it is zero. Two examples of modular addition are shown in Fig 4.23 for illustration.

The modular multiplication using radix 2 is based on Brickell's method with the difference that the operands are in redundant binary form in the present case. The operation in each step is given as

$$P_j = (P_{j+1}.2 + A.b_j) \bmod Q \qquad (4.16)$$

The doubling mod Q of P_{j+1} is done by first left shifting and then adding $-Q$ as described before by examining the MSBs. Next, depending on whether b_j is -1, 0 or 1, -A obtained by negating A or zero or A is added. Next, modulo reduction after addition is done. The method can be extended to radix 4 modular multiplication as well. In this case, however, the expression in (4.16) changes to

$$P_j = (P_{j+1}.4 + A.b_j') \bmod Q \qquad (4.17)$$

s_n

T_{n+1}	t_n / t_{n-1}	-1	0	1
1	0	-	1'	1'
1	1	1'	0	0
0	1	1'	0	0
0	0	0	0	0
0	1	0	0	1
1	1	0	0	1
1	0	1	1	-

(a)

t_v/q_i	0	1
Neg	0	1
Zero	0	0
Pos	1	0

v_i, w_i

t_i/u_i	0	1
1'	0 ,1'	0,0
0	0 , 0	1,1'
1	1, 0	1,0

s_i

w_v/v_{i-1}	0	1
1'	1'	0
0	0	1

(b)

Fig 4.22. Computation rule for the second stage of modular addition (a) at the most significant two positions and (b) at other positions (adapted from [Taka92a] 1992 ©IEEE)

```
Augend A    0 0 1 0  1 1 0 1 0 0 1 0   (462)
Addend B + 1 0 1' 1' 1' 0 1 1' 1  0 0  0 (1176) stage 1
T          1 1'0 1'0 1  1'1 1  1' 0 1' 0
-Q      +   1'0 1 1  0 1 0 0  1 1 1 1 (1)  stage 2
 sum S      0 0 1 0  1' 1 0 1' 1  1' 1 0  (438)

Augend A  0 0 1 0 1' 1 0  1 0  0 1' 0 (462)
Addend B + 1'0 1 1 1  0 1' 1 1' 0 0  0 (24) Stage 1
T         0  1'1 0 1 0  1 0  0 1' 0 1' 0
+Q           0 1 0 0 1  0 1  1  0 0 0  0 (0)  Stage 2
 sum S       0 0 1 0 0  0 0  1' 1' 0 1' 0    (486)
```

Fig 4.23. Example of modular addition (Q=1200) (adapted from [Taka92a] 1992 ©IEEE)

c_j, d_j

$B_{2j+1} \backslash b_{2j}$	1'	0	1
1'	1,1	*0,2/1,2	0,1
0	0,1'	0,0	0,1
1	0,1	*1,-2/0,2	1,1'

$d_j \backslash c_{j-1}$	1'	0	1
2'	-	2	1
1'	2'	1'	0
0	1'	0	1
1	0	1	2
2	1	2	-

Fig 4.24. Decoding rule of a multiplier using redundant representation (adapted from [Taka92b] 1992 ©IEEE)

The evaluation of b_j' shall be such that b_j' {0, 1, 2, -1, -2}. This is achieved in two steps as shown in Fig 4.24. Two adjacent bits b_{2i} and b_{2i+1} are first used to generate c_j, d_j using the knowledge about b^{2i-1}. Then, using c_j, d_{j-1}, b_j is evaluated. Note that next in the implementation of $4.P_{j+1}$, the authors recommend modulo doubling twice to first obtain $2\ P_{j+1}$ mod Q and then $4.P_{j+1}$ mod Q. The block diagram of the modulo multiplier is shown in Fig 4.25.

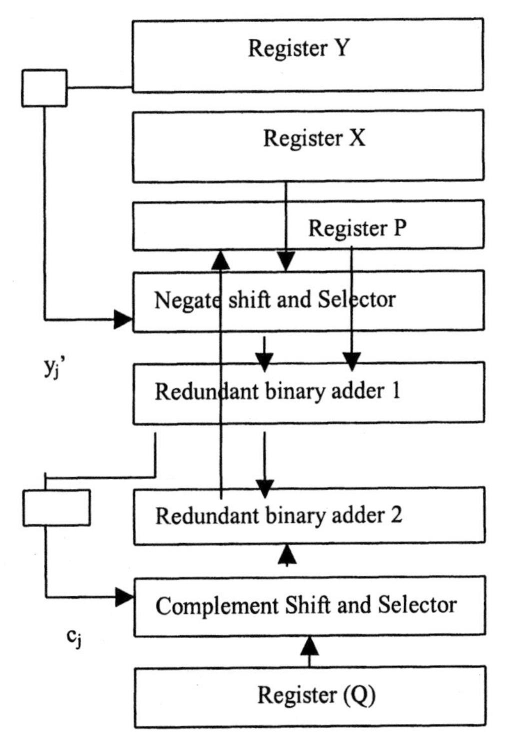

Fig 4.25. Block diagram of the modulo multiplier using redundant arithmetic (adapted from [Taka92b] 1992 ©IEEE)

r_{n+3} , rt_{n+2}

p_{n+1} / p_n	1'	0	1
1'	*1',1'/x	1',0	*0,1'/1',1
0	*0,1'/1',1	0,0	*1,1'/0,1
1	*1,1'/0,.1	1,0	*x/1,1

* p_{n-1} is non- negative/ otherwise.
x never occurs.

(a)
Fig 4.26 (contd)

$ru_{i+1}, rt_i\ (0 \le i \le n+1)$

p_{i-2}/ yx_i	1'	0	1
1'	1',0	*0,1'/1',1	0,0
0	*0,1'/1',1	0,0	*1,1'/0,1
1	0,0	*1,1'/0,1	1,0

*Both p_{i-3} and yx_{n-1} are non- negative/ otherwise.

(b)

$r_i(0 \le i \le n+1)$

rt_i/ru_i	1'	0	1
1'	X	1'	0
0	1'	0	1
1	0	1	X

X never occurs

(c)

$pu_{i+1}, pt_i\ (0 \le i \le n+1)$

r_i/cq_i	0	1
1'	0,1'	0,0
0	0,0	1,1'
1	1,1'	1,0

(d)

$pv := 2(r_{n+3}+cq_{n+3})+(r_{n+2}+cq_{n+2})+pu_{n+2}$

pv must be 1' or 0 or 1.

$pt_i \backslash pu_i$	0	1
1'	1'	0
0	0	1

pu_i =0/1, if c_j is non-positive /otherwise.

(e)

p_{n+1}, p_n

pv/ $p'_{n+1}p'_n$	1'-	01	00	01	1-
1'	x	x	X	1',1'	1',p'$_n$
0	p'_{n+1}, p'_n	p'_{n+1}, p'_n	p'_{n+1}, p'_n	p'_{n+1}, p'_n	p'_{n+1}, p'_n
1	1,p'$_n$	1,1	X	X	X

X never occurs

$p_i = p_i'$ $(0 \le i \le n-1)$.

(f)

Fig 4.26. Tables (a)-(c) used in computational rules for first addition in stages (1)-(2)) and (d) -(f) in second addition stages (1)-(3) (adapted from [Taka92b] 1992 ©IEEE)

Takagi [Taka92b] has described an alternative approach to the radix-4 implementation of modulo multiplier which is briefly outlined next. In their method, the P_j reduction mod Q is preformed by looking at the MSB bits to determine c_j where $c_j = \{ -2, -1, 0, 1, 2\}$ such that
$P'_j = P_j - 4.c_j.Q$ can be evaluated to reduce P_j to a (n+1) bit number. Note that P_j is a (n+4) bit number. They suggest that the top 8 bits of R_j need to be considered with top five bits of 2Q. "Top 6Q" is the most significant 6 digits of Q. The logic used is as follows:

$c_j = -2$ if top $(R_j) <$ - top (6Q)
$c_j = -1$ if - top (6Q) $\leq$ - top $(R_j) \leq$ - top(2Q)
$c_j = 0$ if - top (2Q) $\leq$ top $(R_j) \leq$ top(2Q)
$c_j = 1$ if top (2Q) $\leq$ top $(R_j) \leq$ top(6Q)
$c_j = 2$ if top $(R_j) \geq$- top(2Q)

The reader is referred to [Taka92b] for more details. The addition for the MSBs is different from that for the other bits as shown in Fig 4.26. As explained before, Tables (a)-(c) show the intermediate result computation for all the bits and the addition to yield R_j first. Next, as shown in Tables (d)-(f) in Fig.4.26, the addition of $-4c_j.Q$ is performed. We need a complicated rule for the MSB positions. Note that y x_i is the i th digit of $y_j'.X$ whereas ru_{j+1}, rt_j are the intermediate sum and carry digits at the i th position. Note also that cq_i is the ith digit of $-4c_j.Q$ whereas pu_{j+1} and pt_j are the intermediate carry and sum digits at the i th position. The example due to Takagi is illustrated next in Fig 4.27 to demonstrate the modulo multiplication. The reader is referred to [Taka92c] for modulo multiplication of hybrid numbers i.e. one in normal binary and another in redundant representation.

4.12. CONCLUSION

In this Chapter, several approaches to the design of multipliers using ROMs as well as not needing ROMs have been discussed. In the next Chapter, we consider the important problems of scaling and base extension in detail.

Q= 100101011 (299)
X= 101'1'00101(165)
Y= 1'01100001 (-159=140 mod 299)=Y'=1'22'01

$4P_5$	0 0 0 0 0 0 0 0 0 0 0
$y_4' = 1'$	+ 0 1'0 1 1 0 0 1' 0 1'
R_4	0 0 0 0 1' 1 0 1' 0 0 1' 0 1'
$(c_4=0)$	+ 0 0 0 0 0 0 0 0 0 0 0 0 0
P_4	0 0 0 1' 0 1' 0 0 1' 0 1'
$4P_4$	0 0 0 1' 0 1' 0 0 1' 0 1'
$(y_3'=2)$ +	1 0 1' 1' 0 0 1 0 1 0
R_3	0 0 0 0 1'0 1' 0 1'1 0 1'0
$(c_3=0)$	+0 0 0 0 0 0 0 0 0 0 0 0 0
P_3	0 0 1' 0 1' 0 0 1' 0 1' 0
$4P_3$	0 0 1' 0 1' 0 0 1' 0 1' 0
$(y_2'=-2)$	1'0 1 1 0 0 1 0 1' 0
R_2	0 1' 0 1 0 1' 1 1' 1' 0 0 1' 0
$(c_2=-1)$	+ 0 0 1 0 0 1 0 1 0 1 1 0 0
P_2	0 1' 0 1 1' 0 0 0 1' 1' 0
$4P_2$	0 1' 0 1 1' 0 0 0 1'1'0
$(y_1'=0)$	+ 0 0 0 0 0 0 0 0 0 0 0 0 0
R_1	0 1' 0 1 1' 0 0 1' 1 1 0 0 0
$(c_1=-2)$	+ 0 1 0 0 1 0 1 0 1 1 0 0 0
P_1	1 1' 0 1 1' 0 0 0 0 0 0
$4P_1$	1 1' 0 1 1'0 0 1'1 1'0 0 0
$(y_0'=1)$ +	0 1 0 1' 1'0 0 1 0 1
R_0	1 1' 0 1 0 1'1 1'0 1 1 1 1
$(c_0=2)$ +	1'0 1 1 0 1 0 1 0 0 1 1 1
P_0	0 0 0 1 1' 0 1 1' 1 0 1
$4P_0$	0 0 0 1 1'0 1 1' 1 0 1
$(y_{-1}'=0)$ +	0 0 0 0 0 0 0 0 0 0 0 0 0
R_{-1}	0 0 0 1 1' 0 1 0 1'1 1' 0 0
$(c_{-1}=0)$	0 0 0 0 0 0 0 0 0 0 0 0 0
P_{-1}	1 1' 1' 1 1'0 0 1'1' 0 0

P= 11'1'11'0 0 1'1' (77)

Fig 4.27. An example of modular multiplication (adapted from [Taka92b] 1992 ©IEEE)

5

BASE EXTENSION, SCALING AND DIVISION TECHNIQUES

5.1. INTRODUCTION

Division is an important operation in DSP applications such as filtering. Division by a constant is denoted as scaling. Scaling can usually be by one of the moduli or product of few moduli in a chosen RNS. Once scaling is done in the reduced moduli set, it is necessary to find the residues of the scaled value with respect to the scaling moduli. This operation is known as base extension. Thus, scaling and base extension need to be done together. Several techniques have been suggested in literature for this purpose which will be considered in detail in this Chapter. We first deal with base extension and then discuss scaling in the following sections.

5.2. BASE EXTENSION AND SCALING TECHNIQUES

5.2.1. Szabo and Tanaka Technique

In Residue Number systems, it is often required to find residues corresponding to a modulus or moduli other than those in the RNS. Szabo and Tanaka [Sza67] suggested a technique based on MRC which is considered next. This technique is basically sequential in nature and consumes (n+r) cycles for a n moduli set with r extended moduli to produce the residue in the extended r moduli set. The Szabo and Tanaka method is basically the MRC technique yielding the Mixed Radix Digits a_0, a_1, a_2, ... a_{n-1}, a_n, ... a_{n+r-1}. Note that for the additional moduli, we start with zero residues and process them till the last MRC digit is obtained. Considering the new extended moduli system, the MRC digits corresponding to new moduli a_n, ...a_{n+r-1} shall be zero, since the given number is less than the original dynamic range.

An example is presented to illustrate the technique for a three moduli RNS {2, 3, 5} with base extension needed for modulus 7. The residues considered are {1, 1, 3}. We consider finding the residue of a given number 13 corresponding to base 7. The procedure is sketched below.

Moduli	2	3	5	7
Residues	1	1	3	0
Subtract $a_0 = 1$	1	1	1	1
	0	0	2	6
Multiply by (1/2) mod m_i		2	3	4
		0	1	3
subtract $a_1 = 0$		0	0	0
		0	1	3
Multiply by (1/3) mod m_i			2	5
			2	1
subtract $a_2 = 2$			2	2
			0	6
Multiply by (1/5) mod m_i				3
				4

The value of a_3 can be found as satisfying

$$(x.(1/2.1/3.1/5) \bmod 7 + 4) \bmod 7 = 0 \tag{5.1}$$

which yields x =6. Note that 4 in (5.1) is the last MRC digit which would have been zero, if we had started with the correct residue. Szabo and Tanaka observe that a simplification is possible by terminating one step above i.e. the division by 5 is not needed. This can be seen by rewriting 4 in (5.1) as 6.(1/5) mod 7. Thus, we need to solve (x.(1/2.1/3.1/5) mod 7 +6.(1/5)) mod 7 =0. Multiplying by 5 both sides of this equation, we have (x.(1/2.1/3) + 6) mod 7 = 0 or x = 6.

5.2.2. Shenoy and Kumaresan Base Extension and Scaling Technique based on CRT

Shenoy and Kumaresan [Shen89a] suggested the use of CRT for the above base extension problem so as to reduce the number of cycles needed for base extension. They suggest the use of a redundant modulus for the purpose of evaluating the number of multiples of M in the CRT expansion. Their procedure is briefly described next.

It is known that the application of CRT involves mod M reduction:

$$X= \Sigma_{i=1}^{n} [(r_i /M_i) \bmod m_i].M_i - r_x.M \quad (5.2)$$

where r_x is an unknown integer and $M_i = M/m_i$ and $(1/M_i)$ mod m_i is the multiplicative inverse of M_i with respect to modulus m_i. For base extension to modulus m_{n+1}, it is evident that X mod m_{n+1} is needed for which r_x shall be known. Evidently, $r_x \le (n-1)$. Now, consider that a redundant modulus m_r is also used, whose residue is made available along with X mod m_1, X mod m_2 etc. From (5.2), taking residue of both sides with respect to the redundant modulus m_r, we have

$$X \bmod m_r = \{ \Sigma_{i=1}^{n} M_i.[(r_i /M_i) \bmod m_i] - r_x. M\} \bmod m_r \quad (5.3)$$

From (5.3), we have

$$r_x=\{[(1/M) \bmod m_r].(\Sigma_{i=1}^{n} M_i.[(r_i/M_i) \bmod m_i]-X \bmod m_r)\} \bmod m_r \quad (5.4)$$

Now, r_x can be calculated since it depends on the known residues and X mod m_r. The value of r_x computed can be used in (5.2) to yield X mod m_{n+1} :

$$X \bmod m_{n+1} = \{ \Sigma_{i=1}^{n} M_i.(r_i /M_i) \bmod m_i - r_x.M\} \bmod m_{n+1} \quad (5.5)$$

An example will be illustrated next.

Consider the {3, 5, 7} moduli set with redundant modulus 8 and base extension needed for modulus 11. The use of CRT for residues {1, 2, 3} yields

$X = 35.1.2 + 21.2.1 + 15.3.1 = 157.$

X mod 8 is assumed to be available as 4. Hence, we have

$$(157 - r_x.105) \bmod 8 = 4$$

or

$r_x = |1/105|_8 = 1.$

Thus, X mod 11 = (157 – 1.105) mod 11 = 8, which can be verified to be true since {1, 2, 3} corresponds to 52. Note, however, that the method outlined is feasible, if the residue corresponding to redundant modulus is available.

Shenoy and Kumaresan suggest a $\log_2 n$ level implementation for obtaining the summation in (5.4) as well as in the residue computation of the summation in (5.5) with respect to the redundant modulus. Their architecture is sketched in Fig 5.1, in which there are two $\log_2 n$ level trees. All the $(M_i.(r_i /M_i)) \bmod m_i. [(1/M) \bmod m_i]$ values for i = r as well as n+1 are stored in PROMs. The value $(X.(1/M)) \bmod m_r$ also is stored in PROM.

Shenoy and Kumaresan [Shen89b] extended the above technique for RNS scaling which is considered next. This method is based on the base extension technique to some extent. A knowledge of r_x is first essential. Next, the scaling is always done by a product of s moduli $\pi_{i=1}^{s} m_i$. Denoting this scaling factor as Q, we intend to find the integer part of X/Q. For this purpose, the residues corresponding to the s moduli are used to calculate the number X' in that residue ring. Then, we have the scaled value exactly as

$Y = (X-X')/Q$ (5.6)

where $Q = m_1.m_2....m_s$. The value of X' given as X mod Q, evidently is obtained using CRT in the smaller scaling residue set $\{m_1, m_2, .., m_s\}$ as

$X' = \Sigma_{i=1}^{s} Q_i.[(r_i/Q_i) \bmod m_i] - r_x'.Q$ (5.7)

where r'_x is bounded by (s-1) (i.e. $r'_x \leq s-1$). Note that $Q_i = Q/m_i$. Substituting for x' from (5.7) in (5.6), and using (5.2), we obtain

$Y = r_x' - r_x.M' + \Sigma_{i=s+1}^{n}(M_i/Q).[(r_i/M_i) \bmod m_i] + \Sigma_{i=1}^{s} [(M_i/Q)(r_i/M_i) \bmod m_i - (Q_i/Q).(r_i/Q_i) \bmod m_i]$ (5.8)

where M'=M/Q. As a first approximation, Shenoy and Kumaresan suggest that r_x' be treated as zero with a correction factor for error equalization namely (s-1)/2 since $r_x' \leq (s-1)$ as mentioned above. This error may not be acceptable in some cases. Hence, they suggest another approach to have an error of at most unity.

We first consider the first term in (5.7) defining X', which can be expressed in terms of one modulus $m_j > s$ as an integer and fractional part with respect to $Q_j = Q/m_j$. Thus, we have

$X' = \{Q_j\Sigma_{i=1}^{s}a_j + \Sigma_{i=1}^{s} [Q_i.(r_i/Q_i) \bmod m_i] \bmod Q_j\} \bmod Q$ (5.9)

Denoting the expression inside the modulo Q operator in (5.9) as X'_e, where the suffix e stands for estimate, we have

$X' = X_e' - \varepsilon.Q$ (5.10)

where ε can be 0 or 1. This can be noted by observing that both the terms in (5.9) are less than Q. Next, we have to evaluate r'_x since so far we have not considered the second term of (5.7). For this purpose, we choose another modulus $m_t > s$ and follow the approach similar to that used in the estimation of r_x (see (5.4) above):

$r_x' = \{[(1/Q) \bmod m_t].[\Sigma_{i=1}^{s} Q_i.[(r_i/Q_i) \bmod m_i] - X' \bmod m_t]\} \bmod m_t$ (5.11)

Since X' mod m_t is not available, we can use from (5.10) in place of X' mod m_t, X'_e value from (5.10) which has an error such that $X'_e = X' + \varepsilon.Q$ with ε being 0 or 1(see (5.9)).

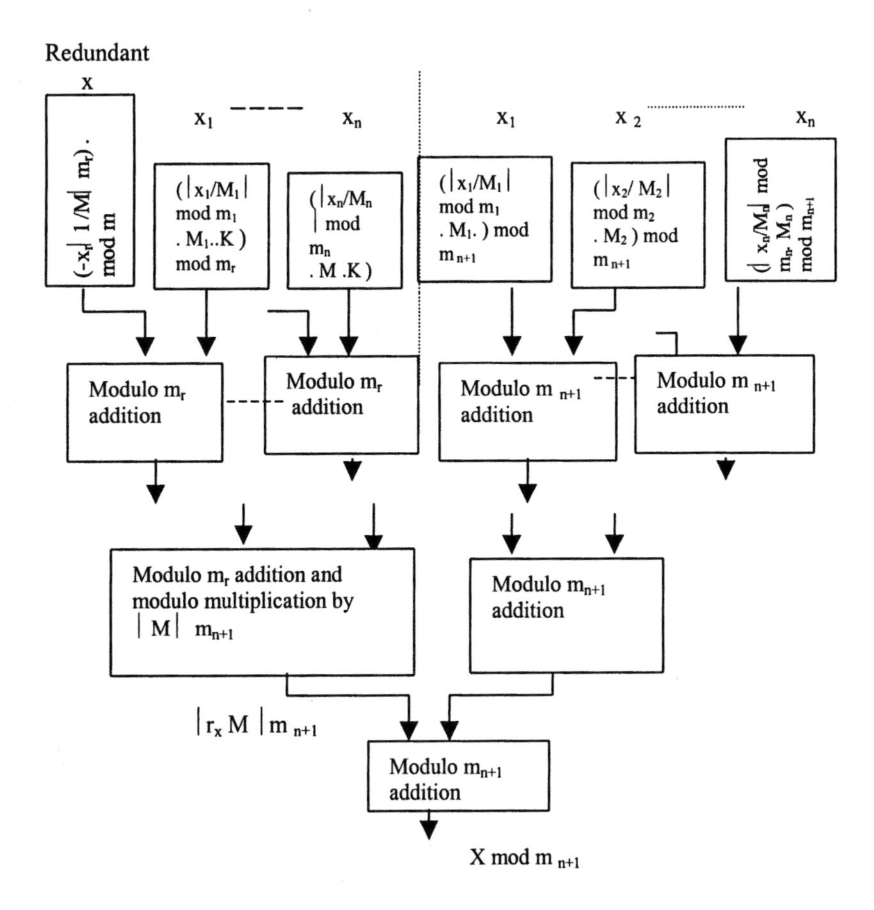

Fig 5.1. Shenoy and Kumaresan Architecture for Base Extension (adapted from [Shen89a] 1989 ©IEEE)

Thus r'_x can be estimated as

$$r_{xe}' = (r_x' - \varepsilon) \bmod m_t. \tag{5.12}$$

The reader is referred to [Shen89b] for a detailed description of this procedure sketched in Fig 5.2. for a nine moduli system.

5.2.3. Shenoy and Kumaresan modular base extension technique

Shenoy and Kumaresan [Shen88] have described an RNS to binary converter using modular look-up tables as mentioned in Section.2.2.7. Here, base extension to the modulus 16 is performed in successive steps so as to obtain

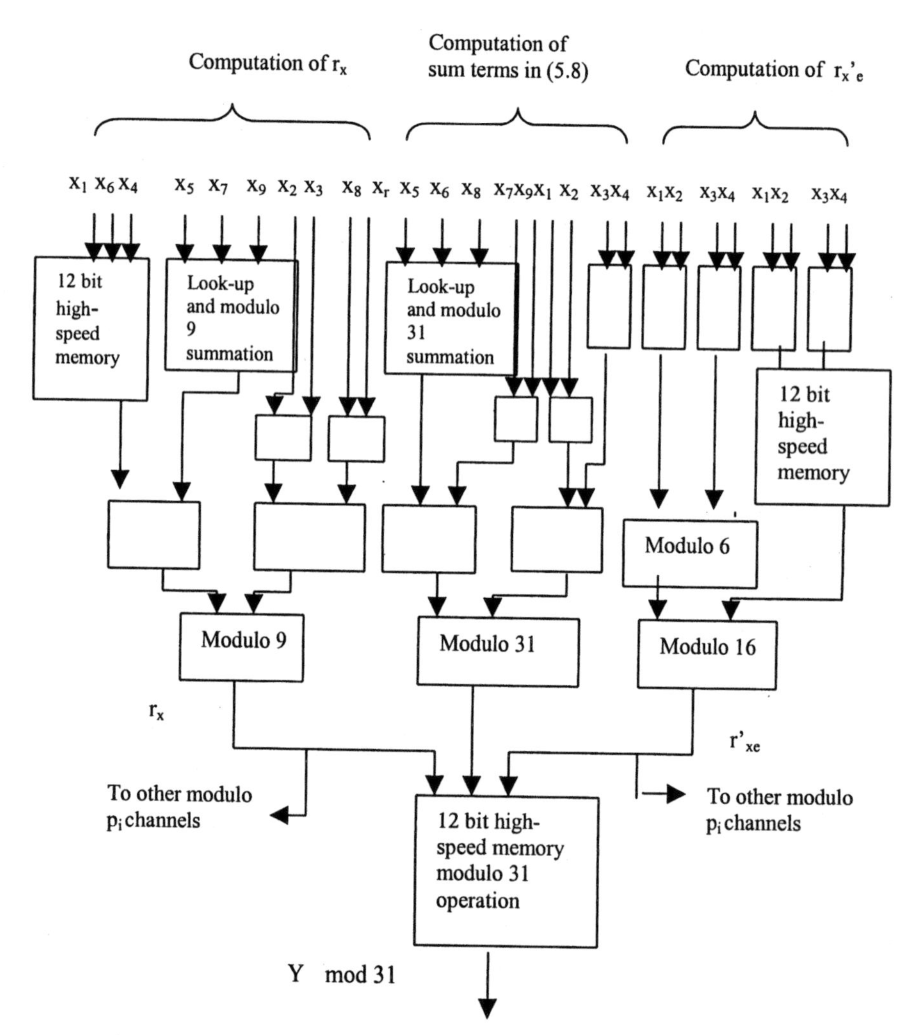

Fig 5.2. Shenoy and Kumaresan Technique for accurate scaling with base extension (adapted from [Shen89b] 1989 ©IEEE)

binary number. First residue mod 16 is obtained by base extension. Then this residue is subtracted and the remaining number is divided by 16 and again base-extended to include the modulus 16. This process is repeated.

An example is presented to illustrate the approach for moduli $m_3 = 19$, $m_2=11$ and $m_1 = 7$. The number 1461 ={17, 9, 5} is assumed to be given as residues. The various steps involved are as follows:

	$m_3 = 19$	$m_2=11$	$m_1 =7$
Residues of X	17	9	5
Step 1: Extend the base to include Modulus 16.			
Subtract r_3	-17	-17	-17
	0	3	2
Multiply by (1/19) mod m_i		*7	*3
		10	6
subtract 10		-10	-10
			3
Multiply by (1/11) mod m_i			*2
			6

Then, performing base extension as described in section 5.2.1, we have $(1/(19.11))\,(X) \bmod 16+11 \bmod 16 = 0 \Rightarrow X_{16}=5=B_1 = b_3b_2b_1b_0 =0101$

Residues of X	17	9	5
Step2: Subtract the LSB and divide by 16			
Subtract (X) mod 16	-5	-5	-5
	12	4	0
Multiply by (1/16) mod m_i	*6	*9	
Residues of Y	15	3	
Step 3a: Extend the base to include The modulus 16	15	3	
Subtract Y_3	-15	-15	
	0	10	
Multiply by (1/19) mod m_i		*7	
		4	

Then, (1/19) mod 16.(Y) mod 16 +7 mod16 = 0 (Y) mod 16=11....$b_7b_6b_5b_4$ = 1011 = B_2

Step 3b: Residues of Y	15	3
Subtract (Y) mod 16	-11	-11
	4	6
Multiply (1/16) mod m_i	*6	*9
	5	5
Residues of Z $\Rightarrow$	5	5

Since Z is less than or equal to (M-1)/256 which is 5, (Z) mod 19 = (Z) mod 16 =5 .....$b_{11}b_{10}b_9b_8$ =0101 =B_3. This completes the conversion process.

The method sketched in Fig 5.3 for three moduli example discussed above, is undoubtedly LUT intensive. Note that the number of base extension cycles and hence the look-up tables can be reduced by selecting a large $m_{n+1} =2^n$ (i.e. number of bits extracted at each step). In general, m_{n+1} can be slightly larger than m_{jmax} so that at each successive step, one modulus can be dropped. This can be understood by noting that at each stage, the number is reducing by 2^n i.e. m_{n+1}.

5.2.4. Jullien's Scaling Techniques

Jullien [Jull78] has suggested an alternative scaling technique which also uses MRC technique, however, with the base extension procedure being slightly different. The base extension is performed by several table look-ups.

The procedure is illustrated for a six moduli RNS {17, 13, 11, 7, 5, 3} next. These moduli are denoted for convenience as $\{m_6, m_5, m_4, m_3, m_2, m_1\}$. The scaling moduli are 3, 5 and 7. The number 44257 corresponding to residues

m_6	m_5	m_4	m_3	m_2	m_1
17	13	11	7	5	3
6	5	4	3	2	1
-1	-1	-1	-1	-1	
5	4	3	2	1	
x6	x9	x4	x5	x2	
13	10	1	3	2	
-2	-2	-2	-2	-2	
11	8	10	1		
x7	x8	x9	x3		
9	12	2	3		
-3	-3	-3			
6	9	10			
x5	x2	x8			
13	5	3	3	3	3
-3	-3				
10	2				
x4	x6				
4	12		6	2	0
-12					
9					
x4					
2			6	1	1
			1	1	1

{6, 5, 4, 3, 2, 1} is considered to be scaled by 105 (product of the three moduli 3, 5 and 7). The MRC yields the residues corresponding to the result as {13, 5, 3} with respect to the three moduli {17, 13, 11}. Next MRC is performed for this number 421 to yield the MRC digits as 2, 12 and 3 respectively. During these steps, the residues of 3 are first looked up in LUTs for finding residues mod m_3, m_2, m_1. Next, upon determining the next

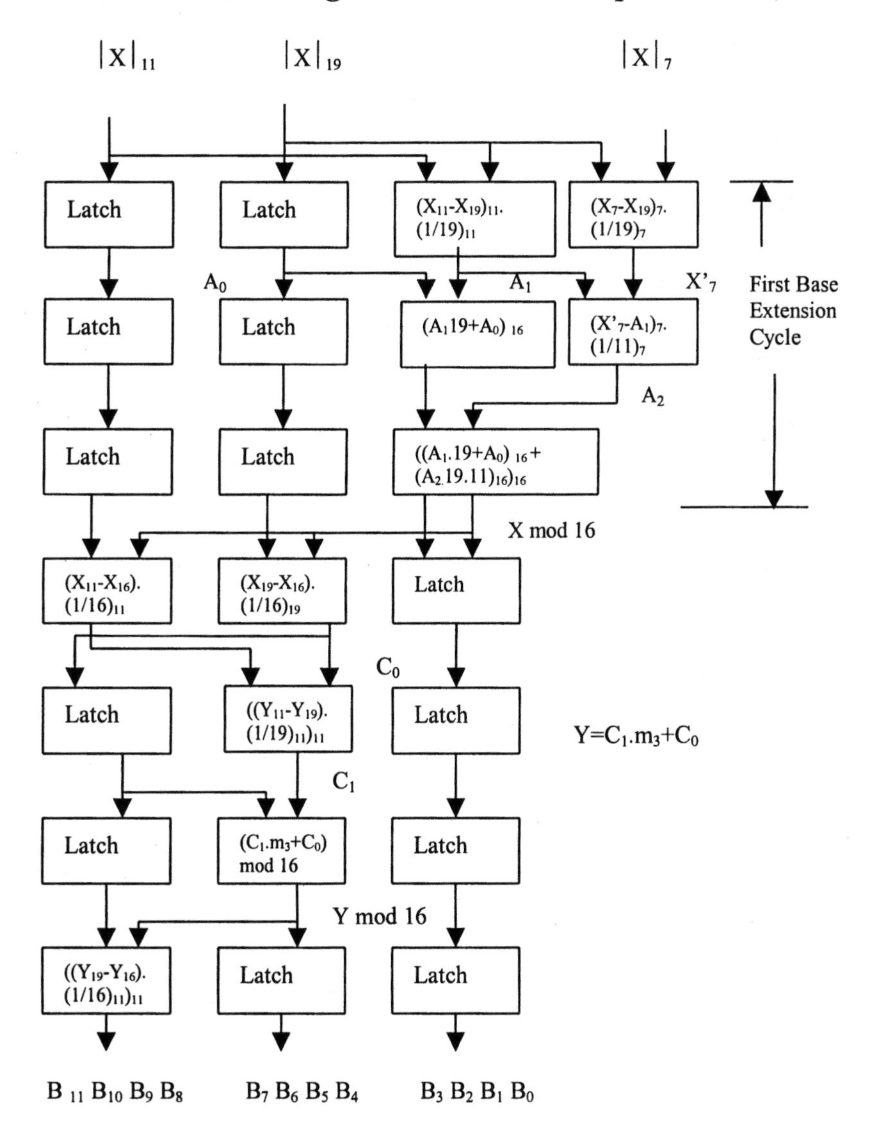

Fig 5.3. Shenoy and Kumaresan RNS to binary conversion using modular look-up tables. (adapted from [Shen88] 1988 ©IEEE)

mixed radix digit 12, again table look-up is performed to find $12.m_4 = 132$ mod m_3, 132 mod m_2 and 132 mod m_1 and added to the previous residues. The next step looks up the LUTs to obtain residues mod m_3, m_2, m_1

corresponding to 2.(143) and the results again accumulated mod m_3,m_2 and m_1 to yield the final result. It is evident that using ROMs, the technique can be implemented easily.

Jullien has suggested an alternative technique based on estimates which is considered next. This is based on CRT. It is known that the decoded binary number is given by

$$X=\Sigma_{i=1}^{n} M_i . [(r_i /M_i) \bmod m_i] \quad (5.13a)$$

where r_i =1, ..., n are the residues of the n moduli system and M_i = M/m_i. Scaling by a product of s moduli Q yields the result

$$Y_E=X/Q=\Sigma_{i=1}^{n} M_i.[(r_i/M_i) \bmod m_i]/Q \quad (5.13b)$$

It can be noted that some of the terms in (5.13b) are integers since M/m_i is exactly divisible by Q while some others are fractions. Jullien suggests individual rounding of all the terms by adding ½ to each term and taking the rounded value. The effect of this operation is that there can be an error in the estimate Y_E of s/2. Next, residues corresponding to Y_E can be estimated by base extension. An example will be in order to illustrate the procedure.

We consider again the same example as before. For this example, the various (M_i).(1/M_i)mod m_i and the quotients corresponding to division by $m_1.m_2.m_3$ are as follows:

(M_i).(1/M_i) mod m_i 195195 157080 46410 145860 51051 170170

(M_i).(1/M_i)mod m_i /105 1859 1496 442 145860/105 51051/105 170170/105

It may be noted that the first three are integers whereas the last three terms are fractions. Multiplying these by the residues and evaluating rounded values by adding ½, we have the corresponding six terms

11154 7480 1768 4167 972 1621.

Note that we have multiplied by the residues viz., {6, 5, 4, 3, 2, 1}. In practice, the complete operations described so far together with their residues mod m_i are obtained by look-up tables. It may be noted that the first three items have only one non zero residue with respect to m_6, m_5 and m_4 respectively whereas the last three terms each have three residues corresponding to m_4, m_5 and m_6. These are as follows:

11154 ⇒ {2, 0, 0}

{7480} ⇒ {0, 5, 0}

{1768} ⇒ {0, 0, 8}

$\{4167\} \Rightarrow \{2, 7, 9\}$
$\{972\} \Rightarrow \{3, 10, 4\}$
$\{1621\} \Rightarrow \{6, 9, 4\}$
Thus adding all these yields the residues of the scaled value as $\{13, 5, 3\}$ which corresponds to 421. Next, base extension of this number is performed as before. Note that the computation can be speeded up by performing several additions in parallel. The schematic of the algorithm as well as fast implementations are sketched in Fig 5.4.

Jullien [Jull78] provided an estimate of the number of cycles as well as Look up Tables needed for both the original and the estimate methods. We denote for this calculation, L_i as the number of look-up-tables and n_i as the number of ROM read cycles. We recall that in the original method three successive computation stages viz., MRC stage, base extension stage and final stage exist. In the estimate method, the first MRC stage is replaced by the estimate stage.
Estimate: $L_1=(N-S).S$ and $n_2 = S$
Original: $L_2=S.(N-((S+1)/2))$ and $n_3=N-S-1$
Mixed-Radix: $L_3=(N-S-1)(N-S)/2$
Final stage: $L_4=S.(N-S-1)$
Total look-up-Table requirement of each method are
Original $L_o= L_2+L_3+L_4$
Estimate: $L_1+L_3+L_4$.
The savings in estimate method is $L_o-L_e = S(S-1)/2$. The number of cycles needed in the estimate method are $n_e=N+n_1-S$ whereas in the original method is $n_o=N$.

5.2.5. Garcia and Lloris modifications of Jullien's technique

Garcia and Lloris [Garc99] suggested three techniques for reducing the size of the look-up tables which are considered next. In the first method shown in Fig 5.5 (a) which is applicable for only small moduli, only two look-up cycles are needed. The first look-up obtains the residues corresponding to moduli m_{S+1} to m_N, whereas the second look-up table uses the outputs of the first look-up table to obtain the residues corresponding to base extension of the scaling moduli. The first LUT is justified by observing that the residues r_j of the moduli m_{S+1} to m_N is dependent only on the residues $r_1, r_2, ..., r_S$

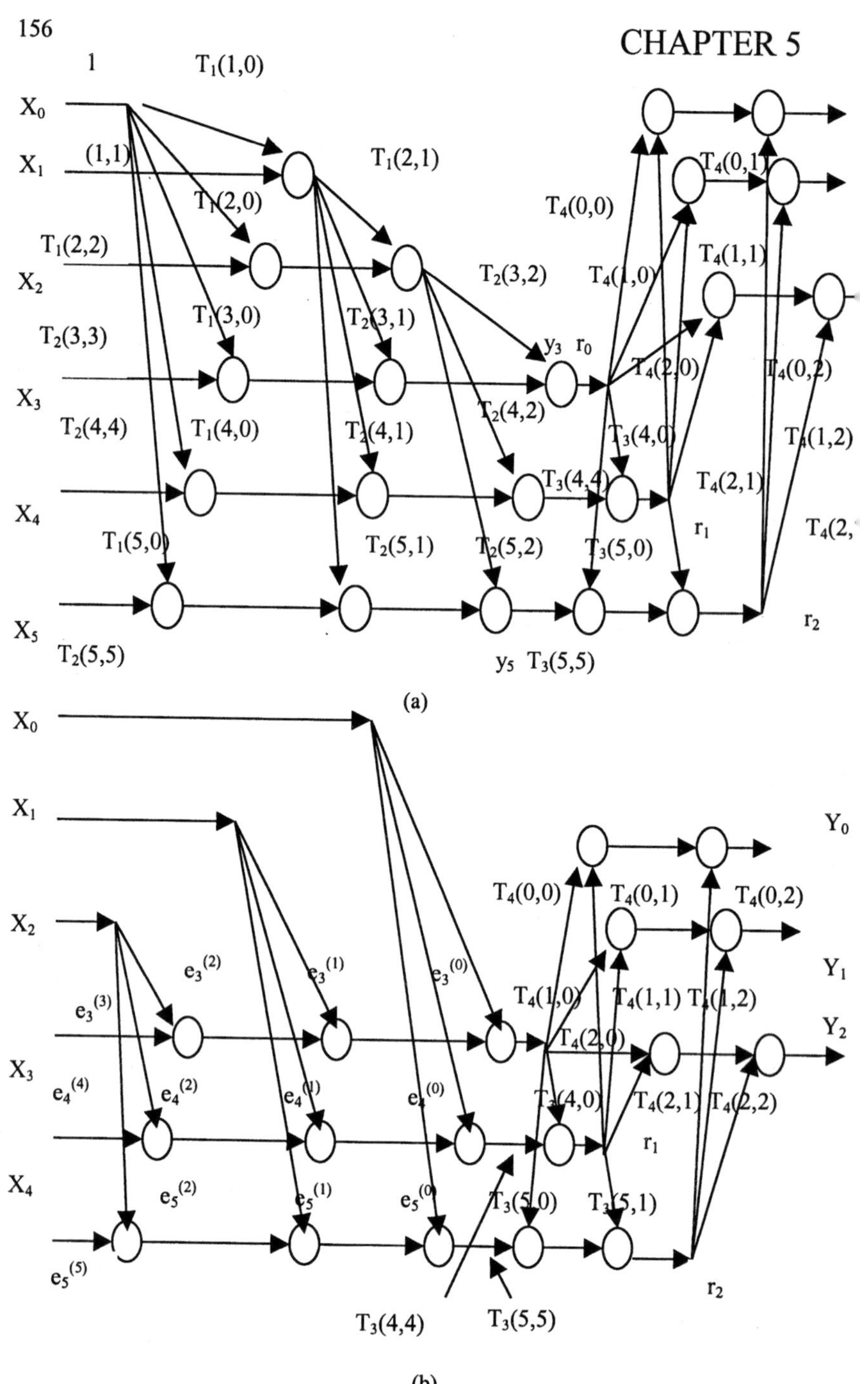
1
$T_1(1,0)$
X_0
X_1
(1,1)
$T_1(2,1)$
$T_1(2,0)$
$T_4(0,0)$
$T_4(0,1)$
$T_1(2,2)$
X_2
$T_2(3,2)$
$T_4(1,0)$
$T_4(1,1)$
$T_1(3,0)$
$T_2(3,1)$
$T_2(3,3)$
y_3 r_0
$T_4(2,0)$
$T_4(0,2)$
X_3
$T_2(4,2)$
$T_2(4,4)$
$T_1(4,0)$
$T_2(4,1)$
$T_3(4,0)$
$T_4(1,2)$
$T_3(4,4)$
$T_4(2,1)$
X_4
r_1
$T_1(5,0)$
$T_2(5,1)$
$T_2(5,2)$
$T_3(5,0)$
X_5
r_2
$T_2(5,5)$
y_5 $T_3(5,5)$
(a)
X_0
X_1
Y_0
$T_4(0,0)$
$T_4(0,1)$
$T_4(0,2)$
X_2
$e_3^{(2)}$
$e_3^{(1)}$
$e_3^{(0)}$
Y_1
$e_3^{(3)}$
$T_4(1,0)$
$T_4(1,1)$
$T_4(1,2)$
Y_2
$T_4(2,0)$
X_3
$e_4^{(4)}$
$e_4^{(2)}$
$e_4^{(1)}$
$e_4^{(0)}$
$T_3(4,0)$
$T_4(2,1)$
$T_4(2,2)$
r_1
X_4
$e_5^{(2)}$
$e_5^{(1)}$
$e_5^{(0)}$
$T_3(5,0)$
$T_3(5,1)$
$e_5^{(5)}$
r_2
$T_3(4,4)$
$T_3(5,5)$
(b)

Fig 5.4 (contd)

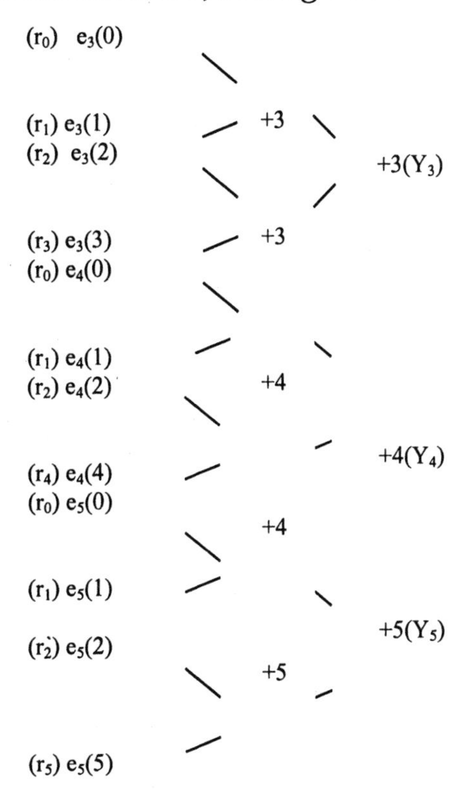

(c)

Fig 5.4 (contd)

and r_j. Note that the number of address inputs to the ROMs in the first and second levels are too large for large number of moduli and large moduli. However, this technique is attractive for small moduli sets e.g. consider {29, 31, 32} to be scaled by any one of the moduli. Evidently, the first level needs 2 ROMs with 10 bit address and 5 bit width. The second level needs another 10 bit address and 5 bit wide ROM. Thus, totally, 15K ROM is needed. This can be compared with that needed for Jullien's technique which needs 3 cycles and 20K ROM (since L_1=2,L_3=1 and L_4=1) for the technique based on estimates.

Garcia and Lloris describe two other techniques shown in Fig 5.5 (b) and (c) which are combinations of Jullien's technique and their first technique of Fig 5.5 (a). In Fig 5.5 (b), the first stage of computing scaled residues

$X_0\ T_1(1,0)$
+1
$X_1\ T_1(1,1)$
$T_1(2,1)$
$X_0\ T_1(2,0)$
$T_1(2,1)$
+2
$X_2\ T_1(2,2)$
$T_2(3,1)$
$T_2(3,2)$
$X_0\ T_2(3,0)$
+3
(y_3)
+3
+3
$X_3\ T_2(3,3)$
$T_2(4,2)$
$X_0\ T_2(4,0)$
$T_2(4,1)$
+4
(y_4)
+4
+4
$X_4\ T_2(4,4)$
$X_0\ T_2(5,0)$
$T_2(5,1)$
$T_2(5,2)$
+5
+5
+5
(y_5)
$X_5\ T_2(5,5)$
Final stage

(d)

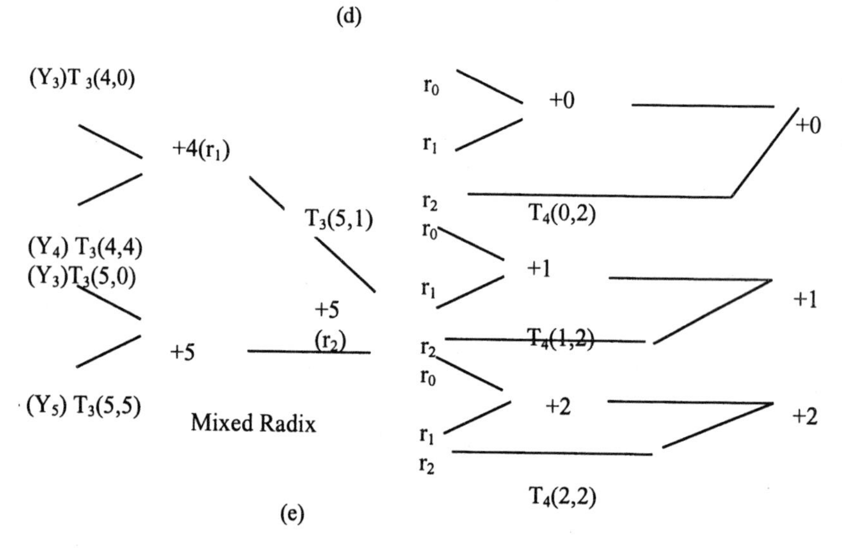

(e)

Fig 5.4. Jullien's Architectures for RNS scaling (a) using look-up tables, (b) using estimates, (c) and (d) High speed realization of the initial stage of estimate and original scaling algorithms, and (e) High speed realization of base extension algorithm (adapted from [Jull78] 1978 ©IEEE)

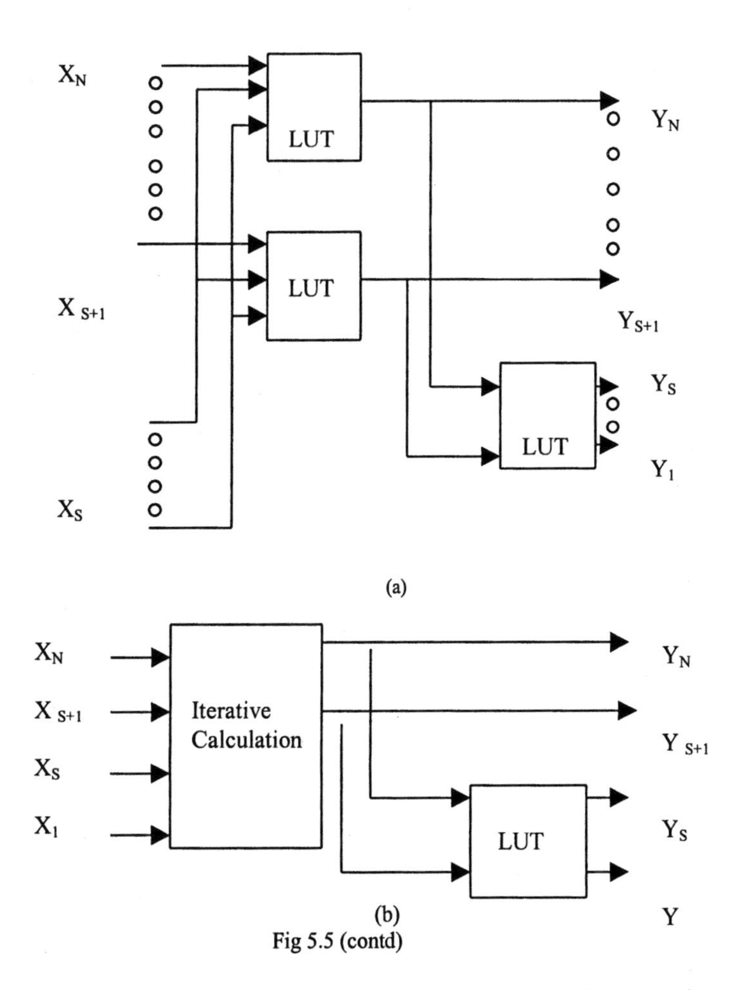

(b)
Fig 5.5 (contd)

corresponding to m_{S+1} to m_N is same as in Jullien's technique whereas the second stage uses LUTs. In Fig 5.5 (c), the first stage uses LUTs and the second stage uses iterative base extension technique as in Jullien's technique. The authors have shown that one of the three new techniques usually leads to smaller ROM requirement and smaller number of cycles. Table 5.1. illustrates for a variety of modulo sets, the comparative requirements of all the four techniques. Note that the technique of Fig 5.5 (b) is called look-up generation technique of y_1 ... y_S and the technique of Fig 5.5 (c) is called look-up calculation of y_{S+1} to y_N.

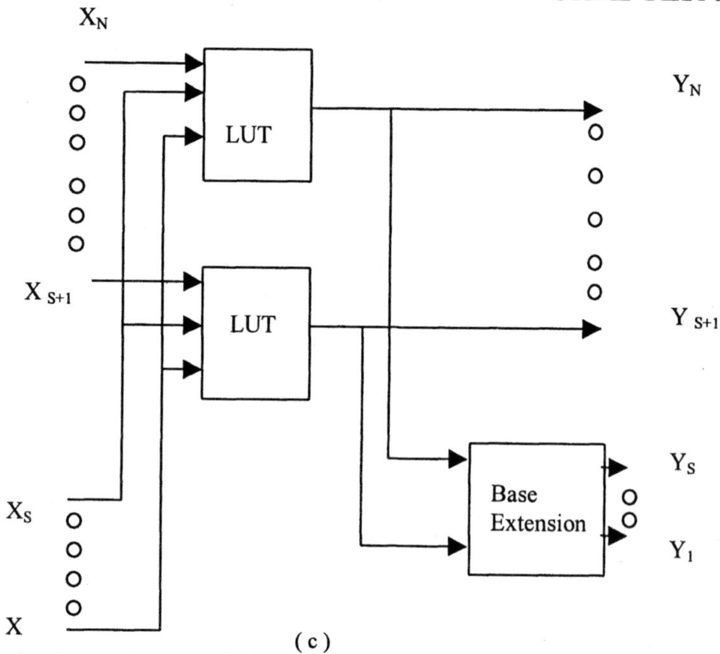

(c)

Fig 5.5. Garcia and Lloris modifications of Jullien's method based on estimates: (a) Two look- up cycle scaling using ROMs (b) Iterative calculation followed by look-up scheme and (c) Look-up followed by iterative calculation (adapted from [Garc99] 1999 ©IEEE)

$\{ m_i \}$	K	Jullien's algorithm n.L memory	Look-up Scheme L(n=2) memory	Look-up generation of $[y_1,y_2,..y_s]$ n,L memory	Look-up generation of $[y_{s+1}, \ldots y_N]$ memory
29,31,32	29	3,4 20K	3 15K	2,3 15K	3,4 20K
29,31,32	29 x32	3,3 15K	3 100K	3,4 10K	2,1 160K
27,29,30, 31	27	4,8 40K	4 175K	2,4 175K	4,8 40K
27,29,30, 31	27 x29	4,8 40K	4 330K	3,6 30K	3,5 335K
25,27,29, 31,32	29	5,13 65K	5 5140K	2,5 5140K	5,13 65K
25,27,29, 31,32	29 x31	5,14 70K	5 800K	3,8 350K	4,10 515K
25,27,29, 31,32	25x 29 x31	4,11 55K	5 10225K	3,9 45K	3,6 10260K

Table 5.1 Speed and Memory use for scaling (adapted from [Garc99] 1999 ©IEEE)

5.2.6. Miller and Polky Scaling method

Miller and Polky [Mill84] describe a scaling algorithm which performs scaling by an arbitrary fixed scale factor in k clock periods with no restrictions on the moduli set. This uses Szabo and Tanaka MRC basically with additional steps being parallelly implemented to evaluate the residues mod all m_i. As an illustration, the five moduli scaling by two residues m_1 and m_2 is shown in Fig 5.6. Note that as soon as a_3 is available, $a_3m_1m_2$ mod all m_j is evaluated. Moreover, instead of waiting for a_2, since a_3 is a function of a_3 and the output of f_{24} block is available, $a_2m_1m_2m_3$ mod all m_j can be evaluated. The last step of calculating a_1 can also be pipelined to similarly obtain $a_1m_1m_2m_3m_4$ mod all m_j. The next step is to add all these mod m_j as soon as two inputs are available. Thus, evidently n cycles are needed for a n moduli RNS since usual MRC needs (n-1) cycles and one last extra step is needed. The authors also observe that scaling of positive and negative values can be accomplished by adding an offset –[M/Q] to the functions g_{ij} since we need to estimate (X-M)/Q. The error introduced in the technique is s/2, since every partial term in the MRC expansion is used to evaluate mod m_j instead of the total summation mod m_j being evaluated. The mappings f_{ij} and g_{ij} are

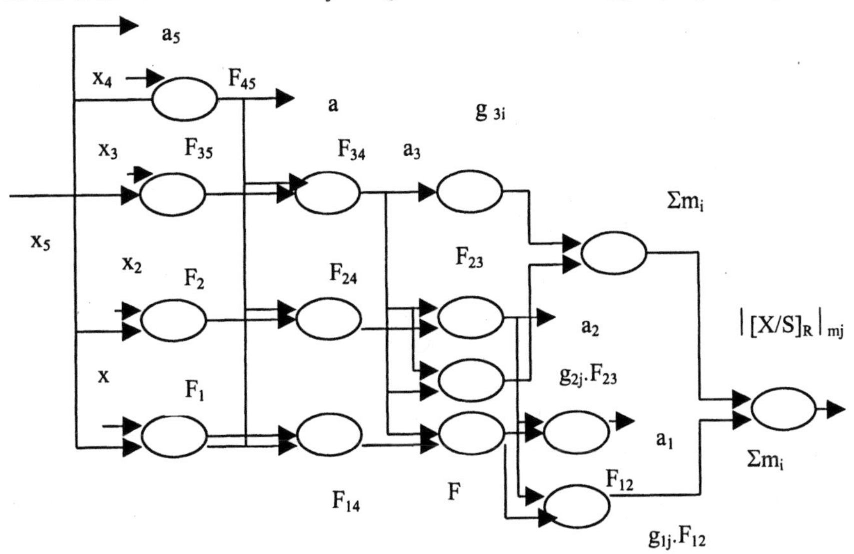

Fig 5.6. Miller and Polky Technique for scaling using modified MRC converter (adapted from [Mill84] 1984 ©IEEE)

as follows:
$f_{ij}:(x,y)=((x-y) \bmod m_i / m_j) \bmod m_i$, $j \neq i$, $g_{ij}:a_i = [(a_i.(m_{i+1}\ldots m_k)/S)_R] \bmod m_j$
where R stands for rounding.

5.3. DIVISION IN RESIDUE NUMBER SYSTEMS

5.3.1. Szabo and Tanaka division Algorithms

Division is one desirable operation and is difficult even in conventional computing. Division of a RNS number by another RNS number has attracted lot of attention. Szabo and Tanaka [Sza67] described two techniques which, however, are difficult to implement as are other methods to be discussed.

In the first method due to Szabo and Tanaka, the nearest powers of two and smaller than the dividend X and divisor Y are first determined. The ratio of these is the partial quotient to be determined. The residues corresponding to all powers of two are stored in LUTs so that Y can be multiplied by the partial quotient and the result subtracted from X. This process is repeated to get the final result. An example will be presented to illustrate the technique.

Consider the RNS {7, 8, 9} and the residues {3, 6, 6} corresponding to X=150. We wish to divide X by Y =11={4, 3, 2}. We first note that $2^7 = 128 < X$ and $2^3=8 < 11$. Hence, the partial quotient Q_0 is $2^7/2^3 = 2^4 = 16$. Therefore, we have in the first step $X_1 = (X - Q_0.Y) = (150- 16.11)$ negative showing that 16 is not the correct choice and hence a power of two less than this shall be chosen. This yields 8 as the Q_0 giving $X_1 = 150-8.11 = 62$. The next steps are as under:
$2^5<62$, $2^3<11$, hence $Q_1 = 4$.
$X_2 = 62 - 4.11 = 18$
$2^4 <18, 2^3<11, Q_2=2$
$X_3 = 18-2.11 = -4$ negative.
Try therefore $Q_2=1$
$X_3 = 18-1.11 = 7$
The result is 8+4+1 = 13.
It may be noted that the multiplication of Q and Y can be in RNS and then RNS to binary conversions can be performed. In the worst case, it may be seen that $\lceil \log_2 (X/Y) \rceil$ steps may be needed to get the result, each comprising

of determination of powers of two which needs several comparisons in the MRC form or in binary form.

We next consider the second technique due to Szabo and Tanaka. This is however approximate. This is based on MRC and is useful only when the divisor Y is such that $Y<Y'<2Y$ where Y' shall be a product of some moduli. Consider $m_1=9$, $m_2=11$, and $Y=4$. Then between 4 and 8 no modulus is lying and hence the method is not applicable. We illustrate the method with an example in the same RNS {7, 8, 9}. Consider division of 150={3, 6, 6} by 61. We can choose Y' such that $61<Y'<122$ and Y' is product of two moduli. Y' can be chosen as 63 or 72. We will consider 72 for illustration. The MRC expansion of 150 yields 150=2.72+0.9+6. We can thus determine Z_1 = integ {150/72}= 2 where { } stands for integer part. Then $X_1 = X_0\text{-}Y.Z_1$ =150-2.61 = 28. We next see that the MRC expansion of 28 yields 28 = 0.72+ 3.9+1. Hence $Z_2 = 0$ terminating the process. The quotient is thus (2+0) =2.

5.3.2. Lu and Chiang Division Technique.

Lu and Chiang [Lu92] have described a division algorithm for RNS. In this technique first signs of the two numbers X and Y are determined which information is saved, to be employed in the end. Next, we find 2^k such that $Y.2^k \leq X \leq Y.2^{k+1}$. Evidently, the quotient lies between 2^k and 2^{k+1}. This value is determined by successive comparisons of the bit positions of the quotient between 2^k and 2^{k+1}. Note that X and Y shall be available as normal binary numbers to carry out this step. An example will be illustrated next using their procedure.

Consider X = 190 and Y =11. It can be seen that $2^k = 16$, since 11.16< 190< 11.32. Thus at most four steps are needed to determine the result. Next, 11.24 can be tested to know whether 11.24 > 190. In case, 11.24 is greater than 190, 11.28 is tried or else 11.20 is tried. In other words, based on the comparison, the successive bit positions in the result are set to 1 or 0. Lu and Chiang however follow slightly different approach. An example will be illustrated to detail their procedure. Denoting the quotient to be obtained as $b_4b_3b_2b_1b_0$, the procedure is as follows:

Set b_4 to 1.

Set b_3 to 1.

11.24 > 190. Reset b_3 to zero.

Set b_2 to 1.
11.20 >190. Reset b_2 to zero.
Set b_1 to 1.
11.18> 190 Reset b_1 to zero.
Set b_0 to 1.
11.17< 190 Procedure is complete.

Note that in each step, the average of the previous attempt and the first decision is tested. As an illustration, 16 is decided as the minimum expected quotient which is fixed and average of 16 and 32, or 16 and 24, or 16 and 20, or 16 and 18 is tried successively.

The crucial problem is the testing needed for comparison of two RNS numbers, sign detçction, as well as detection of overflow in the addition operations above. Lu and Chiang have suggested the use of concept of parity. They suggest the use of one redundant modulus 2 to know whether a number is even or odd. The following rules apply in the case of addition. Overflow occurs when (a) if the sum of two numbers is odd and if both A and B have same parity or (b) when the sum is even and A and B have different parities. If X and Y have same parity, defining $Z=(X-Y)$, $X\geq Y$ if and only if Z is a even number. $X<Y$, if and only if Z is an odd number. Further, if X and Y have different parities, $X\geq Y$, if and only if Z iş an odd number. $X<Y$, if and only if Z s an even number.

Lu and Chiang have suggested the use of CRT technique due to Vu discussed in Chapter II (see Section.2.3.3) to estimate parity of a given RNS number. They first stipulate that all the moduli be odd. Thus, using CRT expansion for residues $\{r_1, r_2, \ldots r_n\}$, we have

$$X = \Sigma_{i=1}^{n} M_i.(r_i/M_i) \bmod m_i - r_x.M \quad (5.14)$$

Evidently, all M_i are odd numbers. Thus, the parity of X is decided by the parities of all terms $(r_i/M_i) \bmod m_i$ and r_x. These in turn are the LSB s of all of these terms. In other words we have the parity P of X given as

$$P = LSB[(r_1/M_1)\bmod m_1] \oplus LSB[(r_2/M_2)\bmod m_2] \oplus \ldots\ldots \oplus LSB[(r_n/M_n)\bmod m_n] \oplus LSB(r_x). \quad (5.15)$$

where $\oplus$ stands for exclusive-OR operation. Note that r_x is still not known in (5.15). The evaluation of r_x needs application of Vu's method to (5.14) after dividing both sides by M:

$$r_x = S_1/M_1 + S_2/M_2 + \ldots. + S_n/M_n - X/M \quad (5.16)$$

where $S_i = (r_i/M_i) \bmod m_i$. Following Vu's method, we observe that the quantities S_i/M_i need few bit precision with integer bits as well. However, since we are interested in the LSB of r_x only, one integer bit is sufficient.

The complexity of Lu and Chiang's method shall have been obvious. The reader is referred to their work for more details.

5.3.3. Gamberger's RNS Division Algorithms

Gamberger [Gamb91] has described an approach to integer division in Residue Number Systems which is next considered. In this technique, basically we wish to find X'/Y' such that it is close to the desired X/Y and also satisfying X'<X and Y'<Y. Repeating the method with the new X'/Y' again, we can obtain the solution in few iterations. The method is illustrated with an example next. We first define D such that (Y.D) = 1 mod M so that

$$X' = \lfloor X.D/M \rfloor \qquad (5.17a)$$

and

$$Y' = \lfloor Y.D/M \rfloor = (Y.D-1)/M \qquad (5.17b)$$

Consider division of 502 by 15 in a system with M=1024.
Iteration 1: X=502, Y=15, D=751 since 751.15 = 11.1024+1. Therefore, X' = $\lfloor(502.751)/1024\rfloor$ = 368 and Y' = (15.751 – 1)/1024 = 11. Note that X' and Y' have decreased.
Iteration 2: X=368, Y=11, D=931 since 931.11 = 10.1024+1. Therefore, X' = $\lfloor(368.931)/1024\rfloor$ = 334 and Y' = (11.931 – 1)/1024 = 10.
Iteration 3: X=334, Y=10, D=751. Since gcd (10,1024) = 2, we have X' = X/2 = 167 and Y' = Y/2=5.
Iteration 4: X=167, Y=5, D=205 since 205.5 = 1.1024+1. Therefore, X' = $\lfloor(167.205)/1024\rfloor$ = 33 and Y' = (5.205 – 1)/1024 = 1.
Iteration 5; Result X=33, Y=1 and hence the quotient is 33.

It is important to note that X.D and Y.D can exceed the dynamic range. Hence, an extended dynamic RNS needs to be employed. Gamberger suggests that additional moduli with product P can be chosen such that P is at least (M-1). The equations (5.17) then get modified as

$$X' = (X.D - (X.D) \bmod P) / P \qquad (5.18a)$$

$$Y' = (Y.D-1)/P \qquad (5.18b)$$

This new moduli set is termed as auxiliary RNS. The additional term -(X.D) mod P = -Z' ensures exact division in (5.18a) by P. This division in RNS can be done in just one step. An example is presented next to illustrate the technique whose flow chart is shown in Fig 5.7.

In this method, first X and Y are converted to the auxiliary RNS in blocks a and b using distinct hardware which needs N steps. There after, the multiplicative inverse D is computed in this RNS and transformed back to the main RNS in R modular steps in the block E. Multiplication of Y by D in the main RNS is executed by block h. In the same block, subtraction of 1 and division by P are also performed. In parallel to the transformation of D to the main RNS, X is directly divided by Y in the auxiliary RNS and the result Z' is transformed to the main RNS blocks d and f respectively. The block g waits for availability of D and calculates X.D. The final value of X' is computed by the block i. Gamberger suggests the use of look-up tables for all the blocks (a)-(i).

Let X=502,Y=16 in moduli {7, 11, 13} i.e. M = 1001 with P=1615 and additional moduli {5, 17, 19}.

m_1=7,m_2=11,m_3=13 p_1=5,p_2=17,p_3=19

X=502 = {5, 7, 8}

Y=16={2, 5, 3}

a X = {2, 9, 8}=502

b Y={1, 16, 16}=16

c Z'={2, 8, 10}=637

d D=101

e D={3, 2, 10}=101

f Z'={0, 10, 0} =637

g X.D={1, 3, 2} =50702

h Y'=[{2, 5, 3}.{3, 2, 10}-{1, 1, 1}]/{5, 9, 3}={1, 1, 1}=1.

i X'= [{1, 3, 2} –{0, 10, 0}]/{5, 9, 3} = {3, 9, 5}=31.

The above technique will not be straight away applicable for the case when gcd (Y,P) =G>1. Normally in this case, we can a straightaway obtain X'/Y' as (X/G) / (Y/G). A variable F is used to indicate whether Y is prime to P. In this case, D is P/P_i where P_i is the modulus having common factor with Y. Then, Z' is defined as 2D. Note that in the method previously described, there are N+R+2 modular steps are needed where R is the number of moduli in the auxiliary RNS. These are indicated in brackets by the side of each block. The reader is referred to [Gamb91] for more details on the number of LUTs needed etc.

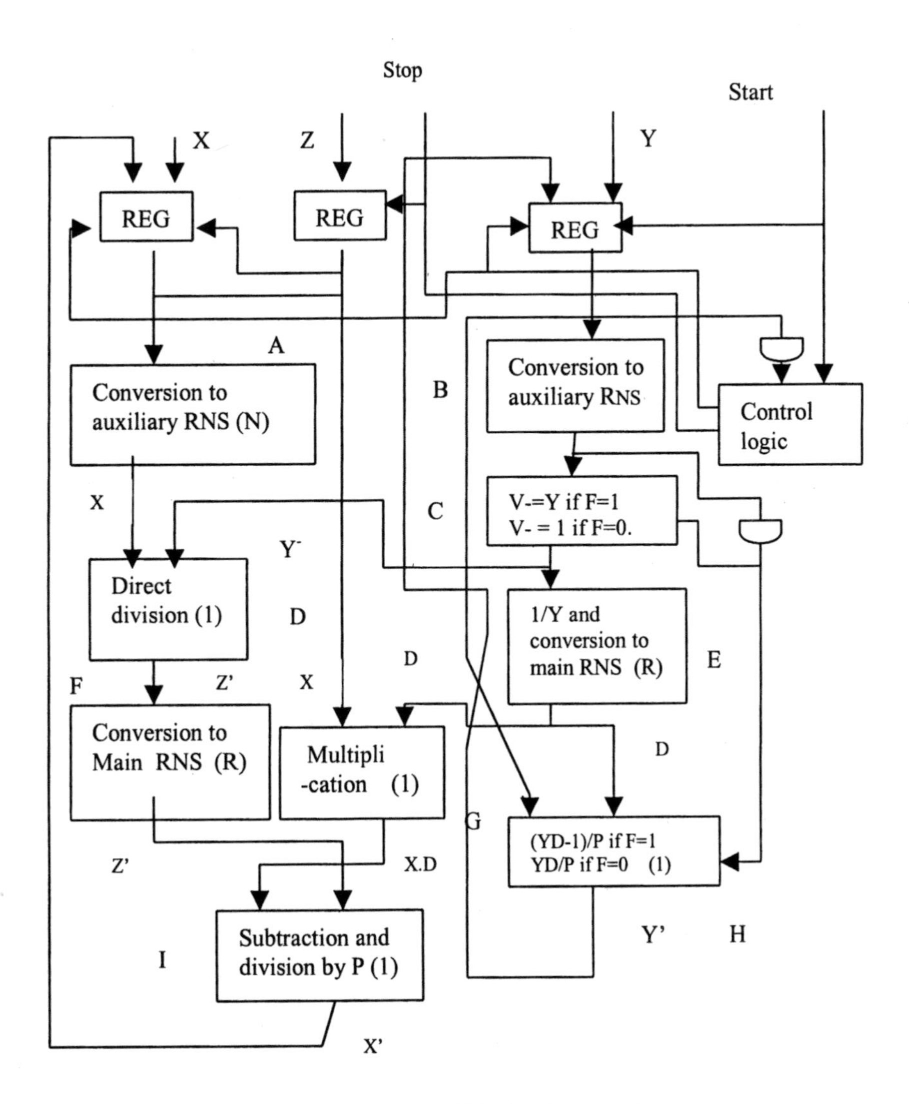

Fig 5.7. Flow chart of Gamberger's RNS division procedure.
(adapted from [Gamb91] 1991 ©IEEE)

5.3.4. Hitz and Kaltofen RNS Division Algorithms

Hitz and Kaltofen [Hitz95] suggest evaluation of (X/Y) mod M as $((X/M).\lfloor(M/Y)\rfloor)$ mod M. They first suggest computation of (M/Y) followed by multiplication by X and then division by M. The operation (M/Y) is denoted as RECIP which is realized using Newton Iteration scheme. The step in Newton iteration scheme is

$$Z_{i+1} = Z_i - (f(Z_i)/f'(Z_i))) \qquad (5.19a)$$

We define the quotient M/Y as Z_1 and $f(Z_1)$ as $((M/Z_1)-Y)$. Hence, it can be easily seen by differentiation of $f(Z_1)$ and from (5.19a) that

$$Z_{i+1}= 2.Z_i - \lceil (Y.Z_i^2/M) \rceil = Z_i.(2M-Y.Z_i)/M \qquad (5.19b)$$

The algorithm starts with Z_i=2. If $(M-Y.Z_2)<Y$, the result is Z_2. Else, the result is (Z_2+1).

An example will be illustrative. Consider M=105,Y=13.
The various steps are as follows:
Step 1 Z_1=0 and Z_2=2
Step2 Z_1=2 and Z_2=2.(2.105 -13.2)/105 =3
Step 3 Z_1=3 Z_2=4
Step 4 Z_1=4 and Z_2=6
Step 5 Z_1=6 and Z_2=7
Step6 Z_1=7 and Z_2=7
Stop.
Since (105-7.13) >13, add one to Z_2 to get the result 8.

The reader is urged to refer to [10] for an estimate on the number of steps needed for estimation of RECIP. Next, the authors suggest multiplication of the result by X to obtain a 2n bit word. This is divided by M to get the final result. Evidently, a comparison is needed in RECIP which can be carried out in RNS by comparing the residue digits for equality. Scaling and base extension are also needed. Equality can be established by bit by bit comparison of the residues and then ANDing the decisions.

A comparison also is needed in the end. This can be achieved by subtraction in the extended residue number system. If X<Y, the result is greater than M. Hence by comparing the residues of Z (=Y-X) with the base extended residues for equality, we can decide whether X>Y or otherwise.

5.4. SCALING IN THE MODULI SET $\{2^n-1, 2^n, 2^n+1\}$

Scaling by 2^n is simpler than scaling by (2^n-1) or (2^n+1). The case of scaling by 2^n is first considered. The RNS to binary conversion procedure described in Chapter III can first be used to obtain the 2n bit MSB word. This corresponds to the quotient whose residues mod (2^n-1) and (2^n+1) can be calculated by summing the n bit LSBs with n bit MSBs mod (2^n-1) and subtracting n bit MSBs from n bit LSBs mod (2^n+1) in the latter case. The residue mod 2^n is n bit LSB word of the quotient itself. As an illustration consider the residues (1, 7, 5} corresponding to moduli {7, 9, 8}. The Piestrak technique gives the MSBs of the converted word as 110 100= $52_{(10)}$. The residues of this word mod 2^n, (2^n-1) and (2^n+1) can be easily computed to be 4, 3 and 7 respectively.

The scaling by any other residue i.e. (2^n-1) or (2^n+1) or product of residues $(2^{2n}-1)$, $2^n(2^n-1)$, $2^n(2^n+1)$ can be done by Mixed Radix conversion as already described in Chapter III. The scaling by any other number can be done in three stages: a RNS to binary conversion, a modulo reduction mod X the divisor using Montgomery algorithm or Binary to RNS conversion techniques described in Chapter II and a binary to RNS conversion of this residue mod X to residues mod 2^n, mod (2^n-1) and mod (2^n+1).

5.5. CONCLUSION

In this Chapter, scaling, base extension and division operations in RNS are studied in some detail. Division by arbitrary number is sill a complicated operation. In the next Chapter, the issue of error detection and correction is considered in detail. This is an important feature perhaps not easily available in the conventional binary number system based architectures.

6

ERROR DETECTION AND CORRECTION IN RNS

6.1. INTRODUCTION

Error detection and correction in RNS has received considerable attention in literature. This was considered to be necessary in ROM based designs where some faults may occur which may be corrected by reconfiguring or bypassing the faulty device using additional residues corresponding to additional moduli. These additional moduli are termed as redundant moduli. The errors evidently can occur in the residues or redundant residues. These redundant residues are mandatory for error detection and correction. Techniques for error detection and correction are considered in detail in this Chapter. The possibility of scaling and residue error correction in a single hardware also is considered.

Single digit error detection can be done using one extra residue whereas the correction needs at least two residues. This can be appreciated by considering an example. This example considers a RNS in the moduli set {3, 5, 7, 11} where 11 is the redundant residue. Consider that the original binary number is 52 which should have been {1, 2, 3, 8} in the given moduli set. Let us consider that an error has occurred and the residues have become {1, 2, 4, 8}. We now define [Jenk83] projections X_j = X mod m_j' where m_j' = $(\Pi_{i=1}^{L+r} m_i) / m_j$ where L is the actual number of moduli and r the number of redundant moduli. In other words, projections are obtained by ignoring one modulus at a time. For the above example, we have the projections {1, 2, 4}, {1, 2,8}, {1, 4, 8} and {2, 4, 8}. Decoding using Mixed Radix digit evaluation, we obtain for these cases, the decoded numbers as follows:
{1, 2, 4} = 4.15+2.3+1 = 67; Moduli set {3, 5, 7}
{1, 2, 8} = 3.15+2.3+1 =52; Moduli Set {3, 5, 11}

{1, 4, 8} = 7.21 + 1.3+1 =151; Moduli set {3, 7, 11}
{2, 4, 8} = 10.35 + 6.5+2 = 382; Moduli Set {5, 7, 11}
It is known firstly that the decoded number shall be less than 105 = (3.5.7). Hence, the last two choices are wrong indicating that there is an error. However, the error may be in 7 or 11. However, the first case also appears to be correct, but we are not definite about an error in 7 or 11. Interestingly, the use of two redundant moduli say 13 also in addition to 11, will enable correct decision to be taken regarding the error as well as facilitate correction as will be shown later.

6.2. SZABO AND TANAKA TECHNIQUE FOR ERROR DETECTION AND CORRECTION

Szabo and Tanaka [Sza67] suggested the error detection technique which is briefly considered next. In this method using redundant residues, the effect of the error is recognized to cause an error in the decoded number by a multiple of $Mm_{r1}m_{r2}/m_i$ where the error is assumed to occur in some residue corresponding to modulus m_i. Note that m_{r1} and m_{r2} are redundant residues. In other words by subtracting $a.Mm_{r1}m_{r2}/m_i$ from X, a systematic search can be conducted exhaustively for all a and m_i such that $1 \leq a \leq m_i$ and which ever yields the value within the dynamic range M is chosen. The method is computation intensive since $\Sigma_{i=1}^{k} (m_i-1).k$ tests have to be conducted, where k is the total number of moduli including two redundant residues. An example will be in order.

Consider the moduli set {3, 5, 7, 11, 13} for which the residues for 52= {1, 2, 3, 8, 0} are assumed to have changed to {1, 2, 6, 8, 0}. We will show for m_i=7, the six tests that need to be done. For the other moduli 3,5,7 and 11 also, 2, 4, 6 and 10 tests need to be done respectively, so as to observe that only one correct result exists. The given {1, 2, 6, 8, 0} corresponds to 2197. Denoting $M_r = M.m_{r1}.m_{r2}$, $M_r/7$ in this case is 2145, whose multiplicative inverse with respect to 7 is 5 i.e. 10725 corresponds to {0, 0, 1, 0, 0}. Hence, we need to determine a i.e. the number of multiples of 10725, we need to subtract from 2197 mod M_r =15015, to get a result within the dynamic range $0<x<105$:

-1.10725 6487
-2.10725 10777
-3.10725 52
-4.10725 4342

-5.10725 8632
-6.10725 12922
-0.10725 2197.

Evidently, 52 is the correct answer obtained by subtracting the error.

6.3. MANDELBAUM'S ERROR CORRECTION TECHNIQUE

Mandelbaum [Mand72] has suggested a computationally- intensive technique which also notes the fact that the effect of errors is to result in a number outside the dynamic range. Mendelbaum observed that the higher order bits are non-zero in the absence of errors, since the legal dynamic range of M permit only $\log_2$ (M) bits. First F the decoded number in the presence of error has to be determined. Denoting the MSBs of F as T (M_r.B/P) and further expressing new quotients Q_1 = T (M_r.B/P) / M_r and Q_2 = (T (M_r.B/P) / M_r)+1, look up tables can be used to obtain the values of B and P. Note that the function T ( ) stand for truncation and M_r is the product of all the moduli including the redundant moduli. P is the modulus corresponding to which error has occurred. B is to be determined. The criterion for selecting B and P is such that M_r.B/P agrees with T (M_r.B/P) to maximum number of decimal places. The next step is to obtain X from F as

$$X = F - M_r.B/P \qquad (6.1)$$

An example will be illustrative to show the complexity of this technique for the moduli set {7, 9, 11, 13, 16, 17} where 16 and 17 are the redundant residues. M_r evidently is 24,50,448 i.e. the product of all moduli. The actual dynamic range i.e. the product of actual moduli is 9009. Consider the number 52 in the moduli set which should be in the extended RNS {3,7,8,0,4,1}. Assume that an error has occurred in the residue corresponding to modulus 11 resulting in residues {3, 7, 0, 0, 4, 1}. We wish to correct this error. The decoded number F is 668356 which can be seen to be 52+668304. Evidently, the latter is represented by a 21 bit number since M_r is a 21 bit number, whereas the actual dynamic range is 13 bits. Thus, the 21 bit word W_A, considering the 13 bit LSB word as zero can be seen to be $2^{19}+2^{17}+2^{13}$, whereas the actual word W is 668304 = $2^{19}+2^{17}+2^{13}+2^{12}+2^{9}+2^{7}+2^{4}$. The 8-bit MSB word Q_1 $0101\ 0001_{(2)}$ can be expressed as a fraction (1/4+1/16+1/256)= 69/256 i.e. the approximated word W_A is expressed as a fraction of 2^x where x is the smallest number such that 2^x is greater than the

M_r. We have therefore $d_1 = 69/256$ and $d_2 = 70/256$. Next, a table can be looked up to find corresponding to each modulus to know which modulus yields the error close to these values.

An example is shown in Table.6.1, where Q_1 and Q_2 are expressed as functions of P viz., $d_1.P$ and $d_2.P$. Evidently, for the modulus 11, we have the error very close to an integer to a maximum number of decimal places and hence is the modulus where error has occurred. The value of B is $d_1.P$ or $d_2.P$ corresponding to this modulus(B=3 in the present case). Next F can be found to be 6683056 – (3/11).(24,50,448) = 52. Undoubtedly, the method is not attractive due to the need for computations in addition to the need of Look-up Tables. A refinement of the method will be discussed in a later section.

$P=m_I$	$0.2695P=d_1.P$	$0.2734P=d_2.P$
7	1.881	1.92
9	2.42	2.73
11	2.959	3.007
13	3.49	3.55
16	4.3	4.37
17	4.57	4.54

Table. 6.1. Values of $d_1.P$ and $d_2.P$ (adapted from [Mand72] 1972 ©IEEE)

6.4. JENKINS'S ERROR CORRECTION TECHNIQUES

It may be noted that even though with two redundant residues, the dynamic range is M_t, only certain integers corresponding to single bit errors are valid in M_t. This aspect can be used to check the occurrence of single error and if so, this error can be corrected. In general, the valid range for an RNS with two redundant residues is 0 to (M-1) and the illegitimate range is bounded by two vacant bands between M and M_t. Considering that the moduli are ordered in ascending order, the vacant bands are M to $M.(M_{r1} - 1)$ and $M_t - 1 - (M_t/M_{r2}) + M$ to M_t-1. As an illustration for the modulo set {3, 5, 7, 11, 13}, the valid and illegitimate ranges and two vacant bands are respectively 0 to 104, 1155 to 13964, 105 to 1154, 13965 to 15014.

Jenkins [Jenk82, Jenk83, Jenk88] has considered error correction based on the MRC technique in great detail. The concept of using redundant modulus for error detection is already introduced together with the definition of projection in Section 6.1. This will be extended to error correction using the same example as in Section 6.1. and uses an additional modulus 13. The various projections for this moduli set {3, 5, 7, 11, 13} are {3, 5, 7, 11}, {3, 5, 7, 13}, {3, 7, 11, 13}, {3, 5, 11, 13 and {5, 7, 11, 13}. Considering the same given number 52, with an error in the residue corresponding to modulus 7, we have the mixed radix digits and decoded numbers for all these cases are as follows:

Moduli {3, 5, 7, 11}; Residues: {1, 2, 4, 8} $\Rightarrow$ 3.105+4.15+2.3+1=382
Moduli {3, 5, 7, 13}; Residues: {1, 2, 4, 0} $\Rightarrow$ 11.1054.15+2.3+1 =1222
Moduli {3, 5, 11, 13}; Residues: {1, 2, 8, 0} $\Rightarrow$0.165+3.15+2.3+1 =52
Moduli {3, 7, 11, 13}; Residues: {1, 4, 8, 0} $\Rightarrow$ 7.231+7.21+1.3+1 = 1768
Moduli {5, 7, 11, 13}; Residues: {2, 4, 8, 0} $\Rightarrow$ 1.385+10.35+6.5+2=767

One important observation is that the correct value among these is distinguished by the most significant MRC digit being zero. Once this test is performed, then base extension needs to be done to evaluate the correct residue corresponding to modulus 7 using the procedures described earlier. It is important to note that the redundant moduli shall be chosen to be larger than the actual moduli.

Jenkins [Jenk82, Jenk88] has observed that the basic MRC structure can be used to determine the various projections and residues by shorting the column and row corresponding to the residue under consideration as shown in Fig 6.1 (a). It may be noted at the outset that direct MRC conversion shall be done first to observe whether any error is present by looking at the MRC digits. If they are non-zero only, the five steps of evaluating MRC digits for projections needs to be done. The architecture is illustrated in Fig 6.1 (b). This structure can take advantage of the fact that the MRC digits already computed can be used to generate the various projections.

Note that pipelining can be possible by using the arrangement of Fig 6.1 (b). Note that the regular MRC converter computes X and X_5 simultaneously, since X_5 ignores the most significant MRC digit of X. The projections occur in the order X_5, X_4, X_3, X_2 and X_1. It may be appreciated that already available MRC digits of each projection can be used to compute the next so that unnecessary re-computation is avoided. As an illustration {x_1, x_2, x_3, x_5}

can be computed, since $\{x_1, x_2, x_3\}$ evaluation is already is available while computing X and X_5.

An alternative technique known as expanded projections also has been suggested by Jenkins [Jenk82] and Jenkins and Altman [Jenk88]. In this technique instead of zeroing the residue and modulus to generate projections, the original moduli set is multiplied by m_i so that $m_i.X$ is considered for MRC. Evidently, this ensures that the residue with respect to m_i is automatically zero thus avoiding errors in that column to decide the result. By evaluating all the projections in $m_i.X$ and observing the most significant MRC digits, one can find whether an error has occurred or not. However, the method is not attractive in the sense that MRC digits have to be found by base extension of X afresh to correct the errors. This may be appreciated noting that MRC digits of $m_i.X$ only have been computed earlier. We illustrate the technique with the same five moduli example considered above. Note that such multiplication by m_j still keeps the number less than M. The implementation needs the same architecture as Fig 6.1 (b), however, with a front-end ROM to multiply the,residues by m_i appropriately in five cycles for each m_i as shown in Fig 6.1 (c).

Consider the residue set {1, 2, 4, 8, 0} =10777, where error has occurred in the residue with respect to 7. The multiplication by 3, 5, 7, 11 and 13 successively of the residue set and conversion to MRC digits yields the following respectively:

x3 $\Rightarrow$ 2301 = 1.1155+10.105+6.15+2.3+0
x5 $\Rightarrow$11570 = 7.1155 + 7.105+1.15+1.3+2
x7 $\Rightarrow$364 = 0.1155+3.105+3.15+1.3+1
x11 $\Rightarrow$13442 = 11.1155+7.105+0.150.3+2
x13 $\Rightarrow$4966 = 4.1155+3.105+2.1+0.3+1

It is known that the result will be less than m_i.105 (where 105=3.5.7 is the original dynamic range). Evidently, the multiplication by 7 has removed the effect of error and hence 7 is the modulus whose residue is in error.

Jenkins and Altman [Jenk83] have observed that the effect of error in any residue in a Mixed -Radix Converter hardware using redundant moduli is same as the error in the input residue in that column. They point out that the errors that occur in hardware can also be corrected by the error correction hardware used to detect errors in the input residue set. The reader is referred to their work for more details.

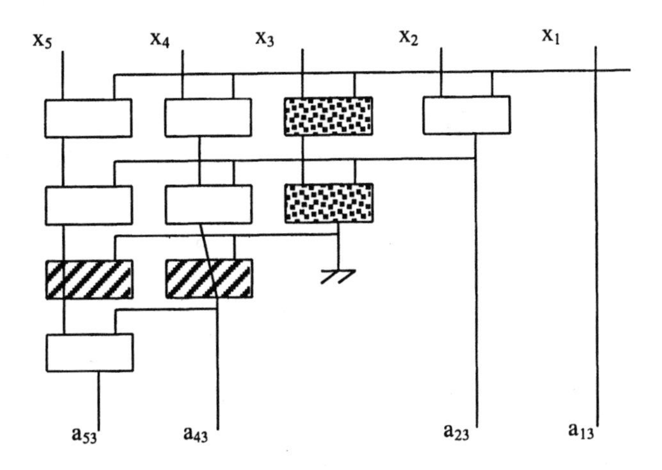

(a)

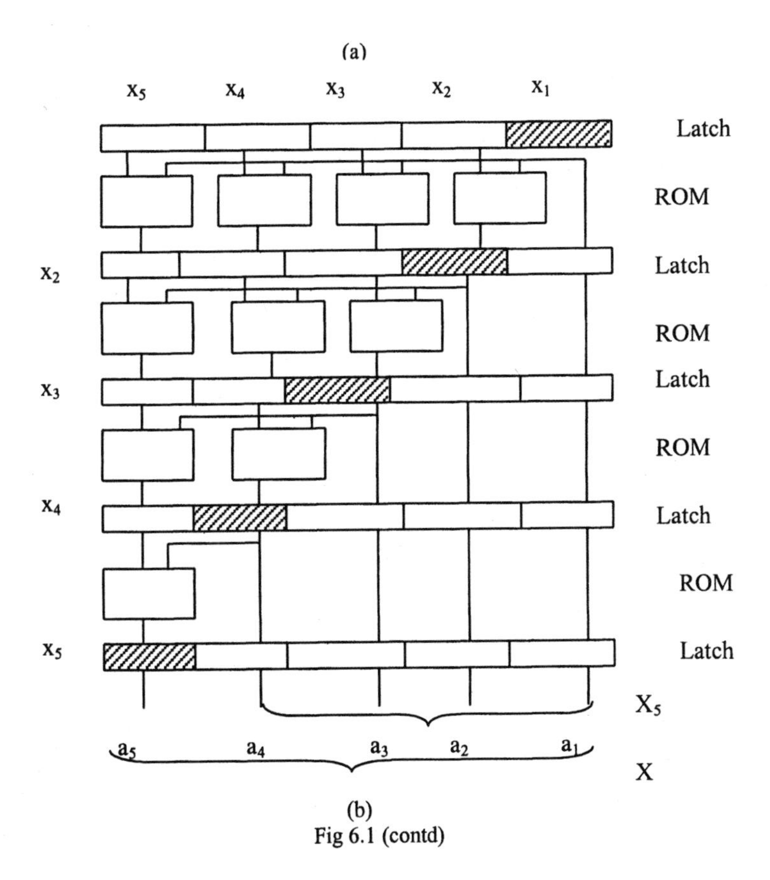

(b)

Fig 6.1 (contd)

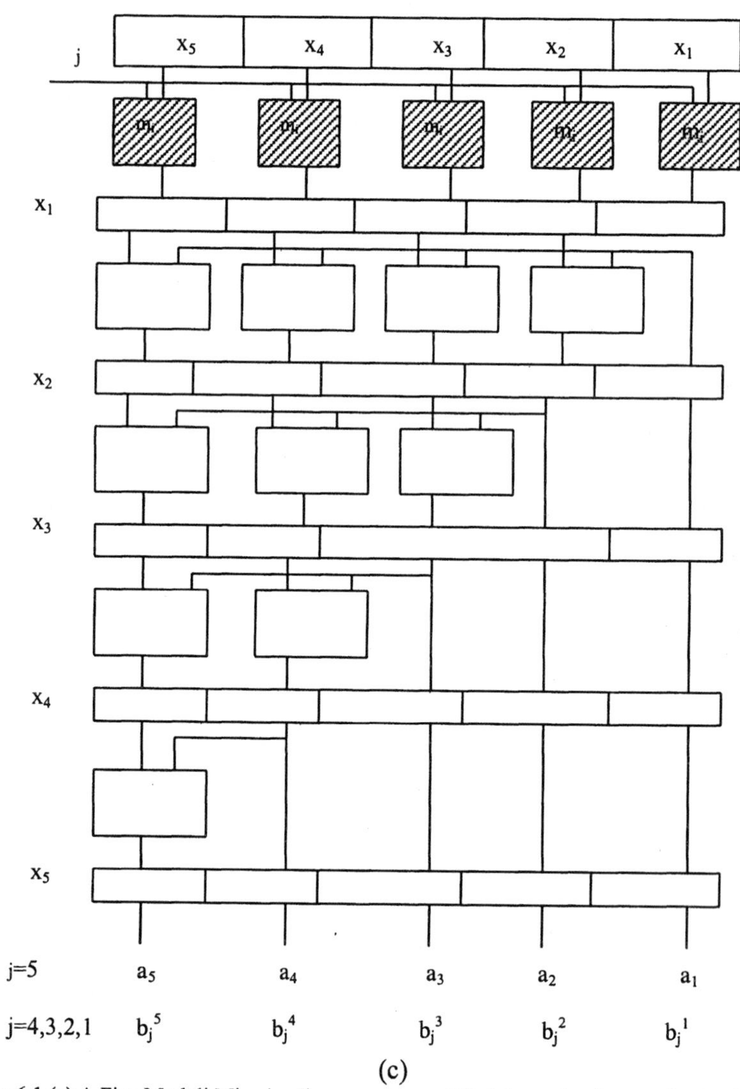

(c)

Fig 6.1 (a) A Five Moduli Mixed radix converter modified for evaluating projection X_3, (b) A pipelined Mixed Radix converter for sequential generation of projections and (c) Modification for obtaining expanded projections ((a) and (b) adapted from [Jenk83] 1983 ©IEEE and (c) adapted from [Jenk82] 1982 ©IEEE)

6.5. RAMACHANDRAN'S ERROR CORRECTION TECHNIQUE

Ramachandran [Rama83] has suggested that fewer tests are needed than those needed in Jenkins and Altman technique for error correction and detection of a single error. For a RNS system with r redundant residues, (2n/r)+2 recombination of residues can correct the error. Two examples are illustrated which use n=6, r=3 and n=7, r=3 respectively. In these cases, six and seven recombinations are needed. In the Jenkins et al technique, even considering that two residues are needed to correct single digit error, eight and nine MRCs corresponding to projections are needed respectively. The basic principle is that in Ramachandran technique, the faulty residue is ensured to be left out at least twice so that the correct value is obtained at least twice. The proof is as follows:

At each attempt, n residues out of $n+r$ residues are chosen leaving r behind. After s recombinations, we leave $s.r$ residues. Since any residue may be in error, we must leave out at least $2(n+r)$ residues so that each residue is left out at least twice. Hence, we require $s.r \geq 2(n+r)$ i.e. the number of recombinations $s > (2n/r)+2$.

The three moduli set with two redundant residues, needs five steps according to the above theory. These combinations for the moduli set {3, 5, 7, 11, 13} could be {3, 11, 13},{11, 15, 13}, {5, 13, 7}, {3, 5, 7} and {5, 7, 11}. Note that we have omitted (5, 7), (3, 7), (3, 11), (11, 13) and (13, 5) in five steps so that each residue is left at least twice. Note that Jenkins et al method also needs five steps of MRC for four moduli RNS, whereas the above needs MRC for three moduli RNS.

6.6. SU AND LO UNIFIED TECHNIQUE FOR SCALING AND ERROR CORRECTION

The one-to-one correspondence between the integers of the original dynamic range and the states of the legitimate range in the redundant Residue Number System (RRNS) can be established by a polarity shift. It may be noted that the actual range in the original RNS is 0 to M-1. In the RRNS, the positive numbers are mapped for odd M from 0 to (M −1)/2 and the negative numbers are mapped from M_t - (M-1)/2 to (M_t- 1). In the case of even M, the positive numbers are mapped from 0 to (M/2)-1 and the negative numbers are mapped from M_t-(M)/2 to (M_t-1). The polarity shift is defined as X' = X+(M/2) for M even and X' = X + (M-1)/2 for M odd. This ensures that since X lies between –M/2 and +M/2 if M is even and –(M-1)/2 to (M+1)/2 for M odd, we have X' between 0 and M-1. Polarity shift therefore needs to be performed before scaling or error correction since X' only belongs to the legitimate range [6.4]. The original encoding in the RRNS and after polarity shift is presented in Fig 6.2 for clarity.

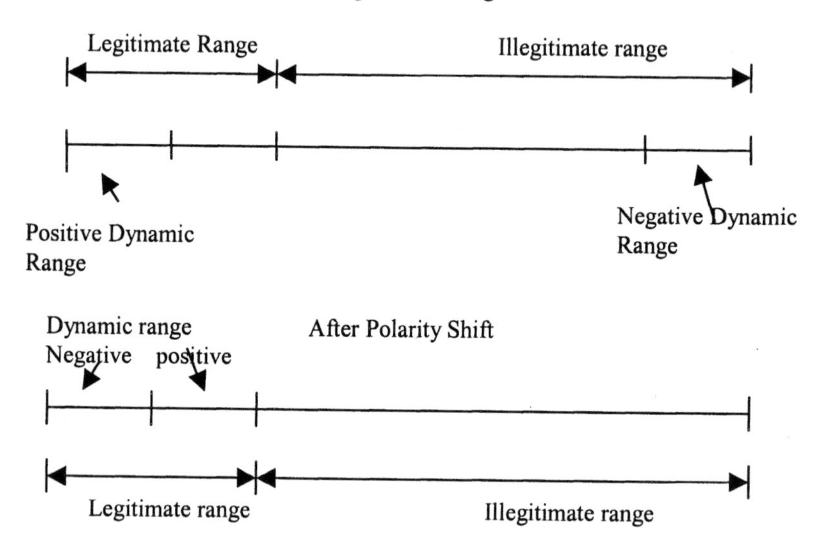

Fig 6.2. Illustration of RNS intervals before and after polarity shift (adapted from [Etz80] 1980 ©IEEE)

The problem of error detection and scaling were treated separately usually. Several scaling techniques have been considered in Chapter V. Su and Lo [Su90] observe that in a single architecture, we can combine both these operations of scaling and error correction. We will consider their approach which uses redundant residues next briefly. In this approach, there are four steps involved as follows:

(a) Compute a polarity shift X' = X+(M/2) for M even and X' = X+(M-1)/2 for M odd.

(b) Compute integer part of $X'/M = a_{l+1} + a_{l+2}.m_{l+1} + a_{l+3}.\ m_{l+1}.m_{l+2} + \ldots + a_{l+r}.\ m_{l+1}\ m_{l+2} \ldots m_{l+r-1}$

(c) Find the error from a LUT by looking into the location corresponding to $a_{l+1}\ a_{l+2} \ldots a_{l+r}$

(d) Correct the error $X = (X' - E_j) \bmod M_t$ where M_t is the product of all the moduli.

Note that the choice of the redundant moduli is governed by certain relationships so that the error table is non-ambiguous. An illustration will aid the understanding of this technique.

Consider the moduli set $\{m_1, m_2, m_3, m_4, m_5, m_6\}$ e.g., $\{2, 5, 7, 9, 11, 13\}$ where 11 and 13 are the redundant moduli. The dynamic range M =630 whereas that including redundant moduli is 90090. We also define $M_r = 143$, product of the redundant moduli. Assume that the given number X is –311 corresponding to {1, 4, 4, 4, 8, 1} and that a single digit error is introduced $e_2 = 4$ corresponding to modulus $m_2=5$ i.e. X' = 53743 ={1, 3, 4, 4, 8, 1}. First, a polarity shift is performed to obtain X' =X + M/2 =54058 = {0, 3, 4, 4, 4, 4}. Decoding using MRC yields [X'/M]=85. A look at Table 6.2 yields the error as $e_2 = 4$. Thus, X can be obtained by subtracting $E_2=\{0, e_2, 0, 0, 0, 0\}$ from residues of X'. Therefore, $X = (X-E_2).\ \bmod M_t =\{1, 4, 4, 4, 8, 1\}$ is the corrected value. Note that $P_j = (w_j.e_j) \bmod m_j$ where w_j is the multiplicative inverse of $(M.m_r/m_i)$.

A look at the table shows that for errors in the redundant moduli, exact high-order MRC digits are obtained (see the last twenty two entries), whereas, the errors in the non-redundant moduli yield a MRC decoded high-order digits which are mixed fractions. Hence two values, one a truncated and the other a rounded value are seen in the fist nineteen entries. The first one is an all-zero high-order MRC digit. Note that CRT has been used to determine the contents of Table.6.2. As an example, $e_1 = \{1, 0, 0, 0, 0, 0\}$ corresponds to $w_i.M_i = (M.M_r/m_i).w_i = 5.7.9.11.13$ where w_i is the multiplicative inverse of $(M.M_r)/m_i$ with respect to $m_1=2$. Thus, dividing by M = 630 yields X/M = (5.7.9.11.13)/630 = 71.5. The table shows this as 71 and 72, two entries. Note that the maximum of the entries evidently can be < 11.13 = 143. No state repeats as well.

Su and Lo determine the conditions under which unique high-order MRC digits can be obtained which are briefly considered next. The reader is

E_i						$[X'/M]$	
e_1	e_2	e_3	e_4	e_5	e_6	$[M_r.P_j/m_j]$	$[M_r.P_j/m_j] + 1$
0	0	0	0	0	0	0	
1	0	0	0	0	0	71	72
0	1	0	0	0	0	57	58
0	2	0	0	0	0	114	115
0	3	0	0	0	0	28	29
0	4	0	0	0	0	85	86
0	0	1	0	0	0	40	41
0	0	2	0	0	0	81	82
0	0	3	0	0	0	122	123
0	0	4	0	0	0	20	21
0	0	5	0	0	0	61	62
0	0	6	0	0	0	102	103
0	0	0	1	0	0	79	80
0	0	0	2	0	0	15	16
0	0	0	3	0	0	95	96
0	0	0	4	0	0	31	32
0	0	0	5	0	0	111	112
0	0	0	6	0	0	47	48
0	0	0	7	0	0	127	128
0	0	0	8	0	0	63	64
0	0	0	0	1	0	26	
0	0	0	0	2	0	52	
0	0	0	0	3	0	78	
0	0	0	0	4	0	104	
0	0	0	0	5	0	130	
0	0	0	0	6	0	13	
0	0	0	0	7	0	39	
0	0	0	0	8	0	65	
0	0	0	0	9	0	91	
0	0	0	0	10	0	117	
0	0	0	0	0	1	11	
0	0	0	0	0	2	22	
0	0	0	0	0	3	33	
0	0	0	0	0	4	44	
0	0	0	0	0	5	55	
0	0	0	0	0	6	66	
0	0	0	0	0	7	77	
0	0	0	0	0	8	88	
0	0	0	0	0	9	99	
0	0	0	0	0	10	110	
0	0	0	0	0	11	121	
0	0	0	0	0	12	132	

Table. 6.2. Correspondence between X'/M and Single Residue digit errors e_i (adapted from [Su90] 1990 ©IEEE)

$[(E_i/K)+(M/2K)]$	$[X'/M]$ $[M_r.P_j/m_j]$ $[M_r.P_j/m_j]+1$
0	0
4536	71 72
3635	57 58
7238	114 115
1833	28 29
5436	85 86
2605	40 41
5179	81 82
7753	12 123
1318	20 21
3892	61 62
6466	102 103
5036	79 80
1032	15 16
6032	95 96
2033	31 32
7038	111 112
3034	47 48
8039	127 128
4035	63 64
1669	26
3307	52
4945	78
6538	104
8221	130
850	13
2488	39
4126	65
5764	91
7402	117
724	11
1417	22
2110	33
2803	44
3496	55
4189	66
4882	77
5575	88
6268	99
6961	110
7654	121
8347	132

Table. 6.3. Correspondence between X'/M and $(E_j/K + M/2K)$ (adapted from [Su90] 1990 ©IEEE)

referred to their work for a rigorous proof. These conditions are (a) $M_r > m_i . m_{l+s}$, $1 \le i \le l$, $1 \le s \le r$ and (b) $M_r > 2m_im_j - m_i - m_j$. These conditions help the detection of a single digit error accurately. It can be checked that the choice of 11 and 13 in the previous example justifies these conditions.

The architecture of implementation of Su and Lo algorithm is as shown in Fig 6.3. Note that the adders at the top row perform polarity shift. Next, the MRC conversion stage yields a_1, a_2, ... a_6. Then, a_5 and a_6 address a LUT to obtain the six errors e_1, e_2 .... e_6. These are next subtracted from input residues to yield the correct result.

Su and Lo observe that scaling by K, product of moduli can be easily accomplished in the same architecture. Herein, the base extension is performed first upon evaluation of MRC digits of the polarity shifted input. Then, the errors are as before evaluated to be subtracted to yield the correct result. Note, however, that since K scales the error as well as the polarity shift M/2, the LUT entries correspond to $(E_j/K+(M/(2K)))$ where K is the scaling factor (a product of scaling moduli). The LUT shown in Table 6.3 presents these values corresponding to the high order MRC digits. An example is next illustrated.

Consider the residue set {1, 0, 5, 2, 4, 3} corresponding to X =205. Assume an error of e_3=1 to yield the modified residues {1, 0, 6, 2, 4, 3}=25535. The polarity shift yields next 25850 ={0, 0, 6, 2, 0, 6}. The MRC process yields the high order digits as X'/M = 41 whence the Table 6.3 yields $[E_j/K +(M/(2K))\}$ of 2605 corresponding to a scale factor of 10 (product of two moduli 2 and 5). The residues of 2605 are {1, 0, 1, 4, 9, 5} which when subtracted yield the scaled number as {0, 0, 1, 7, 2, 6} = -20.

6.7. ORTO ET AL TECHNIQUE FOR ERROR CORRECTION AND DETECTION USING ONLY ONE REDUNDANT MODULUS

The use of only one redundant residue for error detection and correction has been studied in detail by Orto et al [Orto92]. They suggest the use of a power of two redundant modulus larger than all the other moduli. They

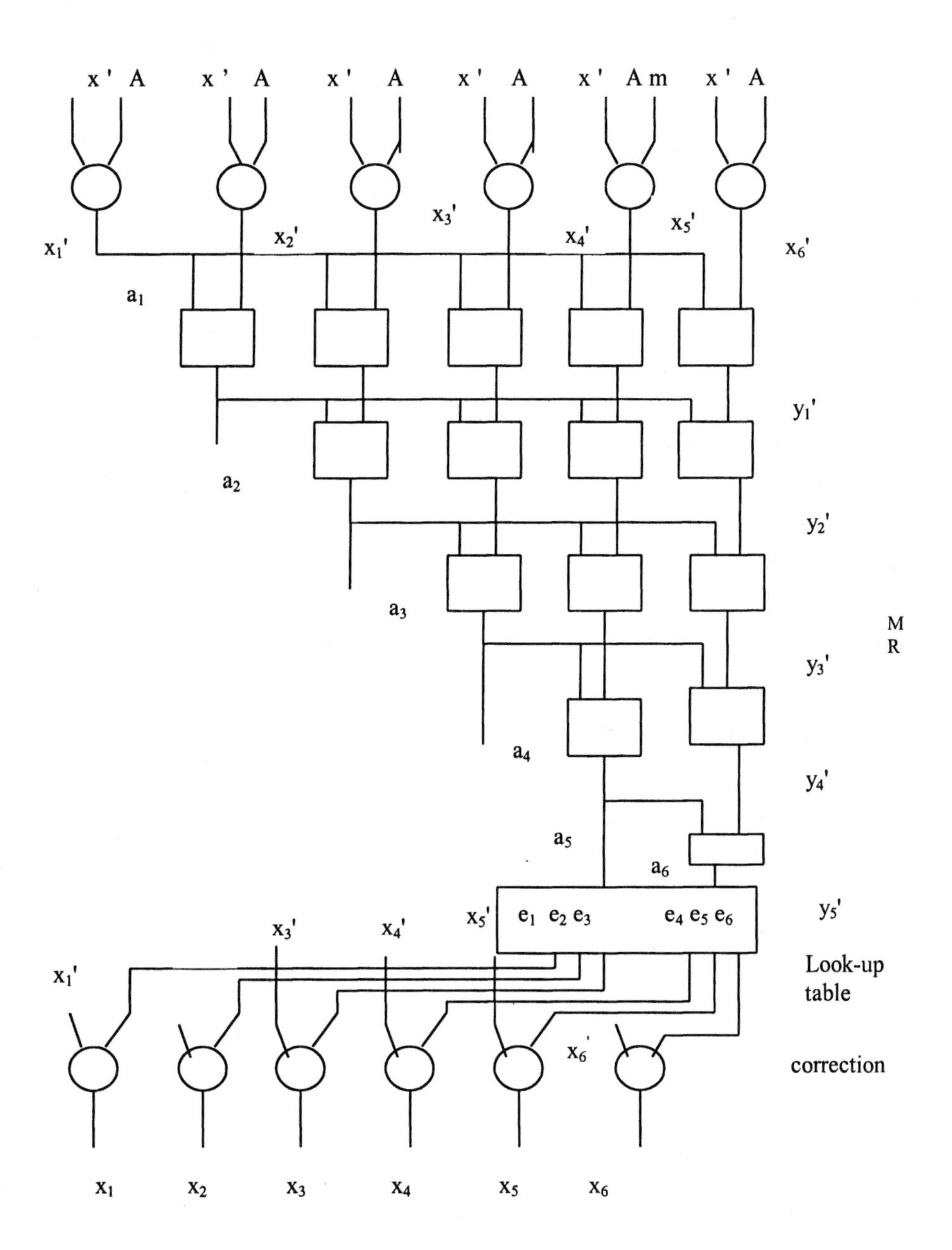

Fig 6.3. Su and Lo architecture for error correction
(adapted from [Su90] 1990 ©IEEE)

suggest the use of MSBs of the scaled result of application of CRT for determining the existence of error. Their technique is briefly as follows: Defining the residues $\{r_1, r_2, r_3,\ldots, r_n, r_r\}$ where $r = n+1$ and $r_r = 2^k$, the CRT expansion yields $[\ \Sigma_{i=1}^{\ r} (M/m_i).\{(r_i//M_i) \bmod m_i\ \}] \bmod M$ where M is the product of all the moduli including the redundant modulus. This can be several times larger than M. We instead evaluate $[\ \Sigma_{i=1}^{\ r} (S/m_i).\{(r_i//M_i) \bmod m_i\ \}] \bmod 2^k$ which is obtained by multiplying the original result by (M/S). Orto et al observe that S can be chosen suitably so as to detect the presence of errors by looking at the MSBs of the result. If the MSB word is all '0' or all '1' s, there is no error.

An example due to them is illustrated. Consider x = 114 = {12, 0, 22, 6, 18} in a moduli system {17, 19, 23, 27, 32}. Note that 32 is the redundant residue. The values $c_i = (r_i//M_i) \bmod m_i$ can be calculated as {11, 0, 7, 3, 30}. Next choosing S as 2^9, the S/m_i values can be either rounded or truncated. Considering that rounding is used, we have $[S/m_i]$ as {30, 27, 22, 19, 16}. Note that the actual values are {30.118, 26.97, 22.26, 18.963, 16.000}. Using these rounded values together with the available c_i values, we have $(\Sigma_{j=1}^{\ 5} [2^9/m_j].c_j) \bmod 2^9 = 509 = 11{,}111{,}101$ (binary). The case with an error in residue corresponding to m_1, to make x' = {13, 0, 22, 6, 18} yields $(\Sigma_{j=1}^{\ 5} [2^9/m_j].c_j) \bmod 2^9 = 239 = 011{,}101{,}111$ (binary). The six most significant bits are not all ones or all zeroes thus showing an error. It may be noted that after detection of the error, still we have to use projections to find the error and correct it. The choice of S has to be proper to yield the correct answer whether error has occurred. The advantage claimed for this technique is that small look- up-table based multipliers can be used since $[2^9/m_i]$ is 5 bits and residues are also 5 bit words. In other words, x_j can directly look into ROM to get the terms $[2^9/m_j].c_j$. The reader is, however, urged to verify the accuracy of this method.

6.8. CONCLUSION

The use of one or more redundant moduli for error correction has been discussed in detail in this Chapter. The use of the same architecture to perform scaling as well as error correction also has been considered. The next Chapter deals with yet another topic in the theory of RNS viz., Quadratic Residue Number Systems which have application in complex signal processing.

7

QUADRATIC RESIDUE NUMBER SYSTEMS

7.1. INTRODUCTION

Complex signal processing can be handled by RNS in a manner similar to standard complex number operations such as addition, subtraction, multiplication etc. However, under certain special cases of choice of moduli, the complete decoupling of computation of real and imaginary parts of the result is feasible. Nussbaumer [Nuss76] suggested that Fermat primes of the type 4k+1 have this property. Later, this advantage has been extended to any primes of the type (4k+1) using the Quadratic Residue Number System (QRNS) [Jull87, Jenk87, Kris86a]. The QRNS also has been shown to be applicable for composite numbers which have (4k+1) as a factor and also to the case of composite numbers having $(2^{2n}+1)$ as a prime factor. The primary disadvantage of the QRNS is the restriction on the type of moduli. Another technique which allows any modulus but with increase in number of multiplications has also been found known as Modified Quadratic Residue Number System (MQRNS) [Sode84b]. This will also be discussed in detail in this Chapter.

7.2. BASIC OPERATIONS IN QRNS

For primes m of the form (4k+1), -1 is a quadratic residue meaning that $(x^2+1) = 0 \bmod (4k+1)$. This has two solutions j_i for x such that $j_i^2 = -1 \bmod (4k+1)$. Thus, the root j_i has a multiplicative inverse of $-j_i$. A given complex number (a+jb) represented as an extension element (A, A*) where

$A = (a+j_ib) \bmod m$ and $A^* = (a-j_ib) \bmod m$. The addition rule applies and multiplication of two complex numbers reduces to only two multiplications unlike the normal complex multiplication. This can be easily seen as follows.

7.2.1. Addition in QRNS

The addition operation is as follows. Consider the two complex numbers $(a+jb)$ and $(c+jd)$ to be added where $j = \sqrt{(-1)}$. The two-tuples corresponding to these are

$(a+jb) \Rightarrow (a+j_ib) \bmod m, (a-j_ib) \bmod m$ (7.1a)

$(c+jd) \Rightarrow (c+j_id) \bmod m, (c-j_i.d) \bmod m$ (7.1b)

The sum is

$(a+c) +j(b+d) \quad A=(a+c) +j_i(b+d) \quad A^*= (a+c) -j_i(b+d)$ (7.2)

The actual sum is obtained from A and A* using the relationship

$Q = (A + A^*)/2 = a+c$ true (7.3a)

$Q^* = (A - A^*)/(2j_i) = (b+d)$ true (7.3b)

An example will be illustrated for $m_i = 13$. j_i can be seen to be 8. Consider addition of $(1+j2)$ and $(2+j3)$. We have the sum as $(3+j5)$ using the conventional method whereas in QRNS the mapping is as follows:

$(1+j2) \Rightarrow 4 \quad 11$

$\underline{(2+j3)} \Rightarrow \underline{0 \quad 4}$

$3 +j5 \Rightarrow 4 \quad 2$ (addition mod 13)

Hence $Q = (4+2)/3 = 3$, $Q^* = ((4-2)/2j_i) \bmod 13 = (1/j_i) \bmod 13 = (-8) \bmod 13 = 5$, which can be verified to be true.

7.2.2. Multiplication in QRNS

Consider the two numbers $(a+jb) \bmod m$ and $(c+jd) \bmod m$ to be multiplied. The element pairs in QRNS corresponding to these are

$A = (a+j_ib) \bmod m \quad A^* = (a-j_ib) \bmod m$ (7.4a)

$B = (c+j_id) \bmod m \quad B^* = (c-j_id) \bmod m$ (7.4b)

Multiplying these yields

$Q_i = (A.B) \bmod m, Q^* = (A^*.B^*) \bmod m$ (7.5)

Note here that A, B, A* and B* are real numbers. The result is obtained by the reverse conversion from the two-tuples Q and Q* to normal CRNS (Complex RNS) using the formula

$$Y_{iR} = [2^{-1}.(Q_i + Q_i^*)] \bmod m_i \quad (7.6a)$$
$$Y_{iI} = [2^{-1}.j_i^{-1}.(Q_i - Q_i^*)] \bmod m_i \quad (7.6b)$$

The proof for multiplication is as follows: The conventional multiplication yields

$$(a+jb).(c+jd) = (ac-bd) +j(bc+ad) \quad (7.7)$$

In QRNS we have

$$(a+jb) \Rightarrow (a+j_ib) \quad (a-j_ib) \quad (7.8a)$$
$$(c+jd) \Rightarrow (c+j_id) \quad (c-j_id) \quad (7.8b)$$

Hence,

$$(a+jb)(c+jd) \Rightarrow Q= (ac-bd) +j_i (ad+bc),\ Q^*=(ac-bd) -j_i (ad+bc) \quad (7.9)$$

whence we can check that the following hold:

$$\text{real part} = (Q+Q^*)/2 = (ac - bd) \quad (7.10a)$$
$$\text{imaginary part} = (Q-Q^*)/2j_i = ad+bc \quad (7.10b)$$

Thus, multiplication of real and imaginary parts separately is sufficient. It must be noted that here j_i is a number.

An example will be in order to illustrate the multiplication operation. Consider $m_i = 13$. It can be verified that $x = 8$ is a Quadratic Residue since $(x^2+1) = 0 \bmod 13$. Consider two complex numbers (2+j3) and (1+j5) to be multiplied mod 13. The conventional complex multiplication gives (0+j0). In QRNS, (2+j3) and (1+j5) are mapped as follows.

$$2+j3 \Rightarrow (2+8.3) \bmod 13 = 0 \qquad (2-8.3) \bmod 13 = 4$$
$$(1+j5) \Rightarrow (1+8.5) \bmod 13 = 2 \qquad (1-8.5) \bmod 13 = 0$$

Thus, the result is $Q_i=0.2=0$, $Q_i^* = 4.0=0$. Then using (7.6), we have the result as $Y_{iR}=0$ and $Y_{iI} = 0$ as desired.

7.2.3. Complex number to QRNS conversion technique

It may be noted that forward and reverse mappings given by (7.8) and (7.10) need to be done. Sousa and Taylor [Sous86] have suggested an architecture for converting complex numbers to QRNS form which is shown in Fig 7.1. Note that in this architecture, the second level accomplishes the function (A+rB) and (A-rB) whereas the first stage is needed to reduce a given binary number to RNS form. The architecture uses the periodic properties of 2^x. mod m_i studied in Chapter II to evaluate the residue mod m_i where alternate words are added and subtracted in case the modulus is of the form 2^n+1.

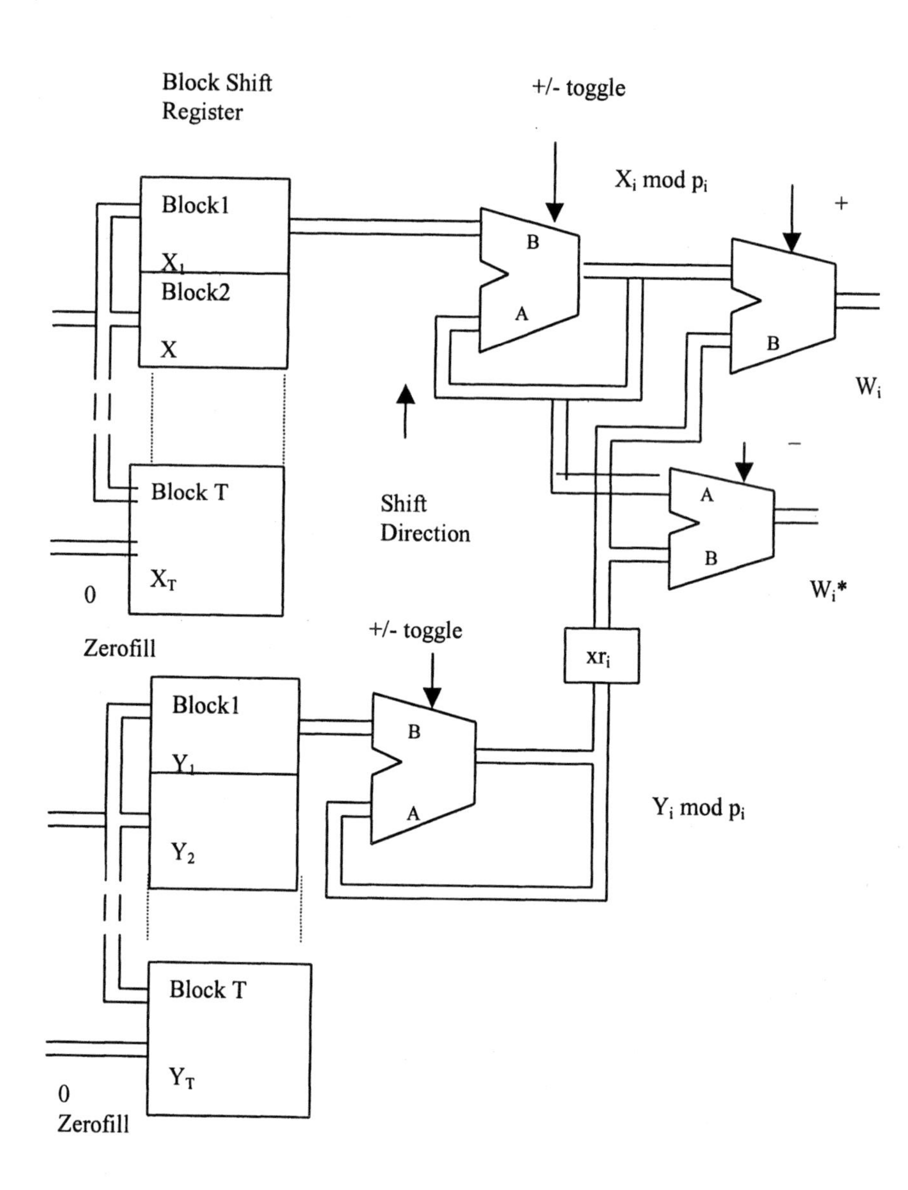

Fig 7.1. CRNS to QRNS integer converter architecture
(adapted from [Sous86] 1986 ©IEEE)

7.3. OTHER QUADRATIC RESIDUE NUMBER SYSTEMS QLRNS AND MQRNS

Soderstrand and Poe [Sode84b] have described a system QLRNS (Quadratic like RNS) which is like QRNS. They suggest that an integer j_1 can be found such that $j_1^2 = -c \bmod p$, $c \neq 1$. Then the mapping a+jb can be seen to be as follows:

$(a+jb) = ((a+ j_1 b\sqrt{c}) \bmod p, j_1 b\sqrt{c} \bmod p) = (z,z^*)$

$a = 2^{-1}(z+z^*)$, $b = (2j_1/\sqrt{2})^{-1}.(z-z^*)$

where the values of z and z* are rounded to the nearest integer. This technique has a disadvantage that the dynamic range shall be greater than $j_1 b\sqrt{c}$.

We next consider the MQRNS system due to Krishnan, Jullien and Miller [Kris86b] in which j_i is a solution of $x^2\text{-}n=0$. While the addition operation is same as in the QRNS case, the multiplication operation gets altered needing additional operations. It can be shown that

$$(a+jb).(c+jd) \Rightarrow (A.B)\text{-}S, \quad A^*.B^* \text{-}S \tag{7.11}$$

where $S = ((j_i^2+1)\, b.d) \bmod m$ where b and d are the imaginary part of A and B and A and B are as defined in (7.4). This is verified as follows:

$$\begin{aligned}(A.B - S) \bmod m &= [(a+j_i b)\,(c+j_i d)\text{-}S] \bmod m\\ &= [ac+j_i\,(bc+ad) + j_i^2.bd - S] \bmod m\\ &= [ac-bd + j_i\,(bc+ad)] \bmod m\end{aligned}$$

where we have used the definition of S.

$$\begin{aligned}(A^*.B^* \text{-}S) \bmod m &= [(a\text{-}j_i b)\,(c\text{-}j_i d) - S] \bmod m\\ &= [(ac\text{-}bd) - j_i\,(bc+ad)] \bmod m\end{aligned}$$

Thus, we have the real and imaginary parts of the result as

$(ac\text{-}bd) = [2^{-1}(A.B+A^*B^*) \bmod m_i - S] \bmod m_i$

$(bc+ad) = [2^{-1} j_i^{-1}(Q_i - Q_i^*)] \bmod m$

Note that several choices of j_i are possible for a given modulus m_i leading to different n values. As an illustration, for m =19, some examples of pairs (j_i, n) are (5, 6), (6, 17), (7, 11), (8,7) etc. It is thus seen that MQRNS multiplication needs three real multiplications, one addition and permits the use of moduli not of the form (4k+1) also.

An example of MQRNS multiplication is presented below for m_i =19, j_i=10 and n=5. Consider multiplication of two complex numbers (2+j3) and

(3+j5). The normal complex multiplication yields the result as 10+j0. The following is the MQRNS evaluation of the same:

$2+j3 \Rightarrow A=13 \quad A^* = 10$

$3+j5 \Rightarrow B=15 \quad B^* = 10$

$A.B = 5 \quad A^*.B^* = 5$

S = bd (n+1) mod 19= 14 since b=3, d=5 and n = 5.

Real Part of the result = (5+5)/2 –14 = 10.

Imaginary part of the result = (5-5)/($2j_i$) =0.

Krishnan et al [Kris86b] observed that for the moduli of the form 2^n-1, the memory requirement for obtaining A and A* can be easily reduced using a single ROM wherein the table is looked for A* by 1's complementing the X value. A table for the modulus 7 is presented in Fig 7.2, for which j' = 6 evidently facilitates the reduction of ROM. As an illustration, corresponding to a=5,b=6, we have ($a+j_ib$) mod m = (5+6.6) mod 7 = 6 whereas ($a-j_ib$) mod m =(5-6.6) mod 7 =4. The later corresponds to (5+1.6) mod 7 since – 6 mod 7 =1. Thus, 1's complementing the x value and looking into the ROM corresponding to this address suffices.

CM

	0(0)	1(6)	2(5)	3(4)	4(3)	5(2)	6(1)	7(0)
0	0	6	5	4	3	2	1	0
1	1	0	6	5	4	3	2	1
2	2	1	0	6	5	4	3	2
3	3	2	1	0	6	5	4	3
4	4	3	2	1	0	6	5	4
5	5	4	3	2	1	0	6	5
6	6	5	4	3	2	1	0	6

CM = compliment or multiplex

(a+jb) mod m = (5+6.6) mod 7 = 6

(a-jb) mod m = (5-6.6) mod 7 = (5+1.6) mod 7 = 4

Fig 7.2. The table used to obtain A and A* in the MQRNS (adapted from [Kris86b] 1986 ©IEEE)

7.4. JENKINS AND KROGMEIER IMPLEMENTATIONS

Jenkins and Krogmeier [Jenk87] in their absorbing paper deal with the particular choice of moduli of the form (2^{2r}+1) which will lead to a simple expression for j' thereby simplifying the implementations. This is next considered briefly. These moduli are called as "Augmented power of two moduli". The mathematical operations mod $2^{2r}+1$ need to handle the case of residue 2^{2r} needing one extra most significant bit only in this case. As a consequence, a diminished-one arithmetic technique [Lieb76], has been suggested in the implementation. In this technique, the state 10000…0 can be used to represent zero. The idea is to represent all non-zero elements by 2r bits so that 10000…0 state can be used to represent zero. This may be appreciated by looking at the two's- complement number mapping to diminished -one number shown in Table.7.1. The lower half corresponds to negative numbers as shown in brackets. Few examples of addition, subtraction and multiplication will be presented next.

(a) Addition:

Consider addition of –5 and 7. The two's- complement addition and diminished one addition are as follows:

	TC	Diminished one
-5	11011	01011
7	00111	00110
2	00010	10001
		0 Complemented MSB bit
		0001 = 2

We consider next addition of two positive numbers 5 and 3.

	Diminished-one
5	00100
3	00010
8	00110
	1
	00111 corresponds to 8

(b)Sign Change:

Changing sign is by complementing the bits except MSB. As an illustration, x=00010 =+3 and –x = 01101 =-3.

(c) Multiplication by 2^k:

Next, we will consider multiplication by 2^k. Consider e.g. (5.2^3) mod 17. Each step involves shifting left and complementing the MSB and adding to LSB. The sequence of operations is as follows:

5 ⇒ 00100 (Diminished)

Step 1 01000
____1 MSB is complemented
01001 Added
Step 2 10010
____0 MSB is complemented
00010 Added
Step 3 00100
____1 MSB is complemented
00101 Added ⇒ 6 =(5.8) mod 17.

Note that in the addition operation MSB needs to be excluded.

Binary	Diminished-1	2's complement	1's complement
00000	1	0	0
00001	2	1	1
00010	3	2	2
00011	4	3	3
00100	5	4	4
00101	6	5	5
00110	7	6	6
00111	8	7	7
01000	9(-8)	8(-8)	8(-7)
01001	10(-7)	9(-7)	9(-6)
01010	11(-6)	10(-6)	10(-5)
01011	12(-5)	11(-5)	11(-4)
01100	13(-4)	12(-4)	12(-3)
01101	14(-3)	13(-3)	13(-2)
01110	15(-2)	14(-2)	14(-1)
01111	16(-1)	15(-1)	15(-0)
10000	17(0)	*****	*****

Table.7.1. Comparison of Diminished one, 2's complement and 1's complement codes (adapted from [Jenk87] 1987 ©IEEE)

(d) Multiplication:

We next consider the multiplication operation, which uses this property. The multiplication shall be first twos complement multiplication followed by a diminished 1 conversion. The next step is to apply the shift property for the MSBs and add to LSBs with carry complementing. Two examples will illustrate this procedure.

(i) 5.6 mod 17

5⇒ 00101 (TC)
6 ⇒ 00110 (TC)
30 00011110 (TC) 000 11101 (diminished one)

$\Rightarrow$ { 1101 + [(0001).2^4]mod 17}mod 17
= {1101 +1111} mod 17 = 11100
0
01100 (-4) mod 17
= 13

(ii) 5 .(-7) mod 17
5 0101 (TC)
-7 1001 (TC)
-35 1101 1101 TC 11011100 (diminished one)
$\Rightarrow$ { 1100 + [(1101).2^4]mod 17 } mod 17
$\Rightarrow$ {1100 +0010} = 01111 (Diminished one)
1 MSB is inverted.
10000 $\Rightarrow$ 16 normal.

The hardware implementation of multiplication is quite involved in diminished arithmetic as can be seen.

The conversion from RNS to QRNS next needs to be considered. It can be seen that j in this case is 2^r corresponding to a modulus of the form $2^{2r}+1$. As an illustration, for r=2 i.e. modulus 17, j = 4 and j^{-1} = -4. The hardware architecture is as shown in Fig 7.3 (a). Similarly, the conversion from QRNS to RNS can be performed by the architecture shown in Fig 7.3 (b). A diminished one multiplier is shown in Fig 7.3 (c) (wherein, first a two's-complement multiplier is used, as demonstrated in the examples above). The upper n bits need to be shifted left and added with inverted carry in few steps to yield the result as explained before. The reader is referred to [Jenk87] for details on the Quantization effects of diminished one arithmetic based operations.

7.5. TAYLOR'S SINGLE MODULUS ALU FOR QRNS

Taylor [Tay85b] has advanced a single modulus ALU which is useful for QRNS applications. This uses a modulus of the form 2^n+1 used in QRNS. He has described the architecture of ALU which can perform addition, multiplication, scaling, negation and hypothesis testing. This uses the ideas described in previous chapters. We will consider only certain new and interesting ideas herein.
Consider an adder modulo 17. Taylor has suggested a first cycle of normal addition and a second cycle in which an offset of 31 is added. As an illustration, let A=B=10. Then

First addition: $S=A+B== 20 >17$.
Second addition: $(S + \text{offset}) \bmod 2^4= 20+31=3$. The overflow is ignored. A multiplexer is used to select the correct result using a comparison block to find whether $S \geq 2^n+1$.

A negator circuit is quite interesting. Here the given number A needs to be complemented mod (2^n+1). Taylor suggests this to be realized as $A' + 1 + 2^n +1$ where $A'+1$ is the two's complement of A and $2^n +1$ is added to reduce the result modulo (2^n+1). The overflow needs to be ignored. The technique for A represented as $a_{n-1}\ a_{n-2} \ldots a_1\ a_0$ is to map the bits according to the following rule:
a_o-a_o', a_1-a_1' exor 1, a_2-a_2' exor a_1', a_3- a_3' exor $a_1'a_2'$ and so on.
Note that the MSB however is exclusive ORed with 1. A hardware schematic for negating 5 mod 17 is presented in Fig 7.4 (a).

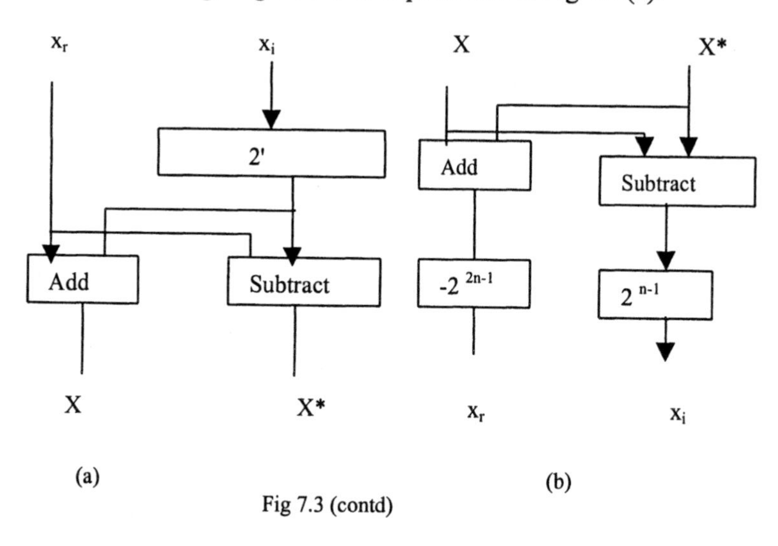

Fig 7.3 (contd)

We next consider a hypothesis testing circuit, which is defined as follows:
$H_0: S=0 \quad H_1: S\geq 0 \quad H_2: S=2^{n-1} \quad H_3: S<0 \quad H_4: S=2^n$.
The logic needed is as shown in Fig 7.4 (b).
The scaling needed in complex ALU by 2^{-1} and $(2j)^{-1}$ mod p can be easily implemented due to the special properties of the modulus 2^n+1. Defining $X = X_{HI}.2^{n/2} + X_{LO}$, the following are the results:
Scaling by $j_1 = 2^{n/2}$ mod p: $j_1X = (2^{n/2}.X_{LO} - X_{HI}) \bmod p$
Scaling by $2^{-1} = (2^{n-1}+1)$ mod p: $2^{-1}.X = ((2^{n-1} +1).X_{LO} + X_{HI}) \bmod p$

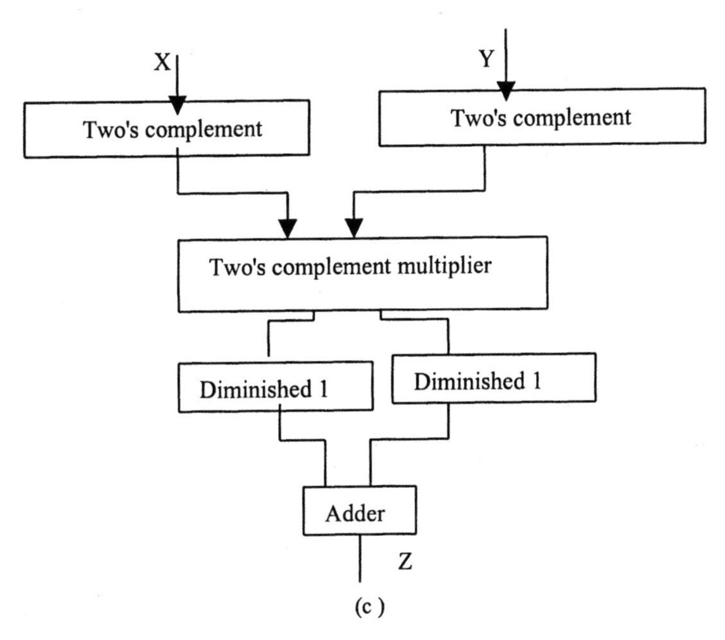

Fig 7.3 (a) A RNS to QRNS converter and (b) A QRNS to RNS converter and (c) a diminished one multiplier (adapted from [Jenk87] 1987 ©IEEE)

Scaling by $(2j_1)^{-1} = (2^{(3n/2)-1} + 2^{n/2})$ mod p: $(2j_1)^{-1} X = (2^{n/2-1}.X_{LO} - X_{HI})$ mod p.
The reader is referred to Taylor [Tay85b] for additional information on enhancements needed to meet special signal processing needs.

7.6. CONCLUSION

The use of RNS for Complex signal processing has been considered in some detail in this Chapter. The complete decoupling between real and imaginary processing while using certain types of moduli simplifies the operations. The extension to other moduli even though has been considered, it is believed to be quite complex for implementation. Extensions to operations in polynomial rings have also been described, which can be found in [Tay85b].

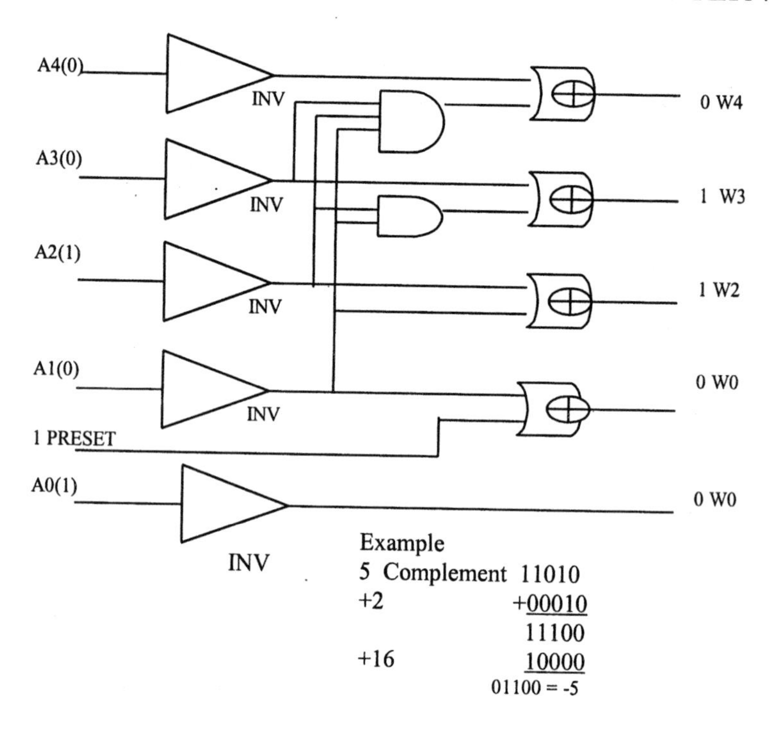

(a)

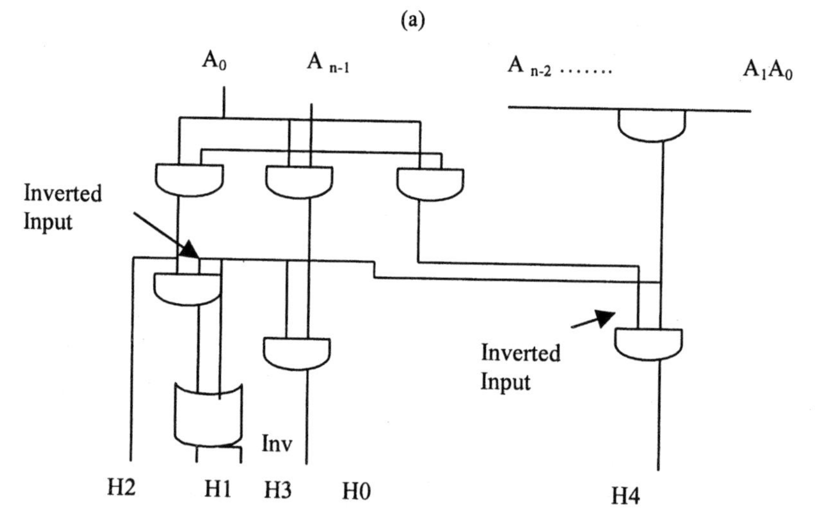

(b)

Fig 7.4 (a) A negator circuit and (b) A hypothesis testing circuit (adapted from [Tay85b] 1985 ©IEEE)

8

APPLICATIONS OF RESIDUE NUMBER SYSTEMS

8.1. INTRODUCTION

In this Chapter, some applications of Residue Number System described in literature are reviewed so as to illustrate the various possibilities. Specifically, Digital to Analog Converters, FIR filters, IIR filters, Adaptive filters, 2-D FIR filters and Digital Frequency Synthesis are considered.

8.2. DIGITAL ANALOG CONVERTERS

Perhaps the earliest work on RNS dealt with this subject. The output of RNS based Digital Signal processor is available as a residue set. A straight-forward method to obtain an analog signal is to use the techniques described in Chapters 2 and 3 to obtain a binary word first. Then, this word is converted into an analog form using a conventional linear D/A converter. As against this approach, Jenkins [Jenk78a] suggested the direct approach presented in Fig 8.1. Note that in this method first, MRC digits $a_0, a_1, a_2, ..$ a_{n-1} are obtained using the techniques described in Chapter 2. Next, these digital words are converted into analog form using n D/A converters, however, of coarser resolution. These are weighted using appropriate resistors to realize the final digital word given by

$$B=a_{n-1}\, m_n .\, m_{n-1}......m_2.m_1 + ...+ a_2.m_1+a_1 \tag{8.1}$$

The sign bit of a_{n-1} can be used to add a fixed dc voltage corresponding to -M/2.

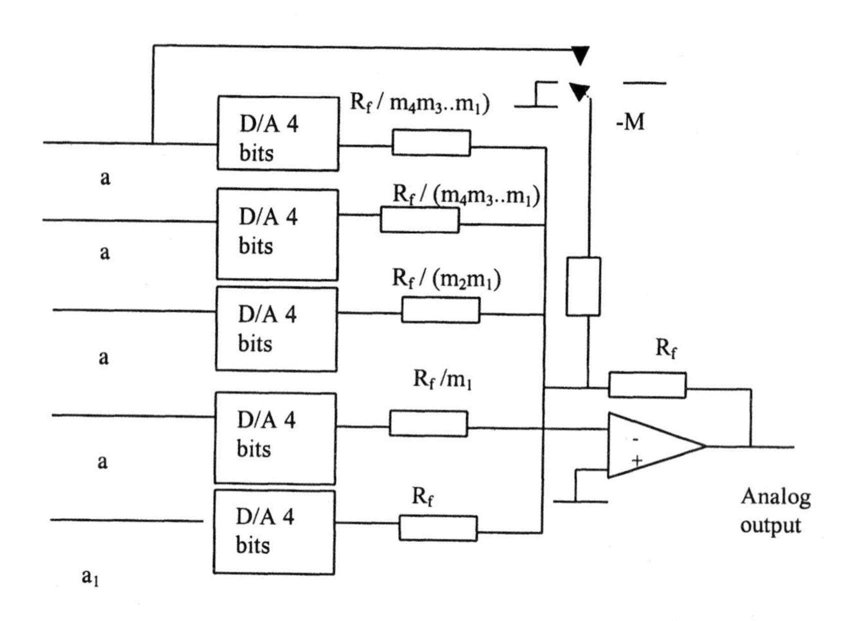

Fig 8.1. Direct RNS Digital to Analog Conversion due to Jenkins. (adapted from [Jenk78a] 1978 ©IEEE)

8.3. FIR FILTERS

8.3.1. Jenkins and Leon FIR filters

The FIR filter implementations due to Jenkins and Leon [Jenk77] use ROM based multipliers. A typical example of their implementation is very instructive. Consider the first-order filter described by the equation

$$y(n) = a_o. u(n) + a_1.u(n-1) \tag{8.2}$$

where $a_o = 127$, and $a_1 = -61$ with an input $u(n) = 30$ and $u(n-1) = 97$. The output shall be -2107. They chose a moduli set {19, 23, 29, 31} with a dynamic range of 392,863 of about 18 bits word length. The coefficients in this RNS can be seen to be $a_o = \{13, 12, 11, 3\}$ and $a_1 = \{15, 8, 26, 1\}$.

The multiplicative inverses of M_i where $M_i = M/m_i$ shall be computed as follows:

$M_1 = M/m_1 = 20677$; $(1//M_1) \bmod m_1 = 4$
$M_2 = M/m_2 = 17081$; $(1//M_2) \bmod m_2 = 20$
$M_3 = M/m_3 = 13547$; $(1//M_3) \bmod m_3 = 22$
$M_4 = M/m_4 = 12673$; $(1//M_4) \bmod m_4 = 5$

Hence, the modified filter coefficients obtained by multiplying the original filter coefficients by $(1//M_i)$ mod m_i are

a_o'= {14, 10, 10, 15}

a_1' = {3, 22, 21, 5}

Evidently, the u(n) and u(n-1) need to be multiplied by these to get the terms a_o.u(n) and a_1.u(n-1) which when added yield the desired output. Noting that u(n) = 30 = {11, 7, 1, 30}, u(n-1) = 97= {2, 5, 10, 4}

we obtain

y'(n)={8, 19, 17, 5} (8.3)

Evidently, this has to be weighted by M_i and summed to yield the result using CRT.

The bit slice approach needs a table which stores the $\Sigma y_{ij}. 2^{j-1}.M_i$ corresponding to all possibilities of j th bits of y_1', y_2', y_3', y_4' where j =0 to 4 since y'_i are 5 bit words. Thus, since y'(n) is represented as five bit words, five table look-ups followed by summation are needed to obtain the result. Specifically, the y_1'(n), y_2'(n), y_3'(n), y_4'(n) in binary form from (8.3) are

y_1'(n) = 01000

y_2'(n) = 10011

y_3'(n) = 10001

y_4'(n) = 00101.

Thus, first the MSB (i.e. j=4) word 0110 addresses the PROM to obtain 30628 which will be doubled to yield 61256. Next, the word corresponding to 0001 is read as 20677 which is added to 61256. This procedure continues to obtain the final result.

Jenkins and Leon [Jenk77], Jenkins [Jenk78b] and Soderstrand [Sode77] have described a combinatorial digital filter architecture due to Peled and Liu [Pele74] which does not use multipliers but instead uses look-up tables stored in ROMs. The Peled and Liu structures are effective for both recursive and non-recursive filters whenever the coefficients are fixed. The speed/cost ratio of these are higher than those using conventional MACs. Since the word lengths in RNS in multiplier based architectures are small, this approach needs fewer cycles to compute the RNS output of convolution summation needed for FIR/IIR filtering. The designs using RNS as well as Peled-Liu architecture have been termed as hybrid residue-combinatorial architectures [Jenk78b].

An architecture due to Jenkins and Leon [Jenk77] is presented in Fig 8.2 (a) where input coding is done with 256x4 PROMs for the moduli set {7, 9, 11, 13, 16} since the input is assumed to be a 8 bit word. The inputs are held in a circulating shift register storing the new and the previous 63

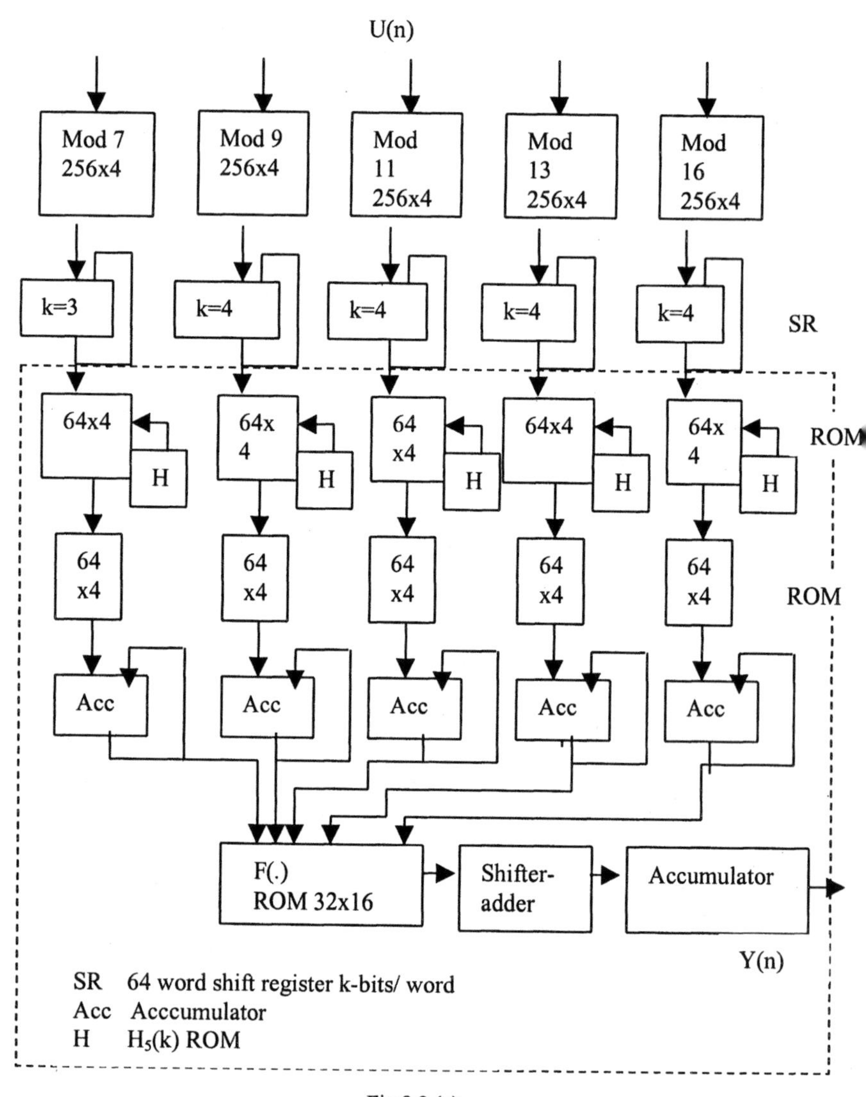

Fig 8.2 (a)

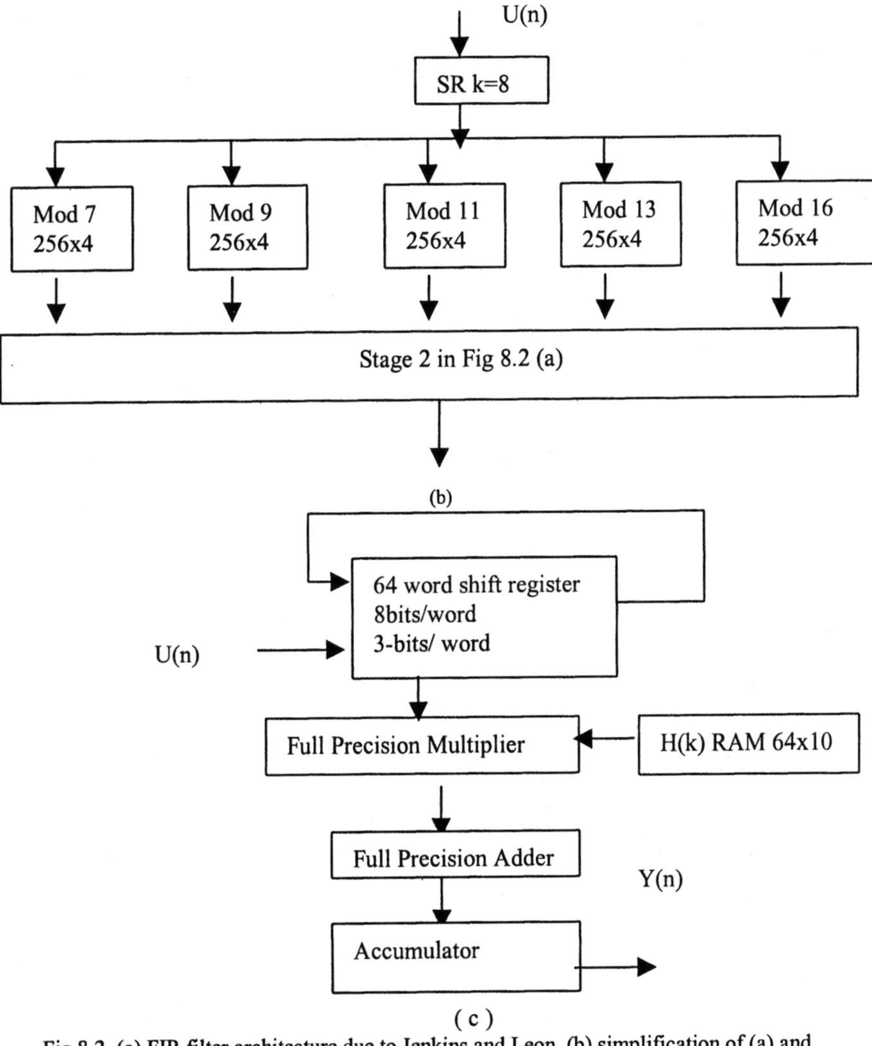

Fig 8.2. (a) FIR filter architecture due to Jenkins and Leon, (b) simplification of (a) and (c) a conventional filter architecture (adapted from [Jenk77] 1977 ©IEEE)

samples to facilitate 64 tap FIR filtering. The filter coefficients are in RAM in residue form so that adaptive filters can be implemented which need updated tap weights. They can as well be in ROM for fixed filter applications. The result of multiplication with the coefficient is available from ROM in the second level and the third level ROM performs the modulo adder function for performing accumulation to yield the convolution sum. The result of all the paths yields $y_i'(n)$ which is fed to the Peled-Liu bit slice implementation (see Fig.8.2 (a)) in f (.) ROM followed by shifter-adder. Note that the shifter by two (i.e. multiplication) is needed, since the memory has only information corresponding to all LSBs of y_i' weighted by M_i. A simplification is possible as in Fig 8.2 (b), where the number of circulating shift-registers have been reduced by relocating the position of input PROMs. As an illustration, a conventional FIR filter implementation using one multiplier and one adder is shown in Fig 8.2 (c).

The important feature of these architecture is that the throughput of the FIR filter is independent of the both the coefficient and data word-lengths. Note that, if desired, the contents of the ROM can be scaled outputs as well, in stead of actual outputs.

8.3.2. Wang's FIR filter architectures

Wang [Wan94] has described a bit-serial VLSI implementation of RNS FIR filters. In his architecture shown in Fig 8.3 (b), for a N tap filter, there are N cells, each cell performing an operation $a_{ij}.x_{n-i,j}$ +S where S is the sum accumulated thus far in the cell above, and a_{ij} are coefficients of the FIR filter and $x_{n-i,j}$ are the delayed input samples and j stands for the j th modulus. Note that the cell accumulates this value and performs the modulo m_i operation. This necessity can be seen from the expression for the desired output sequence

$$y_{nj} = [\Sigma_{i=0}^{N-1} a_{ij}.x_{n-i,j}] \bmod m_j \tag{8.4a}$$

for n=0, 1, 2,... and j=1, 2, ...L. Denoting $x_{n-i,j}$ and a_{ij} as the B bit words

$$x_{n-i} = \Sigma_{b=0}^{B-1} x_{n-i,j}^{b}.2^b \tag{8.4b}$$

$$a_{ij} = \Sigma_{b=0}^{B-1} a_{i,j}^{b}.2^b \tag{8.4c}$$

y_{nj} becomes

$$y_{nj}^{b} = [\Sigma_{b=0}^{B-1} S_{nj}^{b}.2^b] \bmod m_j \tag{8.5a}$$

with

$$S_{nj}^{b} = [\Sigma_{i=0}^{N-1} a_{ij}.x_{n-i,j}^{b}] \bmod m_j \tag{8.5b}$$

In the architecture of Fig 8.3 (b), the bits of x enter serially and are delayed by B+1 clock cycles using delay blocks D_1. The extra clock cycle is required to clear the accumulator addressing the ROM so that fresh evaluation of y_n can commence. Evidently, old samples are multiplied by a_i' and added to the latest sample weighted by a_{i-1}'. Each cell is having an internal architecture shown in Fig 8.3 (a). This comprises of latches at two levels one of which can be eliminated, however. The blocks labeled X are two-input AND gates.

Note that the cells contain $a_i' = a_i \bmod m_j$ rather than a_i. The algorithm used for modulo reduction is interesting and is considered next.

Here, we consider the addition of α and gated β mod m_j. Gated β means that β is allowed or nulled by control bit x_i^b. Wang recommends the evaluation as

$$(\alpha_j+x.\beta_j) \bmod m_j = \langle \alpha_j + x.(\beta_j+r_j)+ x.c.m_j \rangle \quad (8.6)$$

where c is the complement of the carry generated by adding the first two terms, $r_j = 2^B - m_j$, and < > indicates the B LSBs of the adder result.

An example will be in order. For x=1, α=11, β=10 and m_j =13. Then, α_j +x.(β_j+r_j) =11 +(10+3) =24. Evidently, c=0 since carry is generated. Thus, LSBs of the result 24 mean 8 which is the desired answer. As another example, α=11, β=0 for x=1 and m_j=13, we have α_j +x (β_j+r_j) =11+3 =14. Since, c=1, we need to add 13 to yield 27, LSBs of which are 11. Two levels of addition are needed since the knowledge of c is required. Note that the architecture of Fig 8.3 (b) can be pipelined. Note that the accumulation function is also realized in the ROM. The ROM size is $B.2^{2B}$ bits. An alternative architecture is presented in Fig 8.3(c) where the input enters at the top. The circuit, however, has a larger initial delay of (2N+B) clock cycles. Also, the arrangement has bi-directional data flow unlike that in Fig 8.3 (b).

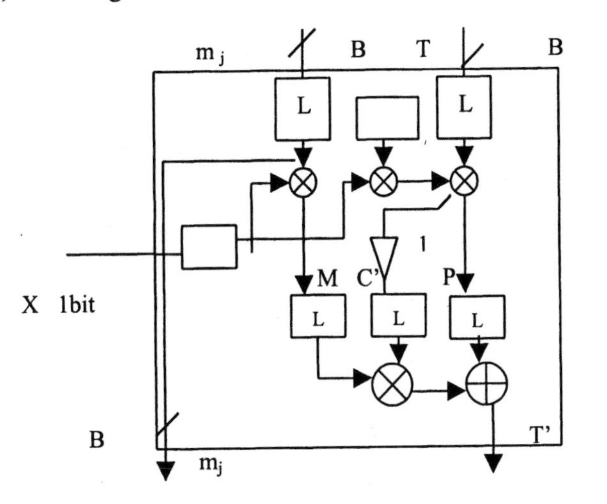

(a)
Fig 8.3 (contd)

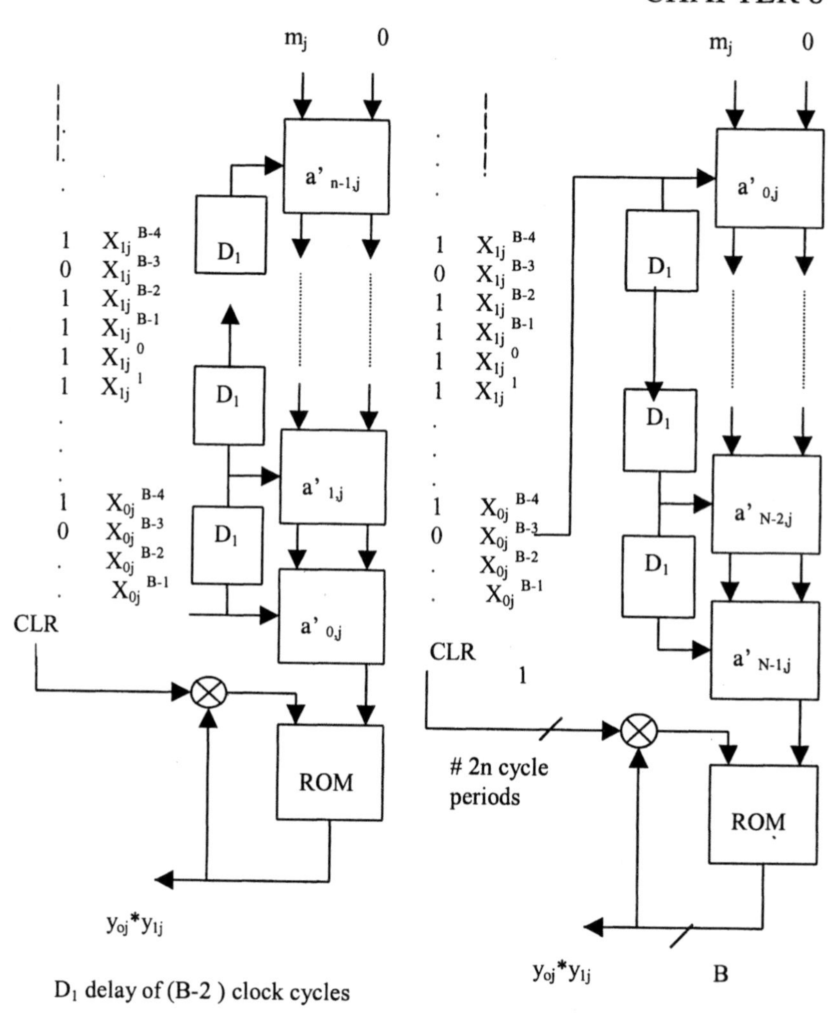

Fig 8.3. Wang's hybrid VLSI FIR filter architectures (b) and (c) and schematic of processor cell in (a) (adapted from [Wan94] 1994 ©IEEE)

8.3.3. RNS based adaptive filters

Miller and Polky [Mill84] described a RAAF (Residue Arithmetic Adaptive Filter) processor. Filtering at real-time throughputs greater than 2MHz has been achieved. They reformulate the LMS algorithm which is faster and flexible than the well-known algorithm. The well-known LMS adaptive filter structure is presented in Fig 8.4 (a), where variable filter weights are used. These filter weights are determined by an error feedback signal generated from the output of the previous iteration. The standard equations of a t-tap LMS adaptive FIR filter have the form

$$y_j = \Sigma_{i=1}^{t} w_{ij}.x_{ij} \tag{8.7a}$$

$$e_j = d_j - y_j \tag{8.7b}$$

$$w_{i(j+1)} = w_{ij} + 2\mu.e_j.x_{ij} \qquad i=1,2,\ldots.t \tag{8.7c}$$

where w_{ij} are the adaptive weights, d_j and x_j are the respective samples of the primary and reference input taken during the j th iteration, e_j is the error feedback signal used as the noise cancelled output, y_i is the filter output used and μ is a scale factor used to control the rate of convergence. Note also that the input samples are shifted by one sample at each iteration. Miller et al observe that the values of w_{ij} shall be scaled after each iteration. This needs a reformulation of the LMS algorithm to allow scaling to proceed in parallel with the main filter realization. The conventional architecture needs four consecutive operations. However, (8.7a) and (8.7c) can be combined as the new equation

$$y_j=\{2\mu\Sigma_{i=1}^{t}x_{i+j-t}(\Sigma_{k=1}^{j} e_k.x_{i+k-t})\}+e_{j-1}.\{2\mu\Sigma_{i=1}^{t} x_{i+j-1}x_{i+j-t-1}\} \tag{8.8}$$

Hence, the computation of y_j needs only the value of e_{j-1}. Hence, as soon as e_{j-1} is available, the y_i can be evaluated. Next, (8.7b) is computed. Hence, eight cycles are sufficient to evaluate y_j. The scaling technique used is already presented in Section 5.2.6.

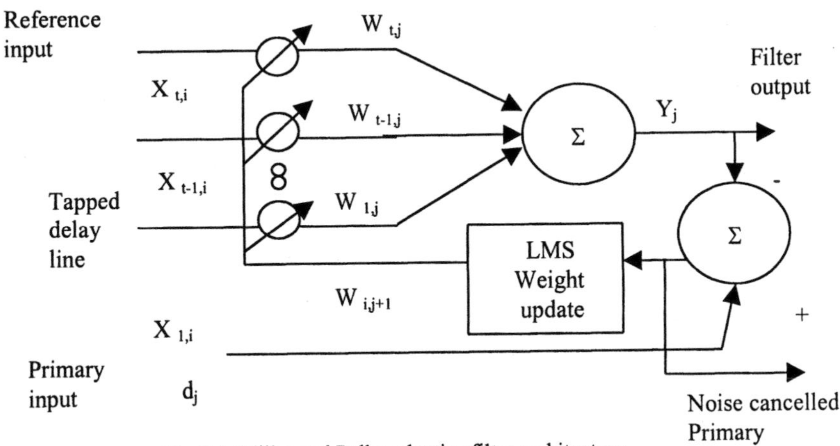

Fig 8.4. Miller and Polky adaptive filter architecture
(adapted from [Mill84] 1984 ©IEEE)

8.3.4. 2-D FIR filters

Huang et al [Hua81] have described a two-dimensional (2-D) digital filter implementation using RNS. This operates on a five- by- five data set of input data using a five-by-five matrix of weighting coefficients. The structure of the filter coefficients is as shown in Fig 8.5 (a) which exhibits certain symmetry. This enables the simplification of the coefficient matrix as shown in Fig 8.5 (b), however, needing addition / subtraction of input samples. The authors interestingly have used the then available microprocessor Intel 8086 using a five moduli set {16, 15, 13, 11, 7}. Mathematically, the operation implemented by the filter is

$$G_{IJ} = \Sigma_{K=1}^{5} \Sigma_{L=1}^{5} H_{KL} \; M_{I-K,J-L} \tag{8.9}$$

The reader is referred to their paper for more details.

F	E	D	E	F
E	C	B	C	E
D	B	A	B	D
E	C	B	C	E
F	E	D	E	F

(a)

F	E	D	E	F
E	C	B	C	E
D	B	A	B	D

(b)

Fig 8.5. Huang's 2-D filter coefficient matrix (a) and its simplification (b) (adapted from [Hua81] 1981 ©IEEE)

Shanbag and Siferd [Shan91] have described a single-chip 2D-FIR filter using residue arithmetic. This realizes a 3x3 2-D filter with linear phase characteristics. The filter coefficients are symmetric as shown in Fig 8.6 (a) and the data window is as shown in Fig 8.6 (b). The operation to be implemented is

$$y(i,j) = A[x(i-1,j) + x(i+1,j)] + B[x(i,j-1) + x(i,j+1] + C[x(i-1,j-1)+x(i+1,j+1)]+D[x(i+1,j-1)+x(i-1,j+1)] + x(i,j) \tag{8.10}$$

Shanbag and Siferd designed the filter using RNS by using modulo multipliers and modulo adders in a six moduli set {13, 11, 9, 7, 5, 4} with a dynamic range of 17.37 bits (=180,180).

The modulo m_i adder and subtractor are shown in Fig 8.7 (a) and (b). In the adder of Fig 8.7 (a), a n-bit ripple carry binary adder (RCBA) is used in first level whose carry out C_f is used to select either the output of RCBA or output of the subtractor SUB which subtracts m_i from the sum output of the

C	A	D
B	I	B
D	A	C

(a)

x(i-1,j-1)	x(i-1,j)	x(i-1,j+1)
x(i,j-1)	x(i,j)	x(i,j+1)
x(i+1,j-1)	x(i+1,j)	x(i+1,j+1)

(b)

Fig 8.6. 2-D Filter coefficients (a) and data window (b)
(adapted from [Shan91] 1991 ©IEEE)

RCBA block. The borrow B_f of the subtractor is used to select the correct output using a 2:1 Mux. As an illustration, consider two examples mod 13. (a) X=6,Y=5. Then $C_f = 0$. Sum of the first adder is 11 which when added to the two's complement of m_i (=13) yields 14 and B_f=0. In this case 11 is selected. As another example, let X=8 and Y=9. Then, the sum is 1 and C_f=1, B_f=0. The result of the subtractor is 4 i.e. after adding two's complement of 13. Thus, the Mux selects the subtractor output yielding 4 as the output.

The modulo m_i subtractor shown in Fig 8.7 (b) uses a first RBBS (Ripple Borrow Binary Subtractor) with borrow B_f to whose output m_i is added in adder ADD block. The valid result is selected by the mux using the B_f output of the first subtractor.

The authors suggest the use of Jenkins and Leon technique [Jenk77] of binary to residue conversion using PLAs. The use of PLAs over ROMs leads to area efficiency. The architecture of a 8-bit binary to RNS converter is shown in Fig 8.8 (a). Writing the given binary word as

$$X = 2^4 . B_M + B_L \tag{8.11a}$$

we have

$$X \bmod m_i = (2^4 . B_M + B_L) \bmod m_i = 2^4 \bmod m_i . B_M \bmod m_i + B_L \bmod m_i \tag{8.11b}$$

The first and second terms are realized using PLAs with only 16 product terms.

The residue to binary converter uses the architecture of Fig 8.8 (b), wherein two residues are combined parallelly to yield via MRC the intermediate result. These are once again combined in two stages to give the final result.

The complete filter architecture is presented in Fig 8.9 (a) and (b), wherein the various stages are as follows: (a) input latches L, (b) Binary to Residue converter BTOR, (c) latches L, (d) FIR RNS filter M9-M13, (e) latches L, (f) RNS to binary converter first stage, (g) latch L, (h) RNS to Binary converter second stage and (i) latches L. The chip is pipelined as evidenced by the positioning of the latches in Fig 8.9 (a). The block diagram of the filter block is shown in Fig 8.9 (b). The correspondence to (8.10) can be clearly seen. The reader is referred to [Shan91] for more details.

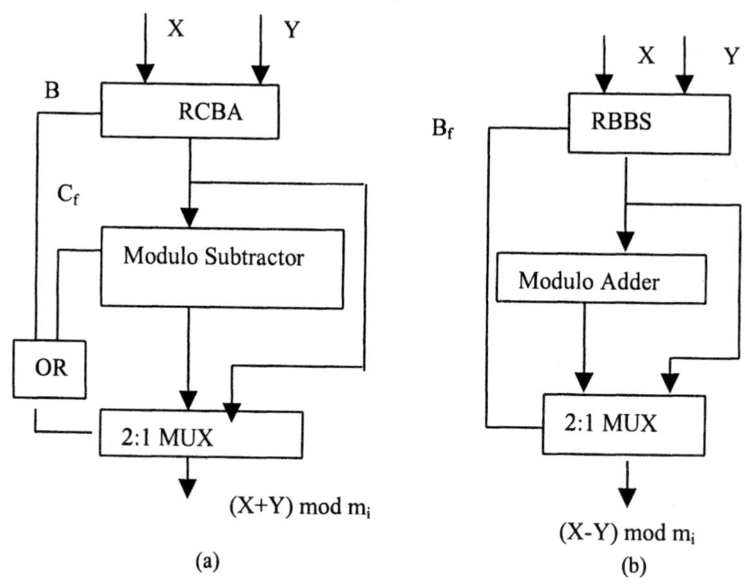

Fig 8.7. Modulo adders (a) and subtractor (b) due to Shanbag and Sieferd (adapted from [Shan91] 1991 ©IEEE)

8.4. RECURSIVE RNS FILTER IMPLEMENTATION

Soderstrand and Sinha [Sode84a] suggested the realization of IIR filters using the RNS, where pipelining was considered so as to increase the throughput of the IIR filter. It was noted that IIR implementations are not easy to be pipelined, since, input needs to be added to a delayed output. A

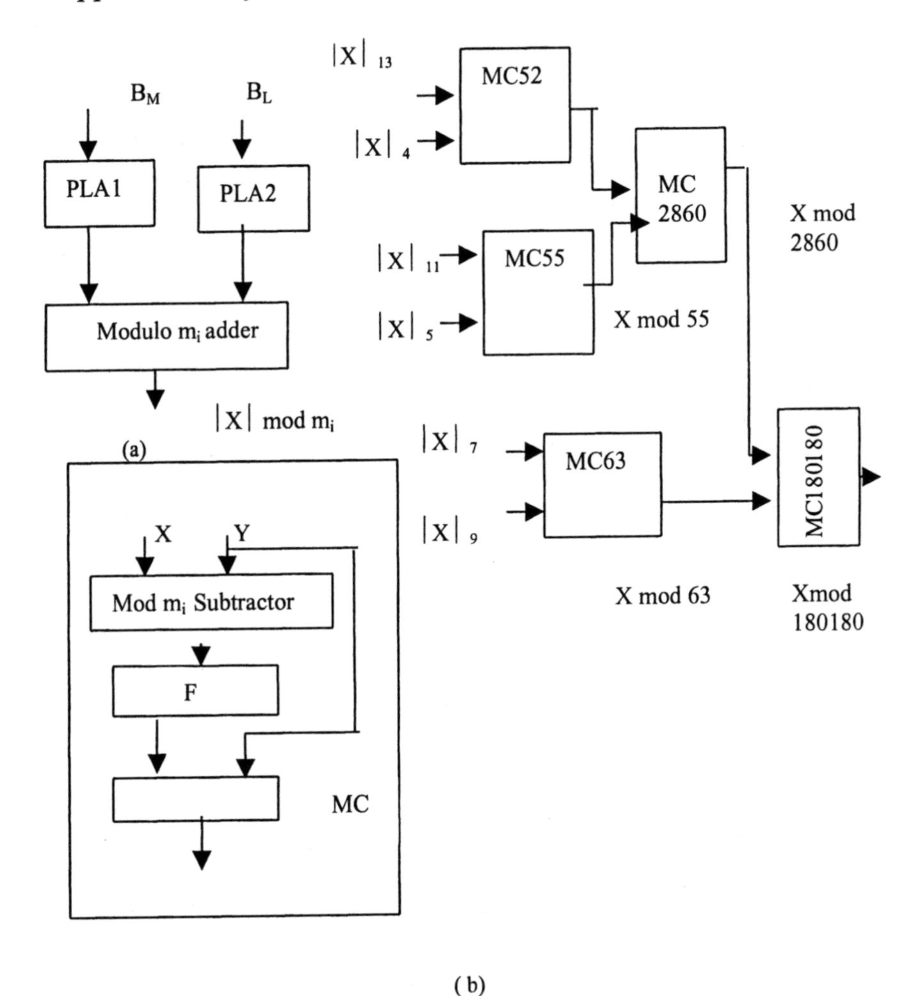

(b)

Fig 8.8. Binary to RNS converter (a) and RNS to Binary converter (b) used in Shanbag and Sieferd's approach (adapted from [Shan91] 1991 ©IEEE)

solution to this problem has been suggested by Sinha and Loomis [Sode84a], which is briefly considered next.

Consider the second order recursion equation of an IIR filter

$$Y(z)=X(z)+b_1z^{-1}.Y(z)+b_2z^{-2}.Y(z) \quad (8.12a)$$

which can be rewritten as

$$Y(z).z^{-1}=X(z).z^{-1}+b_1.z^{-2}.Y(z)+b_2z^{-3}Y(z) \quad (8.12b)$$

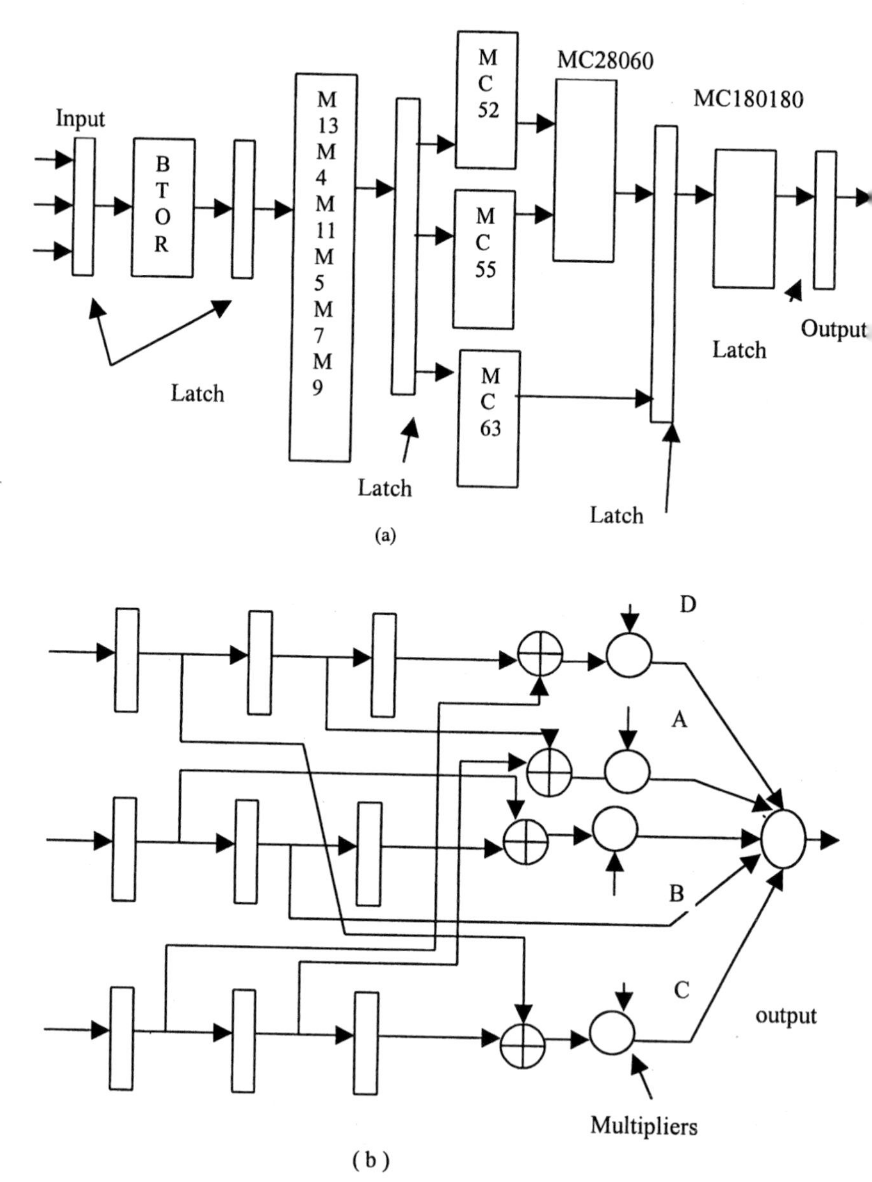

Fig 8.9. Complete filter architecture due to Shanbag and Sieferd (a) and stage 2 (b) (adapted from [Shan91] 1991 ©IEEE)

which, when substituted for the middle term in (8.12a) yields

$$Y(z)=(1+b_1z^{-1}).X(z)+z^{-2}((b_1^{\,2}+b_2)+b_1b_2.z^{-1}).Y(z) \tag{8.13}$$

Equation (8.13) can be expressed in transfer function form as

$$Y(z)/X(z) = (z+b_1) /\{ (z^2-b_1z-b_2).(z+b_1)\} \tag{8.14}$$

Note that the extra pole and zero introduced get cancelled yielding the desired second order transfer function. It is important to ensure the stability of the second-order system. Proceeding in a similar manner, by always substituting for the lower order term in the coefficient of Y e.g. $z^{-2}.(b_1^2+b_2)$. Y(z) term in (8.13), we can obtain high delay pipelined realization. The penalty paid is the need for realization of numerator using a FIR filter which is also of RNS type. The fifth order delay system can be shown to have a transfer function given by

$$Y(z)/X(z) = \{ (z^3+b_1z^2+h_2z+h_3)/(z^5+d_1z+d_2)\} \tag{8.15}$$

where

$h_2 = b_1^2+b_2$,
$h_3 = b_1^3+2b_1b_2$
$d_1=b_1^4+3b_1^2b_2+b_2^2$
$d_2 = b_1^3b_2+2b_1b_2^2$.

The stability issue also needs to be addressed. A typical implementation is shown in Fig 8.10. The internal variables are expressed as follows:

$$V=z^{-1}.X+ z^{-3}.d_2.Y \tag{8.16a}$$
$$V=z^{-2}d_1Y \tag{8.16b}$$
$$z^2Y =V+W. \tag{8.16c}$$

The realized transfer function is

$$Y(z)/X(z) = z^2/(z^5-d_1z-d_2) \tag{8.17}$$

If a general second-order numerator is desired, the overall numerator of FIR filter will be of fifth order (six-tap). The reader is referred to [Sode84a] for more information.

8.5. DIGITAL FREQUENCY SYNTHESIS USING RNS

The use of RNS for digital frequency synthesis (DFS) has been considered by Chren Jr [Chr95, Chr98]. The classical digital frequency synthesizers facilitate generation of sinusoidal output by using the architecture of Fig 8.11 (a). Herein, the 'sine' function is stored in ROM and is sequentially read. The ROM L bit address is generated by a phase accumulator of L bit resolution which repeatedly adds a phase increment of k corresponding to the frequency to be generated. Since, sinewave is symmetric, only sine function for one quadrant(zero to 90 degrees) is stored. The L bit word, however, corresponds to a maximum of 360 degrees. The MSB of this word contains sign information and hence used to generate two's complement of

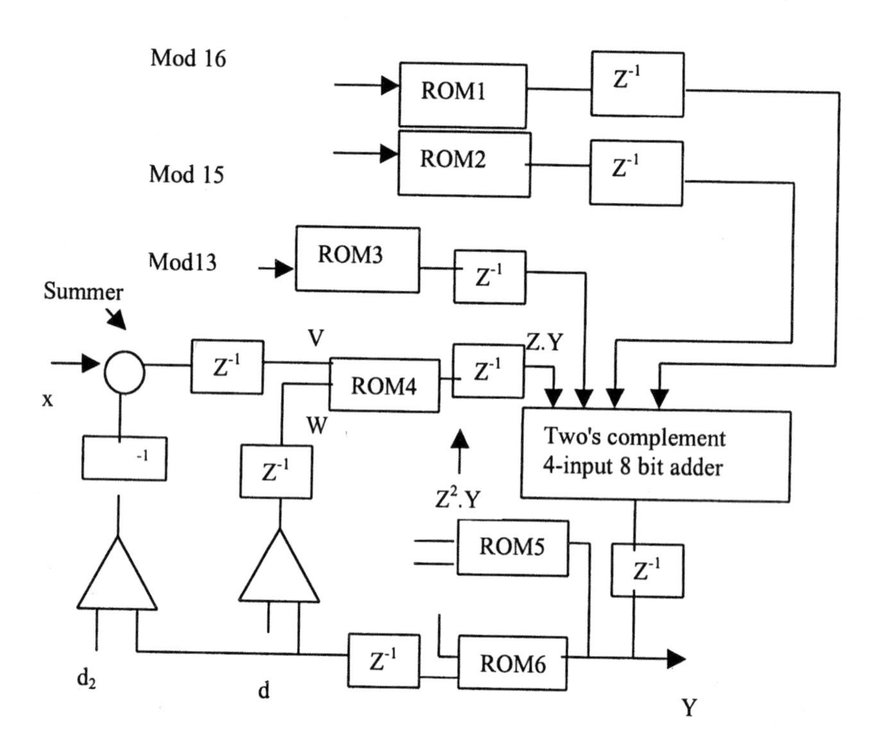

Fig 8.10. Pole forming hardware of the recursive RNS filter implementation due to Soderstrand and Sinha. (adapted from [Sode84a] 1984 ©IEEE)

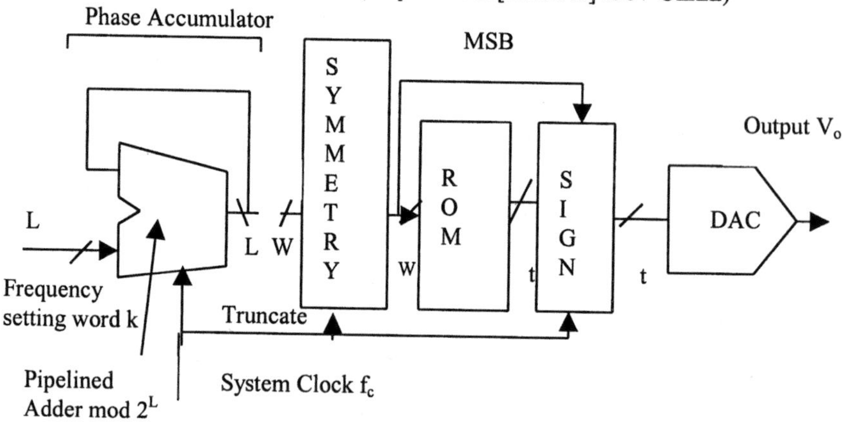

(a)
Fig 8.11 (contd)

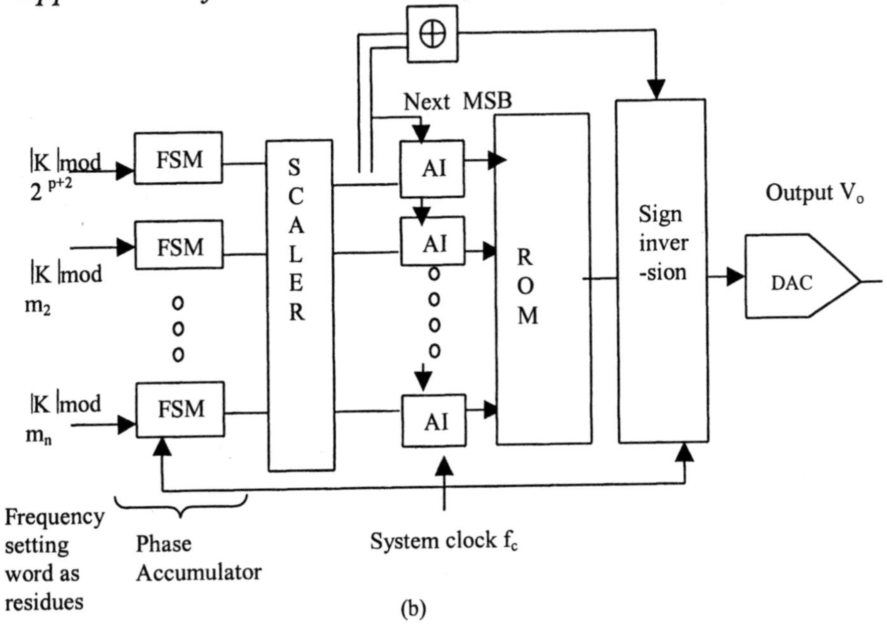

(b)

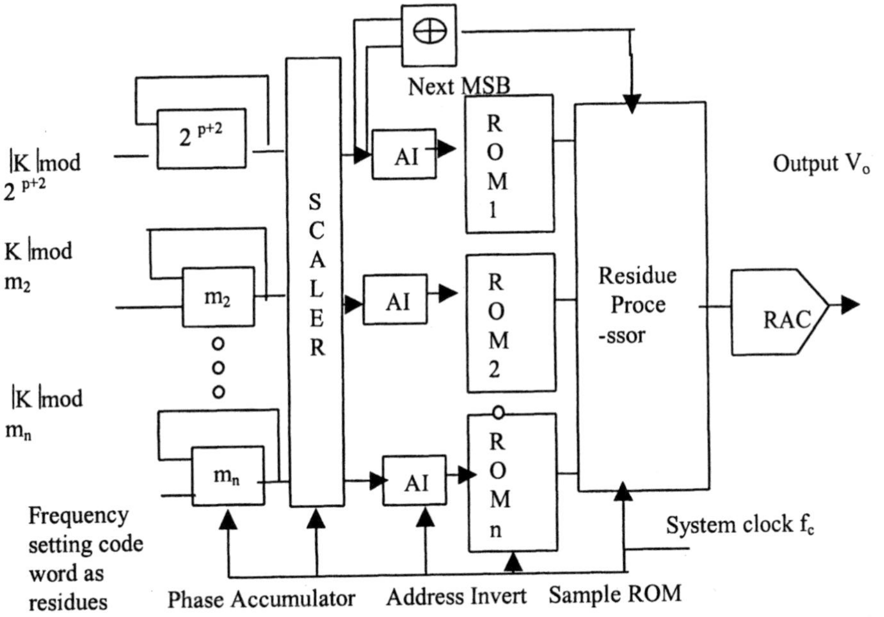

(c)

Fig 8.11 (a) Traditional direct digital frequency synthesizer architecture, (b) Frequency agile direct synthesizer and (c) Reduced area direct synthesizer (adapted from [Chr95] 1995 ©IEEE)

the contents read from the ROM so as to obtain negative value of the sample stored in ROM. In addition, the bit next to the MSB is used to selectively invert the lower order bits of the address by using exclusive-OR gates controlled by the second MSB. This can be appreciated by noting that for the case of values in the second quadrant, complementing the bits yields an address corresponding to the correct value. Assuming a clock frequency of f_c and frequency setting word k, and an L bit accumulator, the frequency of the sinusoid generated is $f_c.k/2^L$. The phase increment is $2.\pi.\ k/2^L$.

The figures of merit of a DFS design are frequency, phase and amplitude resolution. The frequency resolution is evidently $f_c/2^L$ corresponding to k=1. Alternatively, the width of the accumulator is the frequency resolution. Phase resolution is the width W of the ROM address. Note that while phase accumulation can be done with resolution as desired, the restriction on the memory size may need truncation of L bits to W bits. Amplitude resolution is evidently the number of bits used to represent the stored samples in the ROM.

The frequency agile direct Synthesizer due to Chren Jr uses a n moduli Residue number system as shown in Fig 8.11 (b), which is based on the architecture of Fig 8.11 (a) except that the phase accumulator is mod m_i accumulator corresponding to n moduli. The scaler performs the compression of L bits to W bits as in Fig 8.11 (a). The scaling is done by product of (n-r) moduli, since scaling by a product of moduli is easily possible in RNS than scaling by an arbitrary power of two number. One of the n moduli is recommended by Chren Jr to be chosen as 2^{p+2} with the expected advantage that the symmetry and sign inversion was thought to be taken care of easily. Thus, the ROM address for moduli 2^{p+2} is less by 2 bits saving ROM space for this modulus. Chren Jr recommends the use of a finite state machine (FSM) in place of modulo adder (using binary adders and 2:1 multiplexers as described earlier), since the delay can be of only two logic levels. The samples to be fed to the DAC can be obtained by using CRT on the residues. The AI block performs the additive inverse modulo m_i so that the function of symmetry can be achieved.

Chren Jr [Chr95, Chr98] suggests another architecture RADS (Reduced Area Direct Synthesizer) for reducing the area. Herein, the sample values are computed rather than stored. The block diagram is as shown in Fig 8.11 (c). Here, the phase accumulator used adders because they can be smaller.

We first note that in both the schemes of Fig 8.11 (b) and (c), the two MSBs and the remaining LSB word in the first residue (i.e.2^{p+2}) are denoted as a_{p+1}, a_p and Q mod 2^p respectively. Denoting the other residues as Q_i mod m_i, for i=1 to n-1, we need to determine for generating every output sample the accumulated phase value using Chinese Remainder Theorem:

$$\cos(2.\pi.\phi/(2^{p+2}.s)) = \cos(\pi.M_1^{-1}.a_{p+1} + \pi/2.\ M_1^{-1}.a_p + (2\pi M_1^{-1}/m_1.\ \phi) \bmod 2^p + (2.\pi.M_2^{-1}/m_2.\phi) \bmod m_2 + \ldots\ldots + (2.\pi.M_r^{-1}/m_r.\phi) \bmod m_r) \qquad (8.18)$$

where s is the product of all the moduli excluding 2^{p+2}. Next, since the first two terms can be used for exploiting the symmetry and sign inversion, we need to compute only the other terms. Hence, considering that cos $(2.\pi.(M_i^{-1}/m_i.\ \phi) \bmod m_i)$ and sin $(2.\pi.(M_i^{-1}/m_i.\ \phi) \bmod m_i)$ are stored in r locations each corresponding to each m_i, using the basic cosine and sine identities cos (a+b) = cos a.cos b – sin a.sin b and sin(a+b) = cos a . sin b + sin a. cos b , cos(a+b) and sin (a+b) are e computed by a processor P as shown in Fig 8.12. The individual terms in (8.18) in the argument of the cos function are denoted as Q_i. An architecture for computing (8.18) for a four moduli system i.e. for given Q_1, Q_2, Q_3 and Q_4 is shown in Fig 8.13.

The scaler is used to scale the output of the P processor by the product of the overflow moduli. The scaling is needed for fixed-point multiplication. This can be seen by the fact that $(x_1/P).(x_2/P) \Rightarrow x_1.x_2/P$ wherein a fraction is represented by its numerator with the denominator being a fixed product P. Chren suggests that the use of b moduli for base representation and o moduli for overflow representation. The overflow moduli are useful for preventing the numerical overflow during the multiplication operations in the residue processor. The reader is referred to their work for more details on the evaluation of both types of architectures.

An alternative implementation of the above design using a technique known as one-hot residue coding has also been discussed by Chren Jr [Chr98]. Note that this method uses n-1 wires which can be zero or one to represent a number less than n. Note also that only one wire will be active at a time. Thus, the method is applicable for small moduli. As an illustration, addition

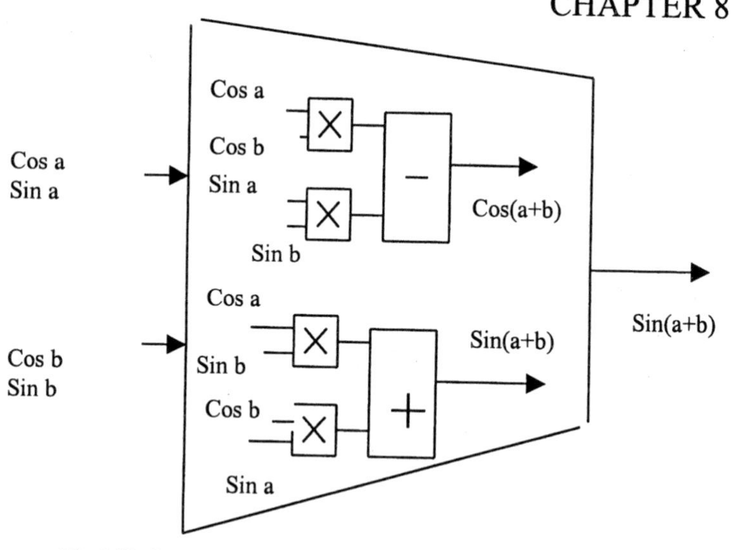

Fig 8.12. Sum Processor P used in Chen Jr's approach. (adapted from [Chr95] 1995 ©IEEE)

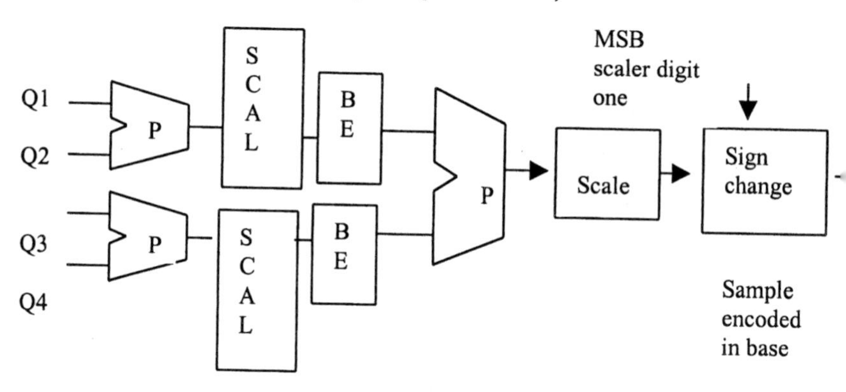

Fig 8.13. Residue Processor architecture for calculation of (8.18) for a four moduli RNS (adapted from [Chr95] 1995 ©IEEE).

in this system amounts to using a barrel shifter as shown in Fig 8.14 (a) and (b). The barrel shifter generates in parallel all possible rotations of the data and selects the appropriate one. The barrel shifters can be realized using pass transistors. Subtractors can be realized in a similar way except that the subtrahend input bus is permuted to generate the additive inverse of the operand. Finally, the multiplication can be realized using the architecture of Fig 8.14 (c) by using wire permutations, barrel shifting and again wire

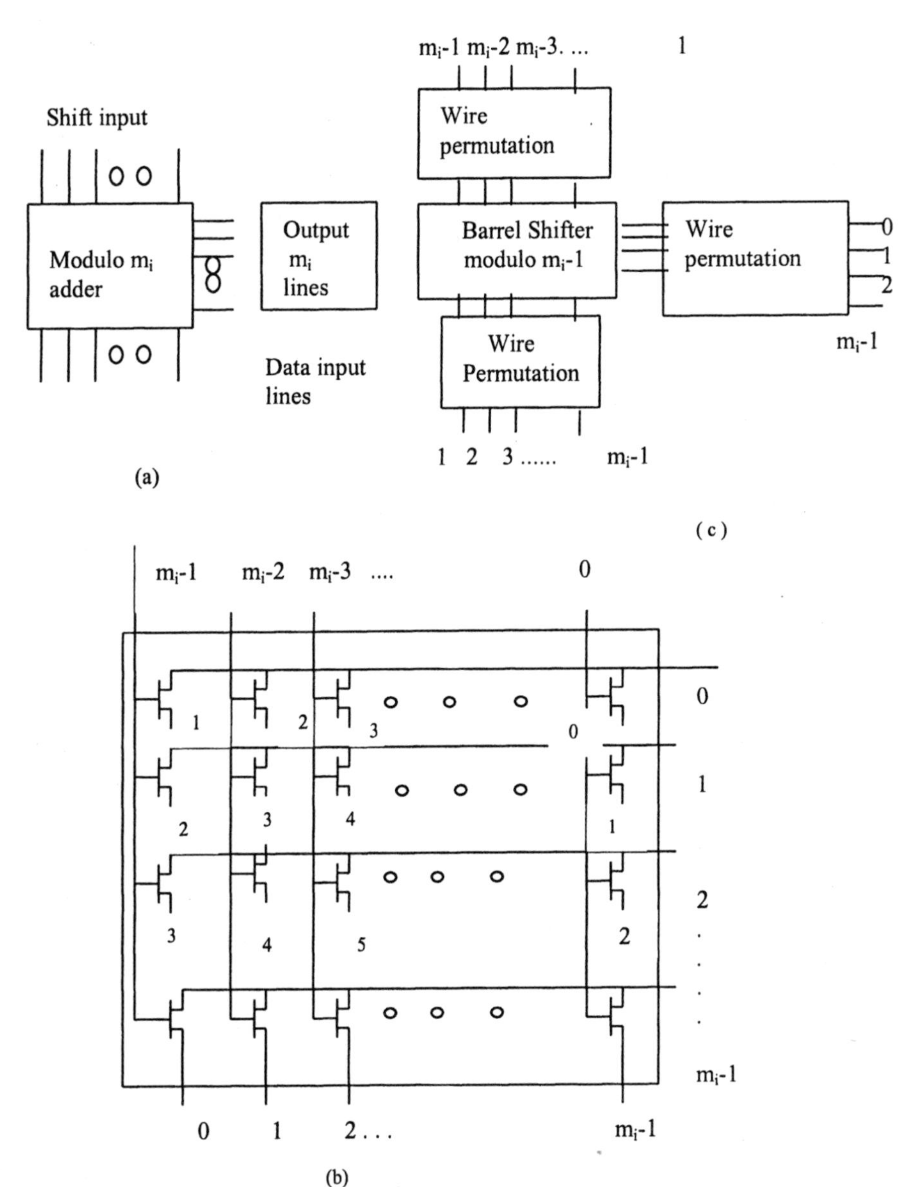

Fig 8.14. Addition, subtraction and multiplication using one-hot residue coding (adapted from [Chr98] 1998 ©IEEE)

permutation. The wire permutation corresponds to use of index calculus. The reader is referred to [Chr98] for more information. This method, however, may not be attractive in practice due to the extensive on-chip wiring needed.

8.6. MULTIPLE VALUED LOGIC BASED RNS DESIGNS

8.6.1. Soderstrand and Escott technique

Soderstrand and Escott [Sode86] have suggested VLSI implementation in multiple valued logic (MVL) of FIR digital Filters. Each signal in MVL can have several levels. Thus a four level MVL signal line can carry two bits of information, while a 8 level signal line can carry 3 bits of information. The attraction of MVL for RNS is that the dynamic range realizable can be increased by using fewer signal lines. Ideally, the levels in the MVL shall be same as the modulus. However, the need for mutually prime moduli restricts the use of moduli to 8, 7, 5, 3 for a four moduli system. Thus, the maximum dynamic range is roughly 840 i.e. <10 bits. If operations such as multiplication need to be done, the input signal shall be less than 5 bits.

Hence, it is preferable to have each modulus represented by few signal lines (more than one) each with at most four levels. This is because of the fact that technology may not be able to support more than four levels. Hence, each residue can be represented by two digits corresponding to each modulus e.g. 9, 25, 49, 64 may be used to yield a 8 level MVL realization. If a dynamic range of 8 bits is desired permitting multiplication to obtain 16 bit words, we can employ the moduli set {25, 49, 64}.

The typical architecture of a MVL based FIR filter is shown in Fig 8.15. The input stage is a binary to RNS converter. Since the RNS is of MVL type, the residues are expressed as two 3 bit numbers for modulus 49 and two 3 bit numbers for modulus 25 each expressed in base 7 and 5 respectively. The binary to RNS converter shown in Fig 8.16 (a) comprises of two stages a ROM and a 3 bit to one line MVL converter. Note that the negative numbers follow the definition of RNS e.g. –10 is represented by $(-10+49) = 39 = 54_{(7)}$. The RNS to binary converter can be based on modified CRT (see section 2.3.2). The architecture is shown in Fig 8.16 (b). The two digits of each residue are converted into normal binary form in two steps. The first step implements the equation

$X/M = X/(64.49.25) = \text{Frac}\,[\,b_{64} + b_{49} + b_{25}]$

where, using the definition of CRT, we have

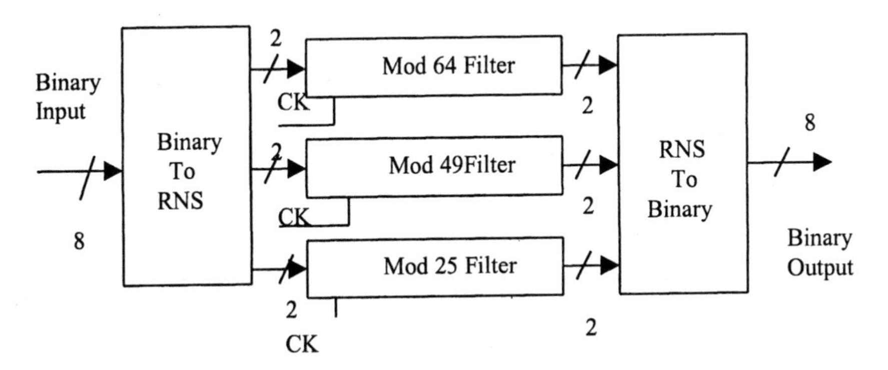

Fig 8.15. Architecture of a MVL based FIR filter (adapted from [Sode86] 1986 ©IEEE).

$b_{64} = (1/64).[(r_{64}/(49.25)) \bmod 64] = (1/64).\ [(57.r_{64}) \bmod 64]$
and similarly
$b_{49} = (1/49).[(23.r_{49}) \bmod 49]$, $b_{25} = (1/25).[(16.r_{25}) \bmod 25]$.
The result is thus larger in dynamic range and needs 3 bits for each modulus to form 3 MVL signal lines. These are converted into normal binary form next.

The filter is a pipelined structure shown in Fig 8.17 (a) which needs multiplication, addition and storage. A MVL implementation of a complete section for one modulus is shown in (b) and its actual implementation (c) is as shown in Fig 8.17 (c). A two digit multiplier is as shown in the dotted lines in Fig 8.17 (c). The reader is referred to [Sode86] for an exhaustive description of the implementation.

8.6.2. Kamayema, Sekibe and Higuchi technique.

Kamayema, Sekibe and Higuchi [Kama89] described multiple valued logic application to RNS. They have suggested a coding scheme which simplifies multiplication operations to shifting and radix 5 arithmetic operations. Herein, a modulus m_i is chosen which has a pseudo-primitive root. This root has a property that $p^0 \bmod m_i$. $p^1 \bmod m_i$ ...$p^{((mi-3)/2)} \bmod m_i$ include every value of 1, 2, 3, ..$(m_i-1)/2$. As an illustration, for $m_i = 7$, p=5, we have $5^0 \bmod 7 = 1$, $5^1 \bmod 7 = -2$ and $5^2 \bmod 7 = -3$. The choice p=5 is applicable to many moduli e.g 7, 11, 17, 19. 23, 37, 43, 53, 59, 73, 79, 83, 97 etc. The residue digit mod m_i can be written as an expression involving powers of p. As an illustration, for $m_i = 7$, a residue can be expressed as
$x = (5^2.x_{i2} + 5^1 x_{i1} + 5^0 x_{io}) \bmod 7$

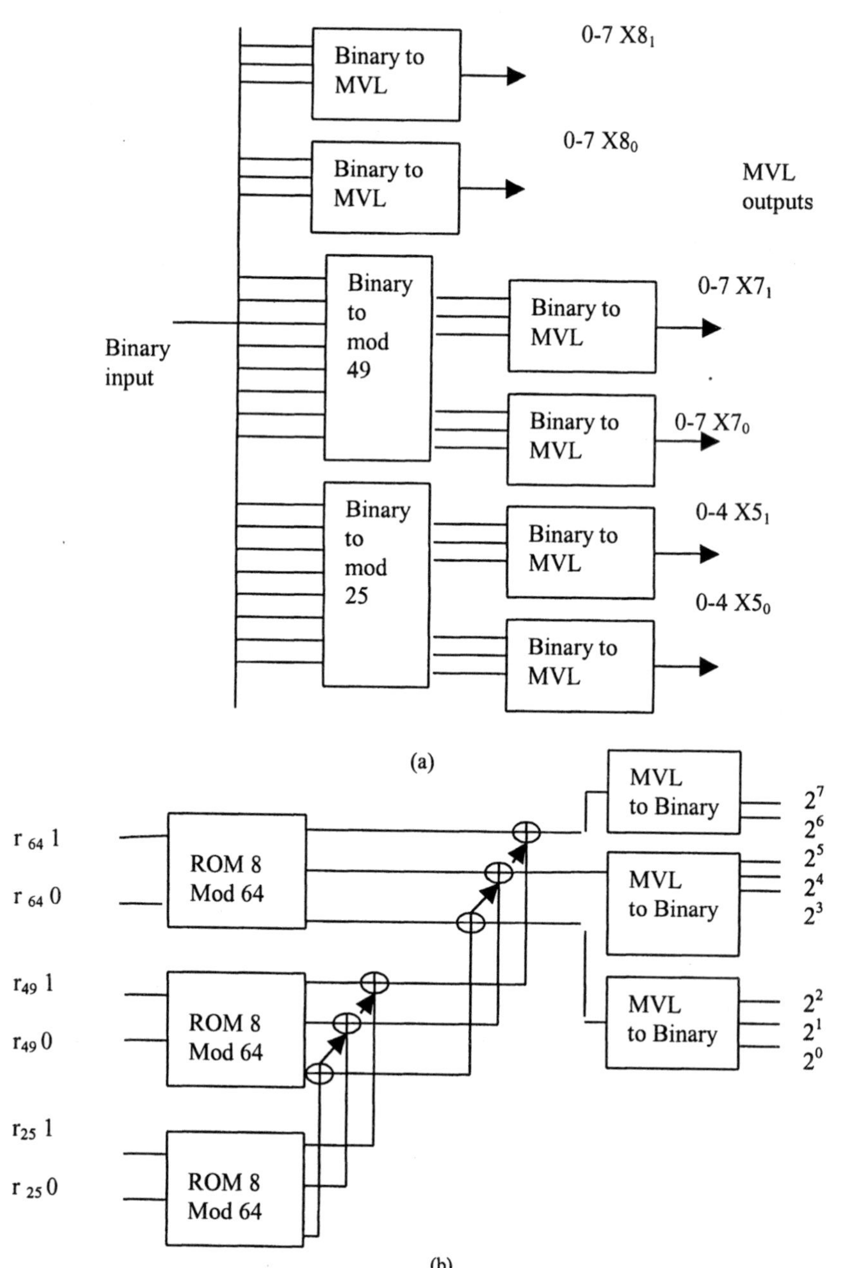

Fig 8.16 (a) Binary to MVL RNS converter and (b) MVL RNS to binary converter (adapted from [Sode86] 1986 ©IEEE)

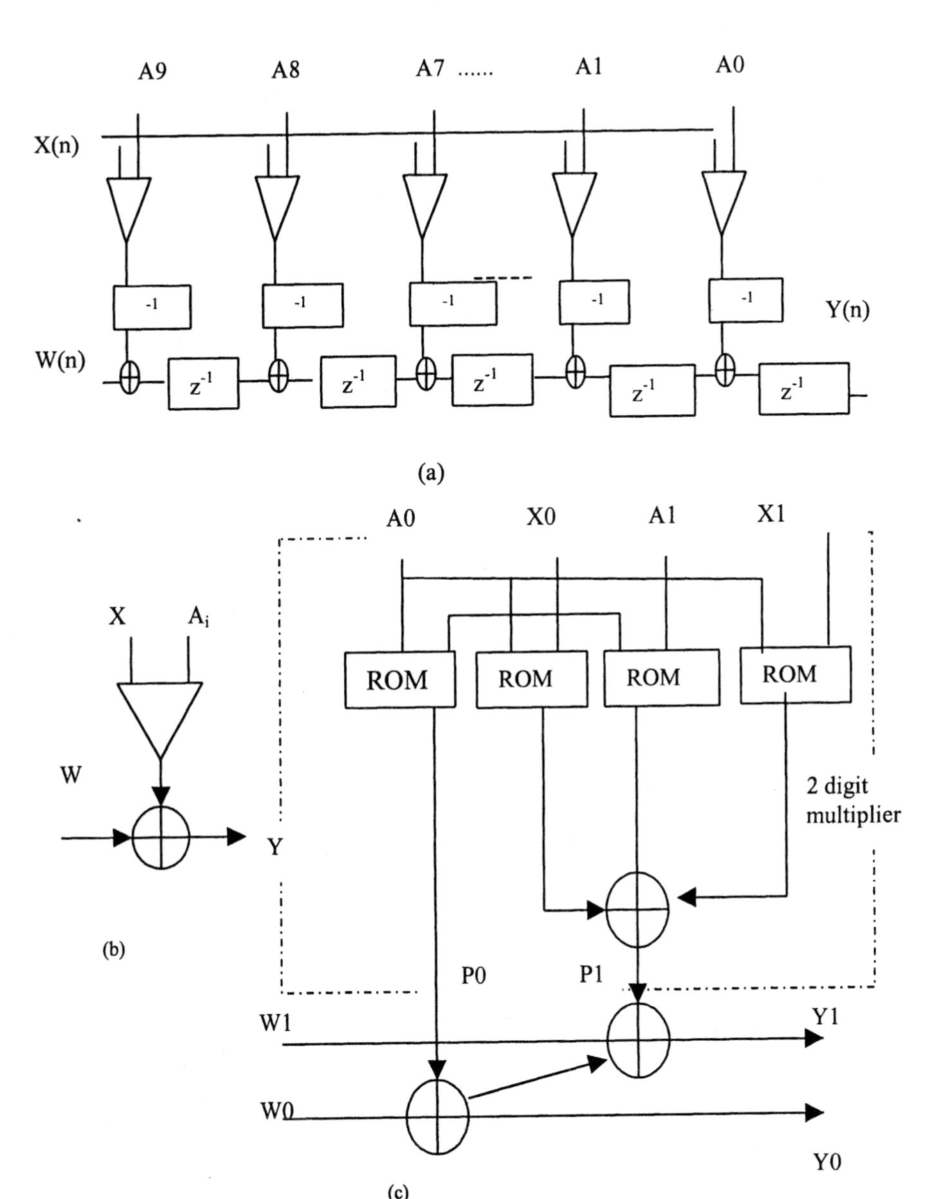

Fig 8.17 (a) A pipelined MVL RNS 10 tap FIR filter structure, (b) structure for one weight and (c) implementation of (b) (adapted from [Sode86] 1986 ©IEEE)

$= (-3.x_{i2} - 2.x_{i1} + x_{io}) \bmod 7.$
All the coefficients thus have magnitudes of 1, 2 and 3. Note that x_i can assume the values {-2, -1, 0, 1, 2}.

We next consider the operation of multiplication by c_i a constant which can be 1 or –2 or –3. The operations performed by c_i are as follows:
c_i = 3 two digit left shift and sign inversion
= 2 one digit left shift and sign inversion
= 1 no operation
= 0 two digit left shift and sign inversion
= -1 sign inversion
= -2 one digit left shift
= -3 two digit left shift
This multiplication by a constant coefficient can be realized by wire strapping.

The mod m_i addition is next considered. In radix -5 SD (signed digit) arithmetic, let x and y be
$x = \Sigma_{j=0}^{n-1} x_{ij}\, 5^j, \;\; y = \Sigma_{j=0}^{n-1} y_{ij}\, 5^j$
Addition of X with Y needs three steps;
$z_{ij} = x_{ij} + y_{ij}$
which can be rewritten as
$z_{ij} = 5c_{ij} + w_{ij}$
where the weight 5 is needed due to the radix- 5 arithmetic. We need thus to determine w_{ij} and c_{ij}. Three cases can be considered as follows:
$w_{ij} = z_{ij} - 5$ and $c_{ij} = 1$ if $z_{ij} > 2$
$w_{ij} = z_{ij}$ and $c_{ij} = 0$ if $-2 \leq z_{ij} \leq 2$
$w_{ij} = z_{ij} + 5$ and $c_{ij} = -1$ if $z_{ij} < -2$.
Also note that
$S'_{ij} = w_{ij} + c_{i(j-1)}$
Since $(5^n) \bmod m_i = 1$ or $(5^n) \bmod m_i = -1$, in the MVL coded residue representation, carry from the MSB can be connected to the LSB as follows:
$s'_{io} = w_{io} + c_{i(n-1)}$ or $= w_{io} - c_{i(n-1)}$
depending on whether $(5^n) \bmod m_i = 1$ or -1. Note that the bounds on the various parameters are
$z_{ij} = \{-6,..0,..6\}$, $w_{ij} = \{-2, -1, 0, 1, 2\}$, $c_{ij} = \{-1, 0, 1\}$ and $s'_{ij} = \{-3, ..0, ..3\}$.
The reader is referred to [Kama89] for details on the hardware implementation.

8.7. PALIOURAS AND STOURAITIS ARCHITECTURES USING MODULI OF THE FORM r^N

Paliouras and Stouraitis [Pali00] recommended the use of a RNS with a modulus of the form r^n. The use of 2^n has been considered earlier by us in Chapter III. The use of moduli of the type r^n can simplify the hardware and can increase the dynamic range significantly. We consider first the design of a multiplier using their approach. Denote the two words A and B to be multiplied as

$$A \bmod r^n = \Sigma_{i=0}^{n-1} a_i r^i. \quad (8.19a)$$

and

$$B \bmod r^n = \Sigma_{i=0}^{n-1} b_i r^i. \quad (8.19b)$$

We first note that a_i and b_i are such that $0 \le (a_i, b_i) \le (r-1)$. The products p_{ij} of the digits a_i and b_j are of the form

$$p_{ij} = a_i.b_j = r.p'_{ij} + p^0_{ij} \quad (8.20)$$

where p'_{ij} is the carry digit and p^0_{ij} is the sum digit. Evidently, if $i+j \ge n$, the carry terms and sum terms do not contribute to the output due to the mod r^n operation. It can also be seen that the carry digit p'_{ij} can be at most (r-2). This can be proved by noting that

$$P_{max} = (r-1).(r-1) = r^2 - 2r+1 = (r-2).r + 1 = d_{carry}.r + d_{sum} \quad (8.21)$$

Thus, two types of cells can be employed one which need to produce a carry to be used everywhere except last column and one which need not produce a carry to be used in the leftmost column. The partial products generated by the pre-processor cells described above need to be added in an array of radix-r digit adders as shown in Fig 8.18 (a) which uses full adder cells. These special full adder cells can produce a maximum carry of 2 for $r>3$ and 1 for $r=3$. This can be seen by noting that the maximum sum of the three inputs to the full-adder is

$$S_r = (r-1) + (r-1) + (r-2) = 3r-4 = 2r+a \quad (8.22)$$

Note that we have used the property that maximum carry from the preceding cell is r-2. Evidently, for $r>4$, an integer a exists such that $r-4=a\ge 0$. If $r=3$, $S_r = 5 = 1.3+2$ showing that the carry digit is unity. A variety of adders can be conceived so as to save the hardware especially more of half-adder type. The adder structure comprises of two stages as shown in Fig 8.18 (b). The corrective logic will evaluate the legitimate carry digit and sum digits.

The binary to RNS conversion can be similar to the one described extensively in Chapter 2. The residues corresponding to powers of two in the

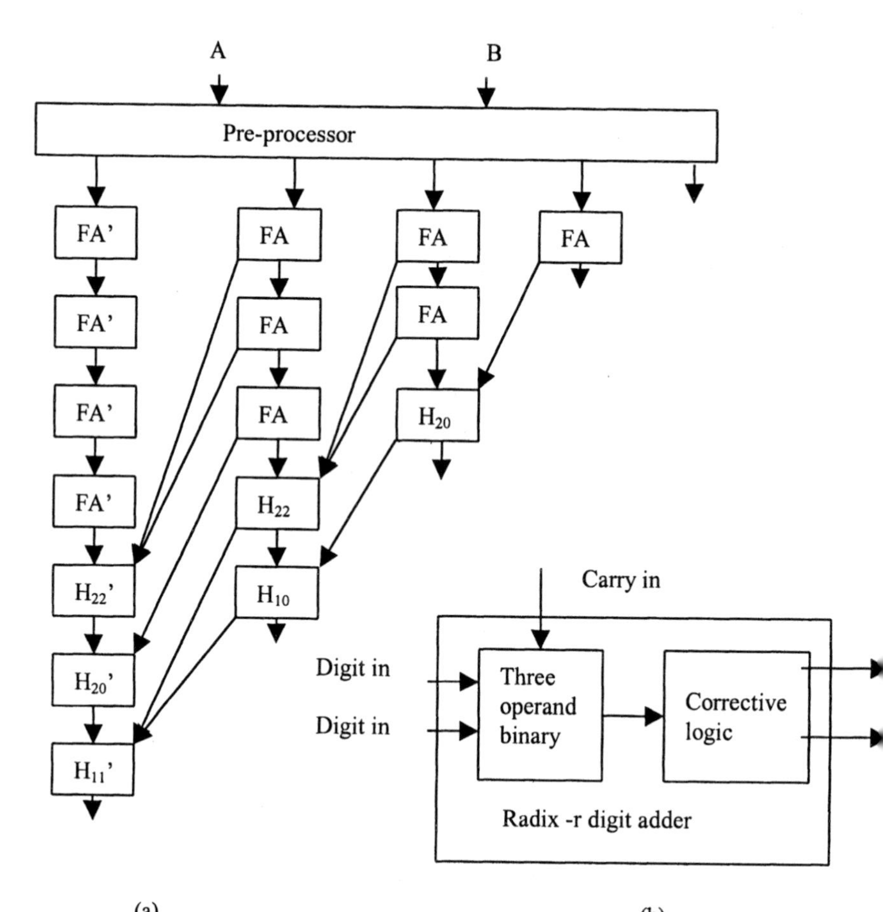

Fig 8.18 (a) Architecture of a high-radix multiplier and (b) general organization of various types of full adders (adapted from [Pali00] 2000 ©IEEE)

given binary number are stored in memory as coefficients of r_0, r_1, ..$r_{n-1}$and then accumulated. As an illustration, 64 can be seen to be 64 = 2.25+2.5+5 showing the corresponding weights in radix r. Next step is to accumulate these corresponding to the given binary number.

The residue to binary conversion can use CRT. As an illustration, for the moduli set $\{2,5,3^3\}$, M=270. The M_i values are 135, 54, 10 respectively. The $(1/M_i)$ mod m_i values are 1, 4 and 19 respectively. Thus, the weights corresponding to modulus 3^3 is 190. The residue corresponding to modulus 3^3 is written as a.9+b.3+c where a, b and c are < 3. Thus, corresponding to the bit positions of a, b and c, corresponding weights can be stored. As an illustration c_0, c_1 correspond to 190, 110, $3b_0$, $3b_1$ correspond to 30 and 60 and $9a_0$, $9a_1$ correspond to 90 and 180.

Paliouras et al compared their realization with those of Di Claudio et al described in Section 4.9.4 and Taheri et al [Tah88] to be described in the next Section. They observe that their design is superior regarding area A, Area.Time A.T, Area.Time ^{2} ($A.T^2$) whereas speed-wise it is slower. The reader is referred to an exhaustive discussion on these to [Pali00].

8.8. TAHERI, JULLIEN AND MILLER TECHNIQUE OF HIGH-SPEED COMPUTATION IN RINGS USING SYSTOLIC ARCHITECTURES

Taheri, Jullien and Miller [Tah88] described high-speed signal processing using systolic arrays over finite rings. The concept of IPSP known in DSP architectures is extended to modulo arithmetic herein. An IPSP (Inner Product Step Processor) basically performs the operations as follows:

$Y_{out} = Y_{in} + A_{in}.X_{in}$ (8.23a)

$X_{out} = X_{in}$ (8.23b)

$A_{out} = A_{in}$ (8.23c)

An IPSP symbol is shown in Fig 8.19 (a). The inputs and outputs are word level or bit level signals. If word level inputs and outputs are considered, the cell inside is an multiplier-Accumulator. The IPSP can be sliced into small bit level blocks as well in which case the architecture simplifies to that shown in Fig 8.19 (b). The use of several of these cells leads to a word level IPSP. Pipelining is possible in bit level IPSPs. An alternative implementation of the same operations in (8.23) can be using totally ROMs as shown in Fig 8.19 (c). The $BIPSP_m$ is needed in RNS to implement (8.23a) modified as

$Y_{out} = [Y_{in} + (A_{in}.X_{in}) \bmod m] \bmod m$ (8.24)

In other words modulo additions and multiplications are needed. The advantage of BIPSP is that the resulting VLSI architectures are modular, regular, homogeneous and have local communication. The $BIPSP_m$ cell performs the operation

$y_{i+1} = y_i \oplus [A_{in}.x_i.2^i]$ (8.25)

where i stands for the i th bit. A possible ROM based implementation is as shown in Fig 8.20 (a) whose inputs are x_i, y_i and the output is y_{i+1}. The cell contains a ROM of size B.m bits and a set of steering switches. The ROM

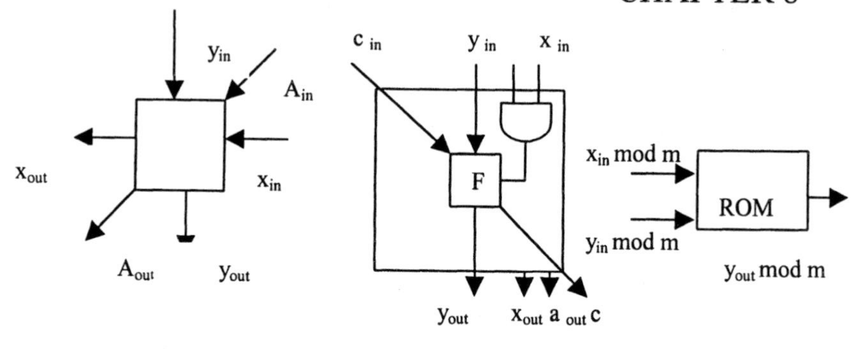

Fig 8.19 (a) An IPSP cell, (b) A BIPSP cell and (c) IPSP for finite ring calculations (adapted from [Tah88] 1988 ©IEEE)

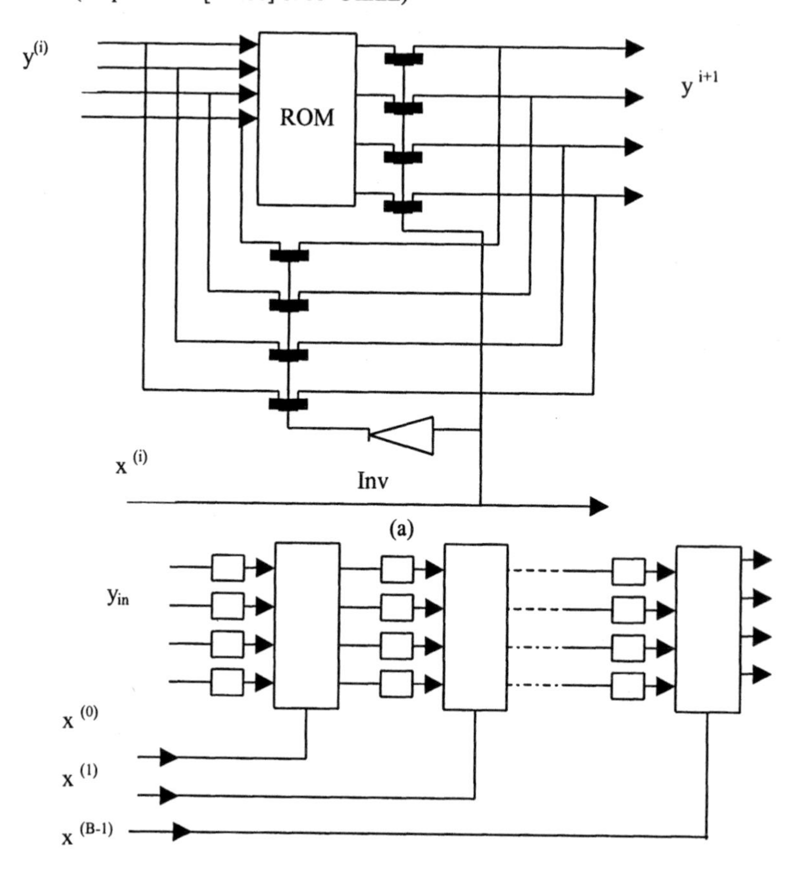

Fig 8.20 (contd)

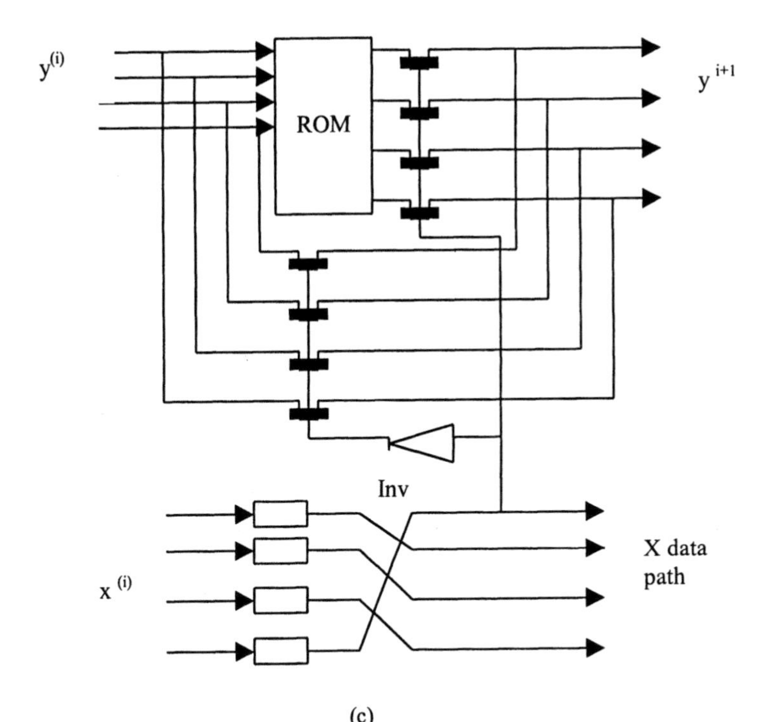

(c)

Fig 8.20. Implementation of $BIPSP_m$ cell (a), $IPSP_m$ cell using an array of cells in (a) and Systolic $BIPSP_m$ cell (c) (adapted from [Tah88] 1988 ©IEEE)

stores the result of the operation $y_i+2^j.A_{in}$. For $x_i = 1$, the memory is read whereas for $x_i=0$, the addition is not performed and the output is just the input itself. In other words, we are realizing the operation

$$y_{out} = y_{in} \oplus [\Sigma_{j=0}^{B-1} A_{in}.x_j.2^j] \qquad (8.26)$$

where all operations are over the ring. We next observe that the operation IPSP m can be realized by the arrangement of $BIPSP_m$ cells as shown in Fig 8.20 (b). Note that the cells are activated by the data input lines x_i.

Taheri et al suggest the modification as in Fig 8.20 (c) so that there is uniformity in the layout. All cells will be routed the x lines but the folding of the data path automatically steers the correct bit to the steering switches in the cell. Note that the small squares are latches. Note that the structure of Fig 8.20 (c) needs $B^2.m$ cells as against $B.m^2$ cells in a single ROM IPSP. The reader is referred to [Tah88] for more information on extensions to input and output conversion and also bit serial implementations.

8.9. RNS BASED IMPLEMENTATION OF FFT STRUCTURES

This topic has received considerable attention in literature [Tay85a, Tsen79, Nag83, Tay90, Bar80]. Fixed point implementation of Fast Fourier Transformation (FFT) algorithms are limited in speed for large word-length applications wherein RNS featuring parallel processing using smaller word length is attractive. Moreover, since Look-up tables can be largely used, pipelining also is feasible thereby further enhancing the speed.

The architecture [Tay85a] of a CRNS (i.e. conventional) radix -4 DFT and its implementation are presented in Fig 8.21 (a) and (b) which need the following operations: 12 real multiplies at level 1, 6 real add/subtracts at level 2, 8 real add/subtracts at level 3 and 8 real add/subtracts at level 4.

The inset in Fig 8.21 (b) shows the details of the CRNS unit. As against this architecture, the QRNS architecture shown in Fig 8.21 (c) needs the following operations: 6 real multiplies at level 1, 8 real add/subtracts and 2 multiplies at level 2 and 8 real add/subtracts at level 3. Pipelined processing is possible in these architectures between various levels. Evidently, the QRNS needs much less hardware. The reader is referred to the references cited above for more information.

8.10. OPTIMUM SYMMETRIC RESIDUE NUMBER SYSTEM

Optimum Symmetric Residue Number System (OSNS) [Pace00] has applications in folding A/D converters and phase sampled direction- finding antennas. The input signal is folded symmetrically in order to reduce the number of comparators needed to analyze the input signal. RNS is used with several moduli one for each channel and the results of all the channels are combined. In this system, corresponding to each m_i there are $2m_i$ states unlike in conventional RNS. This is since the dynamic range consists of increasing segment and a decreasing segment. The period is thus doubled. This is illustrated in Fig 8.22 and in Table.8.1 for $m_1=5$ and $m_2=4$. Note that the residue of natural numbers with respect to the modulus 5 is from 0 to 4 and decreases from 4 to zero and so on. Same is the case with modulus 4 also. Conventional CRT cannot be used for handling the OSNS directly.

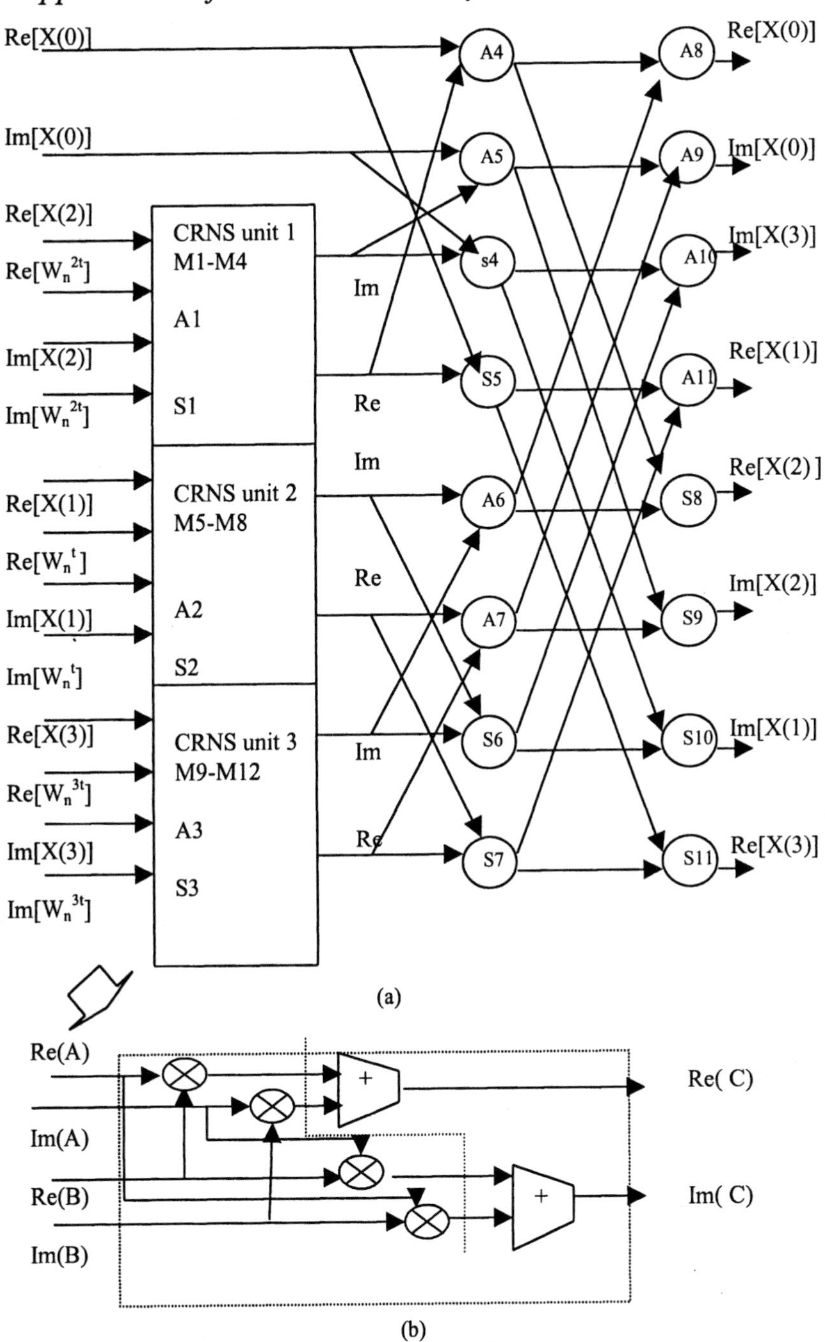

(a)

(b)

Fig 8.21 (contd)

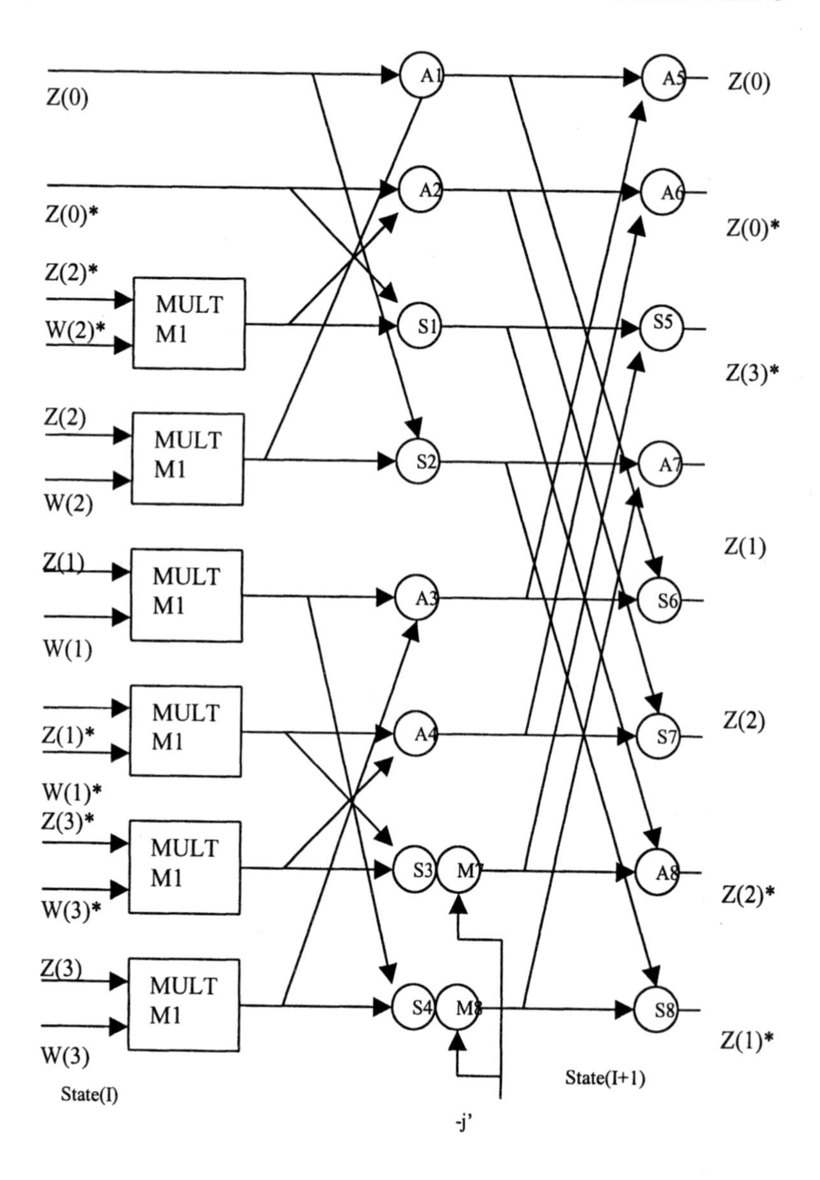

(c)

Fig 8.21 (a) CRNS Radix-4 Butterfly, (b) A CRNS unit and (c) QRNS radix-4 FFT Butterfly. (adapted from [Tay85a] 1985 ©IEEE)

Generalized CRT [Shoc67] needs to be employed.

Note that the theory can be applied to any moduli set. However, due to the advantages of the particular binary set, Pace et al considered the powers of two related moduli set. Given the residues h_1, h_2, h_3 corresponding to the three moduli 2^k-1, 2^k, 2^k+1 respectively, due to the folding, we have the various possibilities:
$h_1 = r_1$ or $(2m_1-r_1-1)$
$h_2 = r_2$ or $(2m_2-r_2-1)$
$h_3 = r_3$ or $(2m_3-r_3-1)$ This can be converted to a standard CRT by dividing first the even residues by two and in the case of odd residues, we need to use $(2m_i-r_i-1)/2$. Next, 1 RNS to binary conversion techniques such as Piestrak and Andraros-Ahmad discussed at length in Chapter 3 can be used. An example will be illustrative.

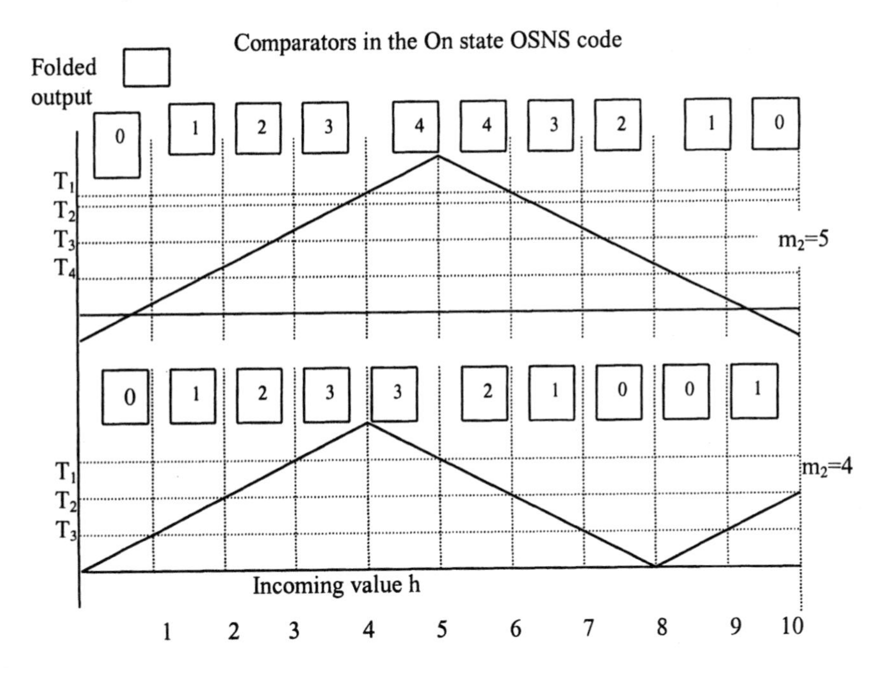

Fig 8.22. OSNS folding waveforms and integer values within each modulus for moduli m_1=5 and m_2=4 (adapted from [Pace00] 2000 ©IEEE)

Unknown incoming value (h)	m_1=5	m_1=4
0	0	0
1	1	1
2	2	2
3	3	3
4	4	3
5	4	2
6	3	1
7	2	0
8	1	0
9	0	1
10	0	2
11	1	3
12	2	3
13	3	2
14	4	1
15	4	0
16	3	0
17	2	1
18	1	2
19	0	3
20	0	3
21	1	2

Table. 8. 1. OSNS vectors for m_1=5 and m_1=4
(adapted from [Pace00] 2000 ©IEEE)

Consider the basic moduli set {65, 64, 63} using OSNS. Assume that we are given $r_1 = 64$, r_2=9 and r_3=46. Then, we have to divide by two using the rule mentioned above to obtain {32, 59, 23}. Using Andraros, Ahmad and Piestrak technique described in Chapter III, the decoded binary word can be obtained as N = 000000 011011 111011. Since $n < M/2$ (i.e. M= 111111 111111 000000), result is 2n = 3574. Otherwise the result is (2M-2n-1). The reader is referred to [Pace00] for an interesting description of PLD implementation of this system. Note that front-end conversion from conventional binary to residues in the OSNS is required as well. Note that there are eight solutions for the problem. But, since the moduli system is $\{2m_1, 2m_2, 2m_3\}$, two of the eight solutions are valid. One corresponds to all even choices of S_i or $2m_i$-S_i and the other corresponds to odd choices.

8.11. CONCLUSION

Several applications of RNS described in literature have been considered in this Chapter. Due to the continued activity in this area of research, it is believed that VLSI architectures using RNS may lead to very fast DSP implementations in the near future.

9
REFERENCES

[Ali84] G. Alia and E. Martinelli, A VLSI algorithm for direct and reverse conversion from weighted binary System to residue number system, IEEE Transactions on Circuits and Systems, Vol. 31, pp 1033-1039, Dec 1984.

[Anan91] P.V. Ananda Mohan and D.V. Poornaiah, Novel RNS to binary converters, IEEE International Symposium on Circuits and Systems, Singapore, June 1991, pp 1541-1544.

[Anan99] P.V. Ananda Mohan, Efficient Design of Binary to RNS converters, Journal of Circuits, Systems and Computers, Vol. 9, pp 145-154, 1999.

[Anan94] P.V. Ananda Mohan, Novel Design for Binary to RNS converters, Proc. International Symposium on Circuits and Systems, London, pp 357-360, 1994

[Anan98] P.V. Ananda Mohan, Evaluation of Fast Conversion techniques for Binary-Residue Number Systems, IEEE Transactions on Circuits and Systems, Vol. 45, Part-I, pp 1107-1109, 1998.

[Anan00a] P.V. Ananda Mohan, On " The Digit Parallel method for fast RNS to weighted number system conversion for specific moduli (2^k-1, 2^k, 2^k+1)", IEEE Transactions on Circuits and Systems, Part-II, Vol. 47, pp 972-974, 2000.

[Anan00b] P.V. Ananda Mohan, Residue Number System based VLSI processors for high-speed signal Processing, Proc. OSEE, 2000.

[Anan00c] P.V. Ananda Mohan, Comments on " Residue to binary converters based on New Chinese Remainder Theorems", IEEE Transactions on Circuits and Systems, Part-II, Vol. 47, p 1541, 2000.

[Anan01] P.V. Ananda Mohan, Comments on "Breaking the 2n-bit carry propagation barrier in Residue to Binary conversion for the [2^n-1, 2^n, 2^n+1] moduli set", IEEE Transactions on Circuits and Systems, Vol. 48, Part-II, pp 1031, 2001.

[And88] S. Andraros and H. Ahmad, " A new efficient memory-less residue to binary converter", IEEE Transactions on Circuits and Systems, Vol. 35, pp. 1441-1444, Nov.1988

[Bar78] A. Baraniecka and G.A. Jullien, On decoding techniques for residue Number system realisation of digital signal processing Hardware, IEEE Transactions on Circuits and Systems, Vol. 25, pp 935-936, Nov.1978

[Bar80] A.Z. Baraniecka and G.A. Jullien, Residue Number System implementations of Number Theoretic Transforms in Complex Residue Rings, IEEE Transactions on Acoustics, Speech and Signal Processing, Vol. 28, pp 285-291, 1980.

[Bars73] F. Barsi and P. Maestrini, Error correcting properties of redundant residue number systems, IEEE Transactions on Computers, Vol. 22, pp 307-315, 1973.

[Bars74] F. Barsi and P. Maestrini, Error detection and correction by product codes in Residue Number Systems, IEEE Transactions on Computers, Vol. 23, pp 915-924, 1974.

[Bay85] M.A. Bayoumi, G.A. Jullien and W.C. Miller, Hybrid VLSI architecture of FIR filters using residue number systems, Electronics Letters, Vol. 21, pp 358-359, 1985.

[Bay86] M.A. Bayoumi, A high-speed VLSI complex digital signal processor based on Quadratic residue Number Systems, VLSI Signal processing II, pp 200-211, S.Y. Kung, R.E. Owen and J.G. Nash, (Eds), IEEE Press, New York, 1986.

[Bay87a] M.A. Bayoumi, G.A. Jullien and W.C. Miller, A VLSI implementation of residue adders, IEEE Transactions on Circuits and Systems, Vol. 34, pp 284-288, 1987.

[Bay87c] M.A. Bayoumi, G.A. Jullien and W.C. Miller, A look-up table VLSI design Methodology for RNS structures used in DSP applications, IEEE Transactions on Circuits and Systems, Vol. 34, pp 604-615,1987.

[Ber85] P. Bernardson, Fast memory-less over 64 bit, Residue to Binary Converter, IEEE Transactions on Circuits and Systems, Vol. 32, pp 298-300, Mar.1985

[Beth89] T. Beth and D. Gollmann, Algorithm engineering for public key algorithms, IEEE Journal of Selected Areas in Communications, Vol. 7, pp 458-466, 1989.

[Bhar98] M. Bharadwaj, A.B. Premkumar and T. Srikanthan, Breaking the 2n-bit carry propagation barrier in Residue to Binary conversion for the [2^n-1, 2^n, 2^n+1] moduli set, IEEE Transactions on Circuits and Systems, Vol. 45, Part-I, pp 998-1002, 1998.

[Bi88] G. Bi and E.V. Jones, Fast conversion between binary and Residue Numbers, Electronics Letters, Vol. 24, pp1195-1197, Sept 1988.

[Blak83] G.R. Blakley, A computer algorithm for calculating the product AB modulo M, IEEE Transactions on Computers, Vol. 32, pp 497-500, May 1983.

[Bren82] R.P. Brent and H.T. Kung, A regular layout for parallel adders, IEEE Transactions on Computers, Vol. 31, pp 260-264, March 1982.

[Bric83] E.F. Brickell, A fast modular multiplication algorithm with application to two-key cryptography, Advances in Cryptography, Proceedings Crypto '82, Plenum, New York, pp 51-60,1983.

[Capo88] R.M. Capocelli and R. Giancarlo, Efficient VLSI networks for converting an integer from Binary System to Residue number system and Vice Versa, IEEE Transactions on Circuits and Systems, Vol. 35, pp 1425-1430, Nov. 1988

[Card88] G.C. Cardaralli, R. Lojocano, G. Martinelli and M. Salerno, Structurally Passive Digital Filters in Residue number Systems, IEEE Transactions on Circuits and Systems, Vol. 35, pp 149-158, February 1988.

[Card00] G.C. Cardiralli, M. Re, R. Lojacono and G. Ferri, A systolic architecture for high-performance scaled residue to binary conversion, IEEE Transactions on Circuits and Systems, Part-I, Vol. 47, 1523-1528, 2000.

[Chak86] N.B. Chakraborthi, J.S. Soundararajan and A.L.N. Reddy, An implementation of Mixed-Radix-Conversion for residue number applications, IEEE Transactions on Computers, Vol. 35, pp 762-764, August 1986.

[Chan92] A. Chandrakasan, S. Sheng and R.W. Brodersen, Low-Power CMOS design, IEEE Journal of Solid-State Circuits, Vol. 27, pp 472-484, 1992.

[Chr90] W.A. Chren Jr., A new Residue Number System division Algorithm, Comp.Math.Appl., Vol. 19, pp 13-29, 1990

[Chr95] W.A. Chren Jr., RNS-based enhancements for direct digital frequency Synthesis, IEEE Transactions on Circuits and Systems, Vol. 42, Part-II, pp 516- 524, 1995.

[Chr98] W.A. Chren, Jr., One-hot Residue coding for low-Delay-Power product CMOS design, IEEE Transactions on Circuits and Systems, Vol. 45, Part-II, pp 303-313, 1998.

[Con99] R. Conway and J. Nelson, Fast Converter for 3 moduli RNS using new property of CRT, IEEE Transactions on Computers, Vol. 48, pp 852-860, 1999.

[Dhur90] A. Dhurkadas, Comments on " An efficient Residue to Binary converter design", IEEE Transactions on Circuits and Systems, Vol. 37, pp 849-850, June 1990

[Dhur98] A. Dhurkadas, Comments on " A High speed realisation of a residue to binary Number system converter, IEEE Transactions on Circuits and Systems, Part II, Vol. 45, pp 446-447, March 1998.

[Di95] E.D. Di Claudio, F. Piazza and G. Orlandi, Fast Combinatorial RNS processors for DSP applications, IEEE Transactions on Computers, Vol. 44, pp 624-633, 1995.

[Dug92] M. Dugdale, VLSI implementation of residue adders based on binary adders, IEEE Transactions on Circuits and Systems, Part-II, Vol. 39, pp 325-329, 1992.

[Dug94] M. Dugdale, Residue multipliers using factored decomposition, IEEE Transactions on Circuits and Systems, Part II, Vol. 41, pp 623-627, Sept 1994.

[Efsta94] C. Efstathiou, D. Nikolos and J. Kalamatianos, Area-time efficient modulo 2^n-1 adder design, IEEE Transactions on Circuits and Systems, Vol. 41, pp 463-467, 1994.

[Eld93] S.E. Eldridge and C.D. Walter, Hardware implementation of Montgomery's modular multiplication algorithm, IEEE Transactions on Computers, Vol. 42, pp 693-699, 1993.

[Ell90] K.M. Elleithy and M.A. Bayoumi, A $\theta(1)$ algorithm for modulo addition, IEEE Transactions on Circuits and Systems, Vol. 37, pp 628-631, 1990.

[Ell92] K.M. Elleithy and M.A. Bayoumi, Fast and Flexible architectures for RNS to Arithmetic decoding, IEEE Transactions on Circuits and systems, Part II, Vol. 39, pp 226-235, April 1992.

[Ell95] K.M. Elleithy and M.A. Bayoumi, A systolic architecture for modulo multiplication, IEEE Transactions on Circuits and Systems, Part II, Vol. 42, pp 725-729, 1995.

[Etz80] M.H. Etzel and W.K. Jenkins, Redundant Residue number systems for error detection and correction in digital filters, IEEE Transactions on Acoustics, Speech and Signal Processing, Vol. 28, pp 538-545, 1980.

[Gall97] D. Gallaher, F.E. Petry, and P. Srinivasan, The digit parallel method for Fast RNS to weighted number System conversion for specific moduli (2^k-1, 2^k, 2^k+1), IEEE Transactions on Circuits and Systems, Part II, Vol. 44, pp 53-57, Jan 1997.

[Gamb91] D. Gamberger, New approach to integer division in residue number systems, IEEE symposium on computer arithmetic, pp 84-91, 1991.

[Garc98] A. Garcia, U. Meyer-Base and F.J. Taylor, Pipelined Hogenauer CIC filter using Field-Programmable logic and Residue Number System, Proc. IEEE Int. Conf. on Acoustics, Speech and Signal Processing, Vol. 5, pp 3085-3088, 1998.

[Garc99] A. Garcia and A. Lloris, A look-up table scheme for scaling in the RNS, IEEE Transactions on Computers, Vol. 48, pp 748-751, 1999.

[Gar59] H.L. Garner, The Residue Number System, IRE Transactions on Electronic computers, Vol. 8, pp 140-147, 1959.

[Hias98] A.A. Hiasat and Hoda S. Abdel-Aty-Zohdy, Residue to Binary arithmetic converter for the moduli set (2^k, 2^k-1, 2^{k-1}-1), IEEE Transactions on Circuits and Systems, Part II, Vol. 45, pp 204-209, Feb 1998.

[Hias00a] A.A. Hiasat, New efficient structure for a modular multiplier for RNS, IEEE Transactions on Computers, Vol. 49, pp 170-174, 2000.

[Hias00b] A.A. Hiasat, RNS arithmetic multiplier for medium and large moduli, IEEE Transactions on Circuits and Systems, Part II, Vol. 47, pp 937-940, 2000.

[Hitz95] M.A. Hitz and E. Kaltofen, Integer division in Residue Number Systems, IEEE Transactions on Computers, Vol. 44, pp 983-989, 1995.

[Hua79] C.H. Huang and F.J. Taylor, A memory compression scheme for modular arithmetic, IEEE Transactions on Acoustics, Speech and Signal Processing, Vol. 27, pp 608-611, 1979.

[Hua81] C.H. Huang, D.G. Peterson, H.A. Rauch, J.W. Teague and D.F. Fraser, Implementation of a Fast Digital processor using the Residue Number System, IEEE Transactions on Circuits and Systems, Vol. 28, pp 32-37, January 1981.

[Hua83] C.H. Huang, A Fully parallel Mixed-Radix Conversion algorithm for residue number applications, IEEE Transactions on Computers, Vol. 32, pp 398-402, April 1983.

[Hun94] C.Y. Hung and B. Parhami, An approximate sign detection algorithm method for residue numbers and its application to RNS division, Comput. Math. with applications, Vol. 27, pp 23-25, 1994.

[Hun95] C.Y. Hung and B. Parhami, Error analysis of approximate Chinese remainder theorem decoding, IEEE Transactions on Computers, Vol. 44, pp 1344-1348, 1995.

[Hwa79] K. Hwang, *Computer Arithmetic: Principles, Architecture and Design,* Wiley, 1979.

[Ibra88] K.M. Ibrahim and S.N. Saloum, " An efficient residue to binary converter design", IEEE Transactions on Circuits and Systems, Vol. 35, pp. 1156-1158, Sept.1988

[Jenk77] W.K. Jenkins and B.J. Leon, The use of residue number systems in the design of Finite impulse response Digital filters, IEEE Transactions on Circuits and Systems, Vol. 24, pp 191-201, April 1977.

[Jenk78a] W.K. Jenkins, Techniques for Residue to analog conversion for residue encoded digital filters, IEEE Transactions on Circuits and Systems, Vol. 25, pp 555-562, July 1978.

[Jenk78b] W.K. Jenkins, A highly efficient Residue combinatorial Architecture for digital filters, Proc. IEEE, Vol. 66, pp 700-702, 1978

[Jenk79] W.K. Jenkins, Recent advances in Residue number techniques for recursive digital filtering, IEEE Transactions on Acoustics, Speech and Signal Processing, Vol. 27, pp 210-221, 1979

[Jenk80] W.K. Jenkins, Complex Residue Number arithmetic for high-speed signal processing, Electronics Letters, Vol. 16, pp 660-661, 1980.

[Jenk82] W.K. Jenkins, Residue Number System error checking using expanded projection, Electronics Letters, Vol. 18, pp 927-928, 1982.

[Jenk83] W.K. Jenkins, The design of error checkers for Self- Checking Residue Number Arithmetic, IEEE Transactions on Computers, Vol. 32, pp 388-396, April 1983.

[Jenk87] W.K. Jenkins and J.V. Krogmeier, The Design of dual-mode complex signal processors based on Quadratic Modular Number Codes, IEEE Transactions on Circuits and Systems, Vol. 34, pp 354-364, April 1987.

[Jenk88] W.K. Jenkins and E.J. Altman, Self-checking properties of residue number error checkers based on Mixed Radix Conversion, IEEE Transactions on Circuits and Systems, Vol. 35, pp 159-167, 1988

[Jenk93] W.K. Jenkins, Finite Arithmetic Concepts, pp 611-676, in S.K.Mitra, J.F. Kaiser (Eds), *Handbook of Digital Signal Processing*, John Wiley and Sons, New York, 1993.

[Jull78] G.A. Jullien, Residue Number Scaling and other operations using ROM arrays, IEEE Transactions on Computers, Vol. 27, pp 325-335, April 1978.

[Jull80] G.A. Jullien, Implementation of multiplication, modulo a prime number with application to Number Theoretic Transforms, IEEE Transactions on Computers, Vol. 29, pp 899-905, 1980.

[Jull87] G.A. Jullien, R. Krishnan and W.C. Miller, Complex digital Signal Processing over finite rings, IEEE Transactions on Circuits and Systems, Vol. 34, pp 365-377, April 1987.

[Kala00] L. Kalampoukas, D. Nikolos, C. Efstathiou, H.T. Vergos and J. Kalamatianos, High-speed parallel prefix modulo 2^n-1 adders, IEEE Transactions on Computers, Vol. 49, pp 673- 679, 2000.

[Kama79] M. Kamayema and T. Higuchi, A new scaling algorithm in symmetric Residue Number System based on Multiple valued logic, Proc. ISCAS, pp 189-192, Tokyo, July 1979.

[Kama89] M. Kamayama, T. Sekibe and T. Higuchi, Highly parallel residue arithmetic chip based on multiple-valued bi-directional current-mode logic, IEEE Journal of Solid-State Circuits, Vol. 24, pp 1404-1411, 1989.

[Kim91] J.Y. Kim, K.H. Park and H.S. Lee, Efficient Residue to binary conversion technique with rounding error compensation, IEEE Transactions on Circuits and Systems, Vol. 38, pp 315-317, 1991.

[Koc96] C.K. Koc, T. Acar and B.S. Kaliski. Jr. Analyzing and comparing Montgomery Multiplication Algorithms, IEEE Micro, Chip, Systems, Software and Applications, pp 26-33, 1996.

[Kris86a] R. Krishnan, G.A. Jullien and W.C. Miller, Complex digital signal processing using quadratic Residue Number Systems, IEEE Transactions on ASSP, Vol. 34, pp 166-177, Feb 1986

[Kris86b] R. Krishnan, G.A. Jullien and W.C. Miller, The modified Quadratic Residue Number System (MQRNS) for complex high-speed signal processing, IEEE Transactions on Circuits and Systems, Vol. 33, pp 325-327, 1986.

[Leu81] S.H. Leung, Application of Residue Number Systems to Complex Digital Filters, Proc.15th Asilomar Conference on Circuits and Systems, pp 70-74, 1981.

[Lieb76] L.M. Liebowitz, A simplified binary arithmetic for the Fermat Number Transform, IEEE Transactions on Acoustics, Speech and Signal Processing, Vol. 24, pp 356-359, 1976.

[Lim83] Y.C. Lim and S.R. Parker, FIR filter design over a discrete powers of two coefficient space, IEEE Transactions on Acoustics, Speech and Signal Processing, Vol. 31, pp 583-591,1983.

[Lin84] M.L. Lin, E. Leiss and B. McInnis, Division and sign detection algorithm for residue number Systems, Comput. Math. Appl., Vol. 10, pp 331-342, 1984.

[Lu92] M. Lu and J.S. Chiang, A novel Division Algorithm for the residue Number System, IEEE Transactions on Computers, Vol. 41, pp 1026-1032, 1992.

[Mand72] D. Mandelbaum, Error correction in Residue Arithmetic, IEEE Transactions on Computers, Vol. 21, pp 538-545, 1972.

[Mee90] S.J. Meehan, S.D. O'Neil and J.J. Vaccaro, An Universal Input and output RNS Converter, IEEE Transactions on Circuits and Systems, Vol. 37, pp 799-803, June 1990.

[Mill84] D.D. Miller and J.N. Polky, An implementation of the LMS algorithm in the residue number system, IEEE Transactions on Circuits and Systems, Vol. 31, pp 452-461, May 1984.

[Mill98] D.F. Miller and W.S. McCormick, An arithmetic free Parallel Mixed-Radix conversion algorithm, IEEE Transactions on Circuits and Systems, Part II, Vol. 45, pp 158- 162, Jan 1998.

[Mont85] P.L. Montgomery, Modular multiplication without trial division, Mat. Comput, Vol. 44, pp 519-521, 1985.

[Nag83] H.K. Nagpal, G.A. Jullien and W.C. Miller, Processor architectures for two-dimensional convolvers using a single multiplexed computational element with finite field arithmetic, IEEE Transactions on Computers, Vol. 32, pp 989-1000, 1983.

[Nuss76] H.J. Nussbaumer, Complex convolutions via Fermat Number Transforms, IBM Journal of Research and Development, Vol. 20, pp 282-284, 1976.

[Orto92] G.A. Orton, L.E. Peppard ad S.E. Tavares, New Fault tolerant techniques for residue number systems, IEEE Transactions on Computers, Vol. 41, pp.1453-1464, 1992.

[Pace00] P.E. Pace, D. Styer and P. Ringer, An optimum SNS to Binary conversion algorithm and pipelined field-Programmable logic design, IEEE Transactions on Circuits and Systems, Part II, Vol.47, pp 735-745, 2000.

[Pali00] V. Paliouras and T. Stouraitis, Novel High-radix Residue number system architectures, IEEE Transactions on Circuits and Systems, Vol.47, Part II, pp 1059-1073, 2000

[Pele74] A. Peled and B. Liu, A new hardware realization of digital filters, IEEE Transactions on Acoustics, Speech and Signal Processing, Vol. 22, pp 456-462, 1974.

[Pies91] S.J. Piestrak, Design of residue generators and multi-operand modulo adders using carry-save adders, Proc.10th Symposium on Computer Arithmetic, France, 1991, pp 100-107.

[Pies94] S.J. Piestrak, Design of residue generators and multi-operand adders using carry-save adders, IEEE Transactions on Computers, Vol. 43, pp 68-77, 1994.

[Pies95] S.J. Piestrak, A high- Speed realisation of Residue to Binary System converter, IEEE Transactions on Circuits and Systems, Part II, Vol. 42, pp 661-663, Dec.1995.

[Pras91] B.S. Prasanna and P.V. Ananda Mohan, Fast VLSI architectures using non-redundant multi-bit recoding for computing exponentiation modulo a positive integer, Proc. ISCAS, Singapore, pp 3054-3057, 1991.

[Pras94] B.S. Prasanna and P.V. Ananda Mohan, Fast VLSI architectures using non-redundant multi-bit recoding for computing A^Y mod N, Proc. IEE, Part G, Vol. 141, pp 345-349, 1994.

[Prem92] A.B. Premkumar, "An RNS to binary converter in 2n+1, 2n, 2n-1 moduli set", IEEE Transactions on Circuits and Systems, Part II, Vol. 39, pp. 480-482, July, 1992.

[Prem95] A.B. Premkumar, An RNS to binary converter in a three moduli set with common factors, IEEE Transactions on Circuits and Systems, Vol. 42, Part II, pp 298-301, 1995.

[Prem98] A.B. Premkumar, M. Bharadwaj and T. Srikanthan," High-Speed and low-cost Reverse converters for the (2n-1, 2n, 2n+1) moduli set", IEEE Transactions on Circuits and Systems, Part II, Vol. 45, pp 903-908, July 1998.

[Radh92] D. Radhakrishnan and Y. Yuan, Novel approaches to the design of VLSI RNS multipliers, IEEE Transactions on Circuits and Systems, Part-II, Vol. 39, pp 52-57, 1992.

[Rama83] V. Ramachandran, Single residue error correction in residue number systems, IEEE Transactions on Circuits and Systems, Vol. 32, pp 504-507, May 1983.

[Ram80] A.S. Ramnarayan, Practical realization of mod p, p prime multiplier, Electronics Letters, Vol. 16, pp 466-467, 1980.

[Raz92] H.M. Razavi and J. Battelini, Design of a Residue arithmetic Multiplier, Proc. IEE Part-G, Vol. 139, pp 581-585, October 1992.

[Shan91] N.R. Shanbag and R.E. Siferd, A single-Chip Pipelined 2-D FIR filter using residue arithmetic, IEEE Journal of Solid-State Circuits, Vol. 26, pp 796-805, 1991.

[Shen88] A.P. Shenoy and R. Kumaresan, Residue to binary conversion for RNS arithmetic using only modular look-up tables, IEEE Transactions on Circuits and Systems, Vol. 35, pp 1158-1162, 1988.

[Shen89a] A.P. Shenoy and R. Kumaresan, Fast base extension using a redundant Modulus in RNS, IEEE Transactions on Computers, Vol. 38, pp 293-297, Feb 1989.

[Shen89b] A.P. Shenoy and R. Kumaresan, A fast and accurate RNS scaling technique for high-speed Signal processing, IEEE Transactions on Acoustics, Speech and Signal processing, Vol. 37, pp 929-937, June 1989.

[Skav92] A. Skavantzos and P.B. Rao, New Multipliers modulo (2^N-1), IEEE Transactions on Computers, Vol. 41, pp 957-961, 1992.

[Sloa85] K.R. Sloan Jr., Comments on " A computer Algorithm for calculating the product A.B mod M", IEEE Transactions on Computers, Vol. 34, pp 290-292, March 1985.

[Smit95] J.C. Smith and F.J. Taylor, A fault tolerant GEQRNS processing element for linear systolic array DSP applications, IEEE Transactions on Computers, Vol. 44, pp 1121-1130, 1995.

[Sode77] M.A. Soderstrand and E.L. Fields, Multipliers for residue-number-Arithmetic Digital filters, Electronics Letters, Vol. 13, pp 164-166,1977.

[Sode77a] M.A. Soderstrand, A high-speed low-cost Recursive digital filter using residue number arithmetic, Proc. IEEE, Vol. 65, pp 1065-1067, 1977.

[Sode80] M.A. Soderstrand and C. Vernia, A high-speed low-cost Modulo P_i multiplier with RNS Arithmetic applications, Proc. IEEE, Vol. 68, pp 529-532, April 1980.

[Sode83] M.A. Soderstrand, C. Vernia and J.H. Chang, An improved Residue Number System Digital-to-Analog Converter, IEEE Transactions on Circuits and Systems, Vol. 30, pp 903-907, December 1983.

[Sode84a] M.A. Soderstrand and Bhaskar Sinha, A pipelined recursive Residue Number System digital filter, IEEE Transactions on Circuits and Systems, Vol.31, pp 415-417, April 1984.

[Sode84b] M.A. Soderstrand and G.D. Poe, Application of Quadratic-like complex residue Number system Arithmetic to ultrasonics, IEEE Inter Conf on ASSP, Vol. 2, pp 28A.5.1-28A.5.4, 1984.

[Sode86] M.A. Soderstrand and R.A. Escott, VLSI implementation in Multiple – Valued logic of an FIR digital filter using Residue Number System Arithmetic, IEEE Transactions on Circuits and Systems, Vol. 33, pp 5-25, 1986.

[Sode86a] M.A. Soderstrand, W.K. Jenkins, G.A. Jullien and F.J. Taylor (Eds), *Residue Number System Arithmetic: Modern applications in digital Signal Processing*, IEEE Press, 1986.

[Soud97] D.J. Soudris, V. Paliouras, T. Stouraitis and C.E. Goutis, A VLSI design methodology for RNS full-adder based Inner product architectures, IEEE Transactions on Circuits and Systems, Part II, Vol. 44, pp 315-318, April 1997.

[Sous86] M. Sousa and F. Taylor, Complex integer to Complex residue encoding, IEEE Transactions on Computers, Vol. 35, pp 648-650, July 1986

[Stou92] T. Stouraitis, Efficient converters for residue and Quadratic residue systems, Proc. IEE, Part G, Electronic Circuits and Systems, Vol. 139, pp 626-634, 1992.

[Stou93] T. Stouraitis, S.W. Kim and A. Skavantzos, Full-Adder based arithmetic units for finite integer rings, IEEE Transactions on Circuits and Systems, Part II, Vol. 40, pp 740-745, November 1993.

[Su90] C.C. Su and H.Y. Lo, An Algorithm for scaling and single residue error correction in residue number systems, IEEE Transactions on Computers, Vol. 39, pp 1053-1064, August 1990.

[Su99] C.Y. Su, S.A. Hwang, P.S. Chen and C.W. Wu, An improved Montgomery's algorithm for high speed RSA Public key Cryptosystem, IEEE Transactions on VLSI Systems, Vol. 7, pp 280-283, 1999.

[Sza67] N.S. Szabo and R.I. Tanaka, *Residue Arithmetic and Its Applications to Computer Technology,* New-York, Mc-Graw Hill, 1967.

[Tah88] M. Taheri, G.A. Jullien and W.C. Miller, High-speed signal processing using systolic arrays over finite rings, IEEE Journal of Selected areas in communications, Vol. 6, pp 504-512,1988.

[Taka85] N. Takagi, H. Yasuura and S. Yazima, High-speed VLSI multiplication algorithm with a redundant binary addition tree, IEEE Transactions on Computers, Vol. 34, pp 789-796, 1985.

[Taka92a] N. Takagi and S. Yajima, Modular multiplication hardware algorithms with a redundant representation and their application to RSA crypto system, IEEE Transactions on Computers, Vol. 41, pp 887-891, 1992.

[Taka92b] N. Takagi, A radix-4 modular multiplication hardware algorithm for modular exponentiation, IEEE Transactions on Computers, Vol. 41, pp 949-956, 1992

[Tay81a] F.J. Taylor, Large moduli Multipliers for signal processing, IEEE Transactions on Circuits and Systems, Vol. 28, pp 731-736, July 1981

[Tay81b] F.J. Taylor and A.S. Ram Narayanan, An efficient Residue to decimal converter, IEEE Transactions on Circuits and Systems, Vol. 28, pp 1164-1169, Dec.1981.

[Tay82a] F.J. Taylor, A VLSI Residue Arithmetic Multiplier, IEEE

Transactions on Computers, Vol. 31, pp 540-546, June 1982.

[Tay82b] F.J. Taylor and C.H. Huang, An auto-scale residue multiplier, IEEE Transactions on Computers, Vol. 31, pp 321-325, 1982.

[Tay83] F.J. Taylor, An over-flow free residue multiplier, IEEE Transactions on Computers, Vol. 32, pp 501-504, May 1983.

[Tay84] F.J. Taylor, Residue arithmetic: A Tutorial with examples, IEEE Computer, pp 50-62, May 1984.

[Tay85a] F.J. Taylor, G. Papadourakis, A. Skavantzos and A. Stouraitis, A radix-4 FFT using complex RNS arithmetic, IEEE Transactions on Computers, Vol. 34, pp 573-576, 1985.

[Tay85b] F.J. Taylor, A single modulus complex ALU for signal processing, IEEE Transactions on Acoustics, Speech and Signal Processing, Vol. 33, pp 1302-1315, 1985.

[Tay86] F.J. Taylor, On the Complex Residue Arithmetic System (CRNS), IEEE Transactions on Acoustics, Speech and Signal Processing, Vol. 33, pp 1675-1677, 1986.

[Tay90] F.J. Taylor, An RNS Discrete Fourier Transform Implementation, IEEE Transactions on Acoustics, Speech and Signal Processing, Vol. 38, pp 1386-1394, 1990.

[Thun86] R.E. Thun, On Residue Number System Decoding, IEEE Transactions on Acoustics, Speech and Signal Processing, Vol. 34, pp 1346-1347, 1986.

[Tsen79] B.D. Tseng, G.A. Jullien and W.C. Miller, Implementation of FFT structures using the residue number system, IEEE Transactions on Computers, Vol. 28, pp831-845, 1979.

[Ulm98] Z.D. Ulman and M. Czyzak, Highly parallel fast scaling of numbers in non-redundant residue arithmetic, IEEE Transactions on Signal Processing, Vol. 46, pp 487-496, 1998.

[Vand90] A. Vandemuelebroecke, E. Vanzieleghem, T. Denayer and P.G.A. Jespers, A new carry-free division algorithm and its application to a single-chip 1024 bit RSA processor, IEEE Journal of Solid-State Circuits, Vol. 25, pp 748-755,1990.

[Vinn94] B. Vinnakota and V.V.B. Rao, Fast conversion techniques for Binary to RNS, IEEE Transactions on Circuits and Systems, Part I, Vol. 41, pp 927-929, Dec.1994

[Vu85] T.V. Vu, Efficient implementations of the Chinese remainder theorem for sign detection and residue decoding, IEEE Transactions on Computers, Vol. 34, pp 646- 651, July 1985.

[Wake76] J.F. Wakerly, One's complement adder eliminates unwanted zero, Electronics, pp 103-104, 1976.

[Walt93] C.D. Walter, Systolic modular multiplication, IEEE Transactions on Computers, Vol. 42, pp 376-378, March 1993.

[Walt95] C.D. Walter, Still Faster modular multiplication, Electronics Letters, Vol. 31, pp 263-264, Feb 1995.

[Wan94] C.L. Wang, New bit serial VLSI implementation of RNS FIR digital filters, IEEE Transactions on Circuits and Systems, Part II, Vol. 41, pp 768-772, November 1994.

[Wan96] Y. Wang and M. Abd-El-Barr, A New algorithm for RNS decoding, IEEE Transactions on Circuits and Systems, Vol. 43, pp 998-1001, December 1996.

[Wan99] Y. Wang, M.N.S. Swamy and M.O. Ahmad, Residue to binary number converters for three moduli sets, IEEE Transactions on Circuits and Systems, Part II, Vol. 46, pp 180-183, 1999.

[Wan00a] Y. Wang, Residue to binary converters based on New Chinese Remainder Theorems, IEEE Transactions on Circuits and Systems, Part II, Vol. 47, pp 197- 205, 2000.

[Wan00b] Z. Wang, G.A. Jullien and W.C. Miller, An improved Residue to Binary Converter, IEEE Transactions on Circuits and Systems, Part I, Vol. 47, pp 1437-1440, 2000.

[Wats67] R.W. Watson and C.W. Hastings, *Residue Arithmetic and Reliable Computer design*, Washington DC, Spartan, 1967.

[Wig90] N.M. Wigley and G.A. Jullien, On modulus replication for residue arithmetic computations of complex inner products, IEEE Transactions on Computers, Vol. 39, pp 1065-1076, 1990.

[Yang98] C.C. Yang, T.S. Chang and C.W. Jen, A new RSA cryptosystem hardware design based on Montgomery's algorithm, IEEE Transactions on Circuits and Systems, Part-II, Vol. 45, pp 908-913, 1998.

[Yass91] H.M. Yassine and W.R. Moore, Improved Mixed radix conversion for residue number system Architectures, Proc. IEE, Part-G, Vol. 138, pp 120-124, February 1991.

INDEX